Digital Printing

Ashok Athalye · Harshal Patil
Editors

Digital Printing

Sustainable Solution to Fashion Industry

Springer

Editors
Ashok Athalye
Department of Fibres and Textile Processing Technology
Institute of Chemical Technology (ICT)
Mumbai, Maharashtra, India

Harshal Patil
CanvaLoop Fibre Private Limited
Surat, Gujarat, India

ISBN 978-981-95-2592-8 ISBN 978-981-95-2593-5 (eBook)
https://doi.org/10.1007/978-981-95-2593-5

This Springer imprint is published by the registered company Springer Nature Singapore Pte Ltd.
The registered company address is: 152 Beach Road, #21-01/04 Gateway East, Singapore 189721, Singapore

Preface

The textile and fashion industry is undergoing a profound transformation driven by the urgent need for sustainability, personalization, and innovation. Among the many advancements reshaping this sector, digital textile printing has emerged as a disruptive and sustainable technology, offering the potential to revolutionize how textiles are designed, printed, and consumed.

This book, *Digital Printing—Sustainable Solution to Fashion Industry*, presents a comprehensive and insightful exploration of digital textile printing as a green alternative to conventional methods. It brings together the collective expertise of academic researchers, technologists, and industry professionals to examine the evolving role of digital printing in fostering eco-conscious, efficient, and flexible textile production.

The book begins with a foundational overview in Chap. 1, introducing readers to the fundamentals of digital textile printing—its evolution, principles, and current relevance. Chapter 2 contextualizes this technology within the broader landscape of textile sustainability, highlighting environmental challenges and the urgent need for cleaner processes.

Moving forward, Chap. 3 delves into the opportunities and challenges of digital textile printing for localized, sustainable production, emphasizing its alignment with modern on-demand and zero-waste business models. Chapters 4 and 5 explore the application of digital printing on various fiber types, including cellulosic, synthetic, and wool-based fabrics, while also showcasing innovations in sustainable inks that enhance both aesthetic appeal and environmental performance.

The science behind digital printing is detailed in Chaps. 6 and 7, focusing on the chemistry of inks, formulation strategies, and the fluid dynamics involved in precise ink-jet application—critical to both performance and sustainability.

A key enabler of digital textile printing is the design process itself. Chapter 8 examines the role of digital designing tools, which allow for greater creative freedom, customization, and reduced material wastage. Complementing the technical discourse, Chap. 9 offers a comparative environmental assessment of digital versus traditional printing methods, reinforcing the ecological advantages of digital solutions.

Finally, Chap. 10 addresses the practical challenges faced during implementation—such as ink compatibility, machine limitations, and substrate behavior—and presents feasible remedies and best practices for smoother industrial adoption.

Together, these chapters provide a holistic understanding of digital textile printing as a transformative force in the fashion industry, bridging creativity with sustainability, and innovation with environmental responsibility.

We hope this book serves as a valuable resource for students, researchers, designers, and industry professionals, inspiring further advancements and collaborations in the journey toward a greener textile future.

Surat, India Dr. Harshal Patil
Mumbai, India Prof. Ashok Athalye

Contents

Chapter 1
Introduction to Digital Printing

Joanna Izdebska-Podsiadły

Abstract This chapter delves into the history and evolution of digital textile printing, outlining its primary applications. It also introduces the fundamental principles of digital printing, providing a concise overview of both direct and transfer printing processes. Furthermore, the study indicates the advantages of digital printing over traditional methods taking into account the aspect of environmental impact.

Keywords Digital printing · Direct printing · Transfer printing · Fundamental principles · Textile

1.1 History of Digital Printing on Textiles

The history of fabric decoration dates back to ancient times, when techniques such as hand painting, stamping, and reserve methods such as batik, tie-dye, and the Chinese zha ran technique were used. Over time, more advanced methods emerged, including woodblock and copperplate printing, which represented a significant step towards the mechanization of the decoration process. A breakthrough moment was the Industrial Revolution, which brought about the implementation of rotary printing machines, enabling the mass production of patterned fabrics (Cahill, 2006).

Since the 1930s, screen printing has been the dominant technique in the textile industry and remains the most commonly used conventional method to this day (Rainert et al., 2024). Digital printing on textiles emerged in the 1990s. Initially, this technology was mainly used for prototyping and small batch production for niche markets. Over time, thanks to technological advances, digital printing gained potential for commercial applications. Recent developments have significantly advanced this technology, paving the way for commercial print runs and product innovation in the industry. Currently, the latest innovations are being implemented in the area of 3D printing (Chakraborty & Biswas, 2020; Xiao & Kan, 2022) and smart fabrics (Marra

J. Izdebska-Podsiadły (✉)
Department of Printing Technology, Faculty of Mechanical and Industrial Engineering, Institute of Mechanics and Printing, Warsaw University of Technology, Warsaw, Poland
e-mail: joanna.podsiadly@pw.edu.pl

A. Athalye and H. Patil (eds.), *Digital Printing*,
https://doi.org/10.1007/978-981-95-2593-5_1

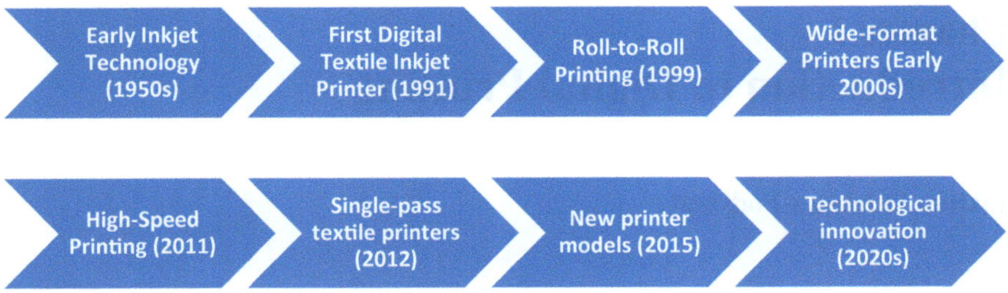

Fig. 1.1 Key milestones in the development of digital textile printing technology

et al., 2021), where printed electronics are used. Key milestones in the development of digital printing technology on textiles are shown in Fig. 1.1.

The beginnings of experiments with digital printing on textiles date back to the 1980s, when the adaptability of inkjet printers for textile printing was tested in laboratory conditions. Challenges included not only adapting the equipment to the specifics of the fabrics (structures, absorbency), but also developing suitable inks and fixing methods. The breakthrough came in 1991, when Stork (now SPGPrints) launched the first industrial piezoelectric printer dedicated to textile printing. In the following years, DuPont developed specialized inks, and Epson unveiled new print heads (Cahill, 2006).

The development of reactive, pigment, acid and sublimation inks made it possible to adapt the technology to different types of fibers, which contributed to the commercialization of digital printing between 2000 and 2009. Dedicated printers from companies such as DuPont, Mimaki and Kornit appeared on the market. In 2005, the introduction of DTG (Direct-to-Garment) technology enabled direct printing on finished garments, which revolutionized the production of personalized clothing and opened up new opportunities for small businesses (Cahill, 2006).

After 2010, digital technology in the textile industry experienced rapid growth. With improved printing quality and increased speed, mass production became possible. At ITMA in 2011, printers reaching speeds of more than 6 linear meters per minute were presented. In 2012, MS Printing Solutions introduced the Lario single-pass printer, printing at 75 m per minute. In 2015, Konica-Minolta unveiled the Nassenger SP-1, reaching speeds of 1,500 to 6,000 linear meters per hour, depending on resolution.

Digital printing has become particularly important in the fast fashion sector, enabling the production of short runs and individual, castomized designs. An additional impetus for development was the growing environmental awareness—digital technology uses much less water and dyes than traditional screen printing, which fits in with the idea of sustainable production. Companies such as Kornit Digital, Epson, Mimaki, EFI Reggiani, HP (Stitch), MS Printing Solutions and Durst began to dominate the market.

The years 2020–2025 will see further expansion of the digitally printed textiles market. According to gminsights.com, Kornit Digital (16%), Seiko Epson Corporation (12%), Mimaki Engineering (8%), as well as Roland DG, Durst Group and

SPGPrints held the largest market share in 2023. Together they accounted for about half of the global market (Gminsights, 2023).

Major growth drivers include on-demand production, waste reduction, the development of new materials and the increasing use of inks that enable printing on cotton, silk or technical fabrics. Artificial intelligence, used for pattern design and integration with 3D software, is also playing an increasingly important role. Shrinking production runs and the need for flexibility are making digital technologies the preferred choice in many market segments. According to API.pl, already 80% of textile production has been digitized (Stawski et al., 2023).

1.2 Textile Digital Printing Technologies

Digital printing is a method of applying designs to fabrics that involves transferring graphic data directly from a digital file to a printing device, without the need for traditional printing forms such as screens. Figure 1.2 shows a diagram of screen printing and digital printing. Unlike conventional techniques such as screen printing, digital printing allows for flexible, fast, and economical production of both short and medium production runs, as well as the personalization of individual products. Table 1.1 shows a comparison of digital inkjet printing on fabrics with traditional screen printing.

Depending on the method of applying the image to the fabric, digital printing is divided into two main categories: direct printing and indirect printing. In the case of direct printing, the image is applied directly to the surface of the fabric or finished product (e.g., a T-shirt), while in indirect printing, the pattern is first printed on a special carrier (e.g., transfer paper or PET film) and then transferred to the fabric using a heat press or calender. Both methods have their applications and are selected depending on the type of material, the expected durability of the print, and the scale of production. The classification of digital printing based on the printing process is shown in Fig. 1.3.

Another classification can be based on printing technology. DOD and piezoelectric technologies are primarily used in fabric printing, but thermal (TIJ) and CIJ

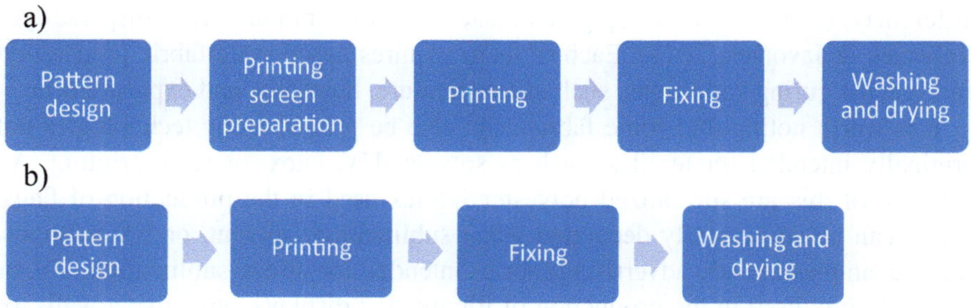

Fig. 1.2 Scheme of printing **a** by screen printing **b** by digital methods

Table 1.1 Comparison of inkjet printing and screen printing

Criterion	Digital inkjet printing	Screen printing (traditional)
Technology	Direct inkjet printing from a digital file	Transferring ink using a printing form (pressing through a screen)
Circulation (profitability)	Profitable for small and medium circulation	Profitable for large circulation
Initial cost	Low (no printed forms, quick implementation)	High (preparation of screens for each color)
Number of colors	Theoretically unlimited (full CMYK, additional colors)	Usually 6–12 colors (each requires a separate screen)
Details and quality	High resolution (photorealistic graphics)	Sharp but limited detail in complex patterns
Production time	Quick start, short lead time	Longer preparation (printing form production)
Personalization	Very easy—each print can be different	Difficult or impossible—each design requires a separate printing form
Ecology	Lower water consumption, less waste	High consumption of water, inks, chemicals
Types of fabrics	Universal (cotton, polyester, silk, etc.)	Also universal, but works better with natural fabrics
Unit cost	Higher with large volumes	Lower with large volumes

technologies can also be used (Vodyashkin et al., 2025). The share of individual printing technologies in the printed textiles market is shown in Fig. 1.4.

From a technological point of view, digital printing on fabrics is mainly based on Drop-on-Demand (DOD) technology (Gminsights, 2023). Piezoelectric printheads are most commonly used, although thermal inkjet (TIJ) technology is also widely used. In some cases, continuous inkjet (CIJ) technology (Tippett, 2002), electrophotography (Khadzhynova et al., 2017), or electrostatic printing (Carr et al., 2000) are also used. The choice of technology depends on the type of ink, quality requirements, and compatibility with the substrate.

Another important criterion for classifying digital printing is the type of dye used. Depending on the type of fiber and the expected performance characteristics, different types of inks are used: pigment, reactive, acid, sublimation and disperse dyes (Javoršek & Javoršek, 2011). Each of them requires appropriate fabric preparation and specific curing conditions, such as temperature, humidity, and exposure time.

It is worth noting that some fabrics can also be printed using technologies not originally intended for textiles, such as solvent, UV, latex, or resin printing. An example of this are specialized polyester fabrics used in the production of flags, which can be successfully decorated using sublimation, solvent, or UV methods. Other examples include advertising fabrics intended for direct, sublimation, UV, or latex printing, used in the production of illuminated displays, advertising walls, or citylights.

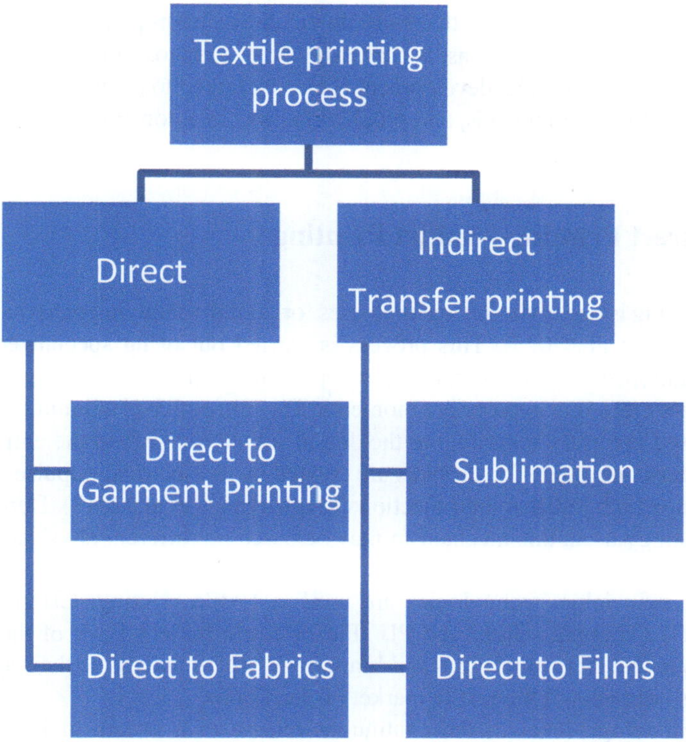

Fig. 1.3 Classification of digital fabric printing

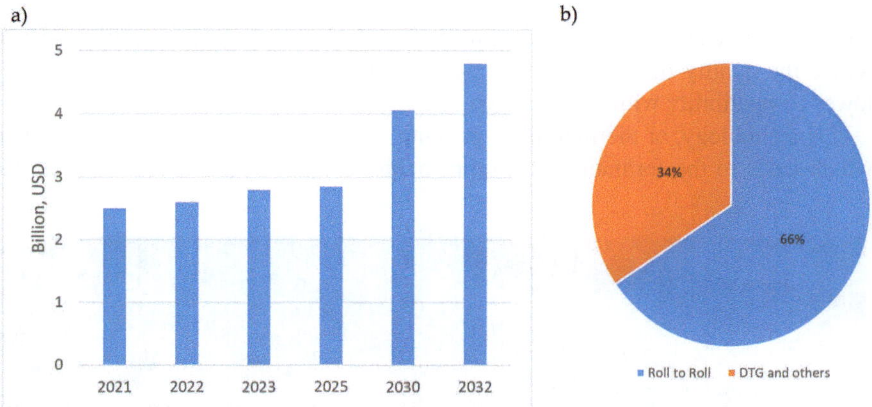

Fig. 1.4 Digital textile printing market: **a** Market size, **b** Market share by printing method (*Source* Based on data from Gminsights, 2023 & Mordor Intelligence, 2025)

Digital printing on fabrics, thanks to its flexibility, high-quality detail reproduction, and customization options, is currently one of the most important tools in the modern textile industry. Its development is closely linked to technological advances in the field of inks, printheads, and production process automation.

1.3 Direct Printing—Inkjet Printing

Inkjet printing is one of the key technologies for digitally applying patterns to fabrics (Vodyashkin et al., 2025). This process is carried out using specialized printing devices equipped with heads with microscopic nozzles (Fig. 1.5), through which ink droplets are ejected at appropriate moments. The ink, which is a liquid form of dye, is dispensed precisely according to the digital source image, and the entire process is computer-controlled. The droplets are created by an electrical impulse sent to the print head, which initiates their ejection (Khadzhynova et al., 2017). Depending on the technology used, this mechanism may be based on different physical principles (Fig. 1.6).

Three main inkjet technologies are used in textile printing: CIJ (Continuous Inkjet), TIJ (Thermal Inkjet) and PIJ (Piezoelectric Inkjet). Each of them has its own unique features, advantages and limitations, which determine their application in various segments of the textile market (Tyler, 2005).

CIJ technology involves the continuous release of a stream of ink through a nozzle. The droplets are electrostatically charged and directed onto the fabric or into a recovery system. The main advantage of this technology is its very high printing speed and the ability to operate in difficult conditions and on uneven surfaces. However, due to its limited precision and lower resolution, CIJ is rarely used in textile printing, where the quality of detail reproduction is crucial. An additional limitation is its lower compatibility with a wide range of textile inks.

TIJ technology, or thermal inkjet printing, is based on heating the ink chamber, which leads to the formation of a vapor bubble and the ejection of a droplet. TIJ

Fig. 1.5 Printhead (*Source* https://www.epson-monnalisa.eu/en_IT/monna-lisa-ml-13000)

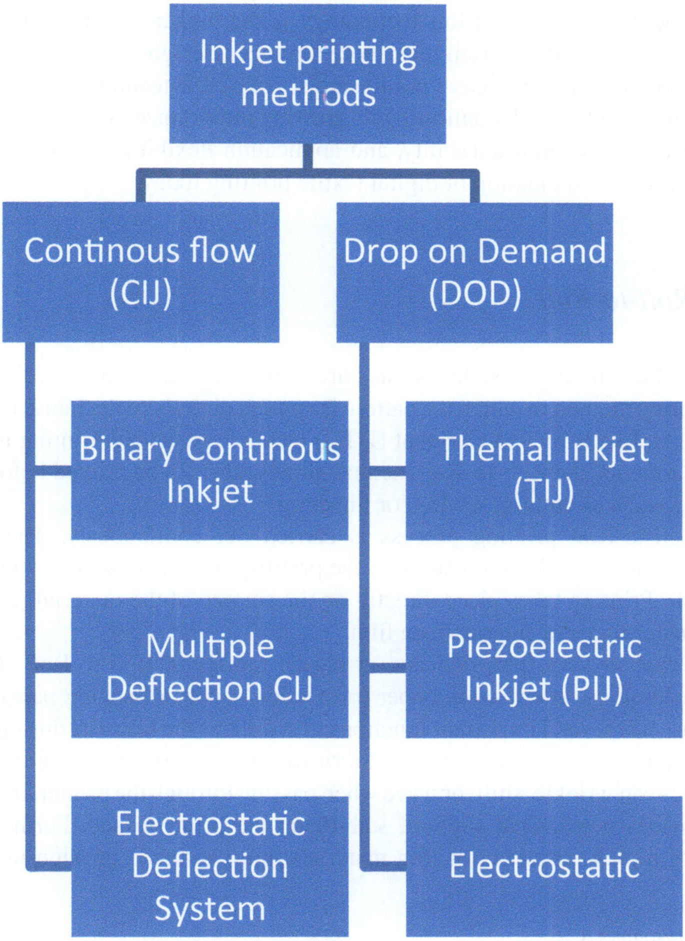

Fig. 1.6 Classification of inkjet printing

is mainly used for printing on cotton and other absorbent materials. Its advantages include high resolution, good print quality, simple printhead design, and low production costs. However, its limitations include shorter printhead life due to heating cycles and limited compatibility with pigment and sublimation inks, which restricts its use mainly to personalized clothing and low-volume production.

The most commonly used and fastest growing technology in professional textile printing is PIJ, or piezoelectric inkjet printing. In this case, ink droplets are pushed out of the nozzle as a result of the deformation of a piezoelectric crystal under the influence of an electric voltage. This technology ensures very high printing precision thanks to excellent control over the size and placement of the drops. Piezoelectric printheads are also characterized by long service life and wide compatibility with various types of inks, including pigment, sublimation, reactive, and acid inks. The only significant limitation of this technology is the higher cost of printheads and printing systems, resulting from their more advanced design.

According to market forecasts (Gminsights, 2023), PIJ technology will grow at a CAGR of 6.4% until 2032, confirming its growing importance in the textile industry. High image quality, print durability, and application flexibility make piezoelectric inkjet printing the foundation of digital textile printing today.

1.3.1 Roll-to-Roll

Roll-to-Roll technology, also known as Direct-to-Fabric (DTFab), is a direct digital inkjet printing method in which the pattern is applied directly to the fabric in roll form (Fig. 1.7). Unlike Direct-to-Garment (DTG) technology, where printing is done on finished garments, the Roll-to-Roll method allows fabric to be printed before further processing, such as cutting, sewing, or finishing.

The Roll-to-Roll printing process is carried out continuously. The fabric is unwound from one roll, passes through the printing zone, and is then wound onto a second roll. Printing takes place directly on the surface of the material, without the use of transfer paper or intermediate film.

In the case of very thin, delicate, or elastic fabrics in Roll-to-Roll (Direct-to-Fabric) technology, interleaving paper (protection paper or backing paper) is often used. Its use has several important functions: it stabilizes the material during printing, protects against ink seepage, facilitates fabric transport, and reduces static electricity. Thin fabrics can wrinkle, shift, or wave when passing through the printer. Interleaving paper provides mechanical support, stabilizing the print surface. Furthermore, in the case of highly absorbent or thin materials, ink can bleed through to the other

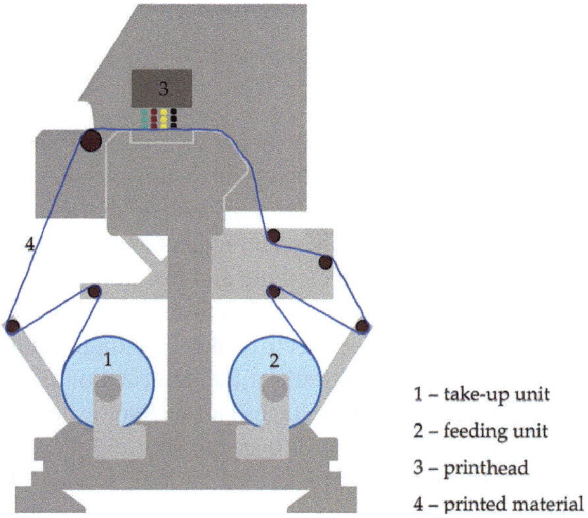

Fig. 1.7 Diagram of a Roll-to-Roll digital textile printer

side. Interleaving paper protects machine components from contamination. Some fabrics are too light to move properly through the roller system. Interleaving paper increases friction and stiffness, making them easier to feed. Synthetic fabrics can become electrically charged, which affects print accuracy. Interleaving paper can help dissipate static electricity.

This technology is used in the production of fashion, decorative, technical, and advertising textiles. The main advantages of Roll-to-Roll technology include: high efficiency and the ability to operate continuously, precise reproduction of patterns and colors depending on the technology and ink used, the ability to print on a wide range of fabrics, and compatibility with industrial production lines.

This technology is particularly valued in mass production and in environments where integration with further stages of textile processing is crucial. Thanks to the ability to print on wide fabric rolls (e.g., 160–320 cm), Roll-to-Roll is also used in the interior textiles sector, such as curtains, tablecloths, upholstery, and acoustic fabrics.

1.3.2 DTG (Direct-to-Garment)

Direct-to-Garment (DTG) technology is a digital inkjet printing method in which designs are applied directly to finished garments such as T-shirts, sweatshirts, bags, and polo shirts. Unlike Roll-to-Roll technology, where printing is done on fabric by the meter, DTG enables personalization and individual production directly on the sewn product.

The DTG printing process involves the precise application of ink directly onto the surface of the material (Fig. 1.8), in accordance with a digital graphic design. Pigment inks are used for printing, consisting of tiny pigment particles suspended in an aqueous solution with stabilizing and fixing additives. This technology works best with cotton fabrics and their blends due to their high absorbency and ability to bind dye to the fiber. Once printing is complete, the product is ready for use or further finishing, such as thermal fixing (Prybeha et al., 2021).

DTG is primarily used in the production of clothing with rich, multicolored, and detailed designs. When printing on dark fabrics, white ink is used as a base coat, forming a covering layer on which CMYK colors are printed. DTG technology enables both single orders and short runs to be fulfilled, making it an ideal solution for the personalized clothing industry, e-commerce, and on-demand production.

The main advantages of DTG include high print quality, durability, and no noticeable layer of ink on the surface of the material. Prints made using this method do not crack or peel off, as the ink penetrates directly into the fiber structure. However, the durability of the print depends on several factors, such as the quality of the fabric, proper surface preparation (e.g., pre-treatment), and how the garment is cared for. Pre-treatment improves ink adhesion, prevents ink from spreading, and increases color durability and intensity.

When printing on polyester fabrics, which naturally do not bind ink as effectively as cotton, it is necessary to apply a special primer beforehand. However, for

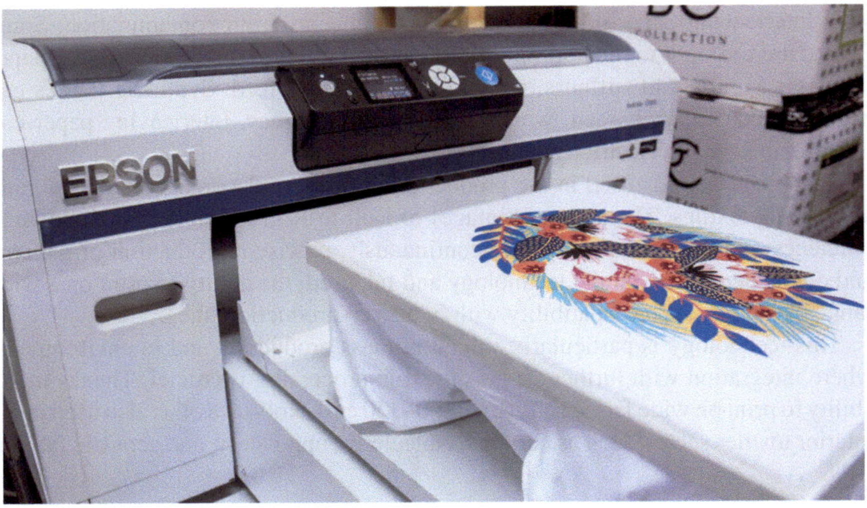

Fig. 1.8 DTG printing (*Source* https://printing-season.com/wp-content/uploads/2020/09/untitled-3-4.jpg)

best results on polyester clothing, it is recommended to use DTF (Direct-to-Film) technology, which offers better compatibility with synthetic materials.

1.4 Transfer Printing

1.4.1 Thermal Transfer

Thermal transfer printing, also known as hot transfer printing, is one of the classic methods of indirectly applying graphics to fabrics. Although this technology has been around for many years, it is still widely used in the production of promotional and personalized clothing, as well as in small and medium-volume projects. Its popularity stems from the simplicity of the process, low technological entry threshold, and the possibility of using it on a variety of textile substrates.

The thermal transfer process involves printing on a special medium—paper or transfer foil—using the appropriate type of ink or toner. Depending on the technology, these can be pigment, latex, solvent inks or laser toners. After printing, the image is transferred to the fabric using a heat press, which ensures the right temperature, pressure, and contact time. Under the influence of heat and pressure, the dye or adhesive layer is activated and bonds with the surface of the material (Rainert et al., 2024).

One of the main advantages of this technology is its material versatility. Thermal transfer printing can be used on cotton, polyester, or fiber blends. What's more, it is possible to print on both light and dark substrates by using appropriate layers of white undercoat.

A specific type of thermal transfer is flex or flock printing. It is intended for designs with relatively simple shapes and a limited number of colors. Special flex or flock films are cut into shapes using a cutting plotter and then heat-pressed into the material under pressure and at high temperature. Flex foil is a foil with a matte or glossy coating, available in solid colors or in a form imitating denim, chrome, or hologram. Flock foil, on the other hand, is a foil with a surface resembling suede or plush. It is thicker than flex foil. It is suitable for use, for example, on fleece sweatshirts, where other printing methods are difficult to use. Flex and flock thermal transfer films can be used as solid colors immediately after cutting (Fig. 1.9) or can be used for solvent, eco-solvent, and latex printing, and only after printing are they dedicated for application on T-shirts, fabrics, and other textile materials.

The transfer process is relatively quick and simple. It only requires a properly prepared print, a heat press, and knowledge of the transfer parameters—typically, the temperature ranges from 150 to 180 °C, the contact time is 10–20 s, and the pressure is medium to high. After the process is complete, the transfer medium is removed—depending on the type of material and film—either hot (hot peel) or cold (cold peel).

The advantages of thermal transfer technology include the ability to personalize individual items of clothing, relatively low equipment costs, wide availability of consumables, and the ability to obtain clear, durable prints. This technology

Fig. 1.9 Flex printed t-shirt

works particularly well in the production of printed T-shirts, promotional clothing, workwear, and in decorating textile accessories such as bags, caps, and aprons.

However, it is worth noting that the durability of thermal transfer prints may be lower than in the case of sublimation or DTF technology, especially with intensive use and frequent washing. This is because the print is deposited on the surface of the fabric rather than penetrating its structure, which can lead to cracking or abrasion over time. Nevertheless, with the right choice of materials and transfer parameters, it is possible to achieve prints with very good durability and aesthetics.

Thermal transfer technology remains an important part of today's textile printing market, offering flexibility, accessibility, and cost-effectiveness, especially in the context of low-volume and personalized production.

1.4.2 Sublimation

Sublimation transfer printing is one of the key technologies for digital decoration of synthetic fabrics, primarily polyester (Rainert et al., 2024). Its operation is based on the phenomenon of sublimation, i.e., the direct transition of a substance from a solid to a gaseous state, bypassing the liquid phase. In the context of printing, this means that the sublimation dye, under the influence of high temperature, turns into a gas and penetrates the structure of the fabric fibers, creating a durable, abrasion-resistant and imperceptible to the touch print.

The sublimation process consists of two stages. In the first stage, a print is made on special transfer paper using water-based sublimation inks. Then, in the second stage, the printed paper is applied to the fabric and subjected to high temperature (usually in the range of 180–200 °C) and medium pressure. The transfer is carried out using a flat press (Fig. 1.10) or a roller calender. As a result of the heat, the dye sublimates and permanently bonds with the polyester fibers. The entire process usually takes 40–60 s, and the transfer paper is removed while hot, in a procedure known as hot peel.

This technology requires the use of fabrics containing at least 80% polyester, although some say that 60% is sufficient. Natural fabrics such as cotton, linen, and viscose are not compatible with sublimation dyes because they do not have the appropriate chemical properties to allow the dye to bond permanently with the fiber. Furthermore, sublimation works best on light-colored fabrics because sublimation dyes are transparent and do not contain white pigment, which prevents their effective use on dark substrates.

The main advantages of this technology include exceptional print durability, which does not crack, wear off, or fade under UV radiation. Sublimation printing is characterized by high-quality reproduction of details and colors, as well as being completely imperceptible to the touch, which is particularly important in the case of functional clothing. This technology also allows printing on both roll fabrics (Fig. 1.11) and sheets (Fig. 1.12), making it extremely flexible and versatile for start-ups as well as small and larger companies.

Fig. 1.10 Sublimation printing—flatbed thermal press (*Source* https://arteam.com.pl/galeria)

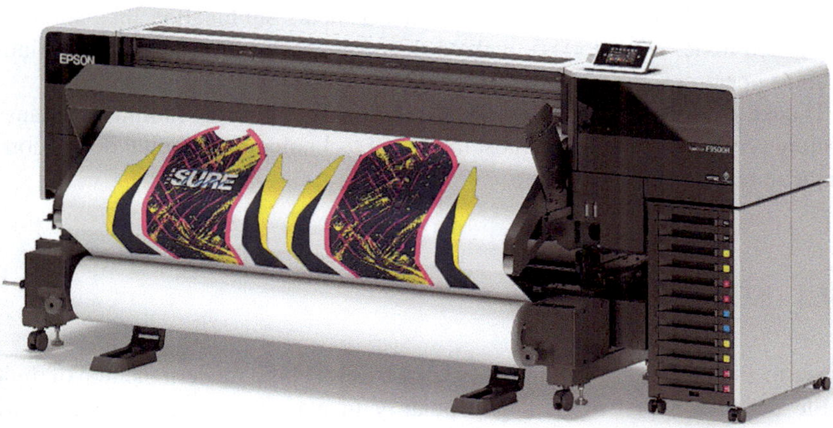

Fig. 1.11 Roll-to-Roll sublimation printer (*Source* https://www.epson.pl/pl_PL/produkty/drukarki/large-format/kompaktowa%2C-64-calowa-drukarka-sublimacyjna-surecolor-f9500h/p/40351)

Sublimation transfer printing applications cover a wide range of textiles, including sports and technical clothing, advertising textiles (such as flags, banners, and exhibition walls), as well as interior decorations, e.g., curtains, cushions, and tablecloths.

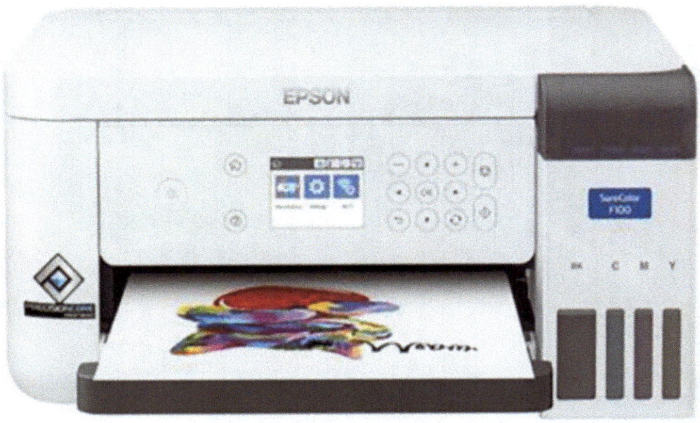

Fig. 1.12 Sublimation sheetfed printer (*Source* https://www.epson.pl/pl_PL/verticals/business-solutions-for-professional-graphics/textile/dye-sublimation-printers#1f093279-ac7d-4a48-94b0-75d5bf081f61)

Thanks to their resistance to weather conditions, sublimation prints are particularly valued in outdoor applications.

It is also worth noting the ecological nature of this technology. The sublimation process does not require the use of water or additional chemicals to fix the print, which significantly reduces its impact on the environment. Sublimation inks are water-based, non-toxic, and odorless, making them safe for both end users and printing machine operators.

Thanks to its quality, durability, and flexibility of application, sublimation transfer printing is currently one of the most important technologies in digital decoration of synthetic fabrics.

1.4.3 DTF (Direct-to-Film)

Direct-to-Film technology, commonly known as DTF, is a modern method of digital textile decoration that is gaining popularity thanks to its versatility, high print quality, and applicability to a wide range of materials. Unlike direct-to-fabric printing methods, DTF uses an intermediate medium—a special PET film—on which the print is made (Fig. 1.13a) and then transferred to the fabric (Fig. 1.13b) using a heat press.

The process begins with the preparation of graphics, which are printed in mirror image on smooth, heat-resistant PET film. First, CMYK colors are applied, followed by a final layer of white ink, which serves as a background and ensures adequate color saturation on dark fabrics. After printing, a special adhesive powder, usually based on thermoplastic polyurethane, is applied to the still wet print. This powder sticks to the moist ink, forming an adhesive layer which, when heated, allows the print to bond permanently with the fabric.

(a) (b)

Fig. 1.13 DTF printing **a** printed film, **b** print transferred from the film to a fabric bag (*Source* https://dtfwarszawa.pl/)

The next stage is to fix the powder. The film with the print and powder is placed in an oven or heating tunnel, where, under the influence of a temperature ranging from 120 to 160 °C for 2–3 min, the powder melts and forms a flexible, thermoactive layer. The transfer prepared in this way is ready for application to the fabric. The transfer process is carried out using a heat press, in which the printed film is applied to the material and subjected to a temperature of 150–160 °C for 10–15 s at medium or high pressure. After the transfer is complete, the film is left to cool completely and then peeled off cold, leaving the print permanently bonded to the fabric surface. To increase the durability and resistance of the print, it is recommended to perform additional ironing using silicone paper or a Teflon sheet.

Although the transfer process in DTF resembles that used in sublimation, it differs significantly in terms of technical parameters. First of all, the transfer takes place at a lower temperature and in a shorter time, and the film is removed only after it has cooled down completely, which distinguishes this technology from sublimation, where hot peeling is used.

One of the biggest advantages of DTF technology is its material versatility. Unlike sublimation, which is limited to polyester fabrics, DTF allows printing on a wide range of materials, including cotton, viscose, polyester, nylon, their blends, and even leather. What's more, this technology works well on both light and dark substrates thanks to the use of white ink as a base layer.

DTF offers exceptionally vivid and durable colors, high precision in reproducing details, and the ability to print complex, multi-color designs. Prints made using this method are characterized by greater durability and better color reproduction than those obtained with DTG (Direct-to-Garment) technology. Additionally, DTF allows for the use of special effects, such as glitter or glossy finishes, making it an attractive option for the production of fashion and promotional clothing. A comparison of DTG and DTF technologies is provided in Table 1.2.

DTF technology is used in the production of various textile products, such as T-shirts, sweatshirts, company shirts, sportswear, backpacks, aprons, flags, and uniforms. Thanks to its ability to handle both small and large print runs, DTF is suitable for both small studios and production plants.

In summary, DTF is a versatile and rapidly developing technology that combines high print quality with flexibility in material applications. Its growing popularity stems from its ability to print on almost any type of fabric, the high durability

Table 1.2 Comparison of DTG and DTF printing

Feature	Direct-to-Garment	Direct-to-Film
Printing process	Direct printing on fabric	Printing on PET film, then transfer to fabric
Type of printed fabrics	Mainly cotton and blends with a high cotton content	Wide range of fabrics: cotton, polyester, nylon, leather
Substrate preparation	Requires pre-treatment	Does not require pre-treatment
Print durability	Very good, especially on cotton	Very good, but may be less flexible
Softness of the print	Very soft, "blended" into the fabric	Sometimes more "rubbery" or noticeable, or even raised
Image rendering	Poorer image rendering, depending on graphics	Better image rendering
Applications	Premium clothing, personalization	Casual wear, gadgets, prints on various materials
Lead time	Longer (pre-treatment, drying)	Faster and more flexible
Cost	Higher for small print runs	Lower unit cost, especially for larger print runs
Ecology	Less plastic consumption, but more water consumption	Less water consumption, but use of PET film

of the prints, and a relatively simple production process that can be successfully implemented in both small and large textile factories.

1.5 Classification of Digital Printing by Type of Ink Used

Direct inkjet printing on fabrics can use pigment and dye inks: disperse, reactive and acid inks. Each type of ink is used for specific types of textiles and requires proper preparation of the material before the printing process and/or its finishing after printing (see Table 1.3). Inks for textile printing should have very good lightfastness, water resistance and good adhesion to the substrate, and should be non-toxic and thermally stable (Khadzhynova et al., 2017).

1.5.1 Pigment Printing

Pigment printing is a universal digital technology that allows patterns to be applied to almost any type of textile fiber. This is due to the peculiarities of pigment inks, which can be applied to any type of fiber if the right binder is used (Fashandi, 2017). In practice, it is most commonly used for printing on cotton, but it is also successfully used on viscose, linen, cellulose fabrics, as well as their blends with polyester (PES)

Table 1.3 Comparison of digital printing on fabrics by type of ink used

Type of ink	Type of fabric to be printed	Treatment before printing	Processes after printing	Print quality	Eco-friendliness
Based on pigment	Almost any kind of fabric	No	Short heat treatment	**	***
Based on reactive dyes	Cotton, viscose, wool and silk	Yes	Steam treatment and washing	***	*
Based on acid dyes	Silk, wool, polyamide	Yes	Steam treatment and washing	***	*
Disperse (for direct printing)	Polyester	Yes	Steam treatment and washing	***	**
Disperse (for sublimation printing)	Transfer paper/ polyester	No	High transfer temperature	**	**

Explanation of labels: ***—very good, **—good, *—average

or polyamide (PA). It is also possible to print on polyester fabrics, although this requires appropriate surface preparation (Rainert et al., 2024).

Unlike dye-based technologies, pigment printing does not involve chemical bonding of the dye to the fiber. Instead, pigments are deposited on the surface of the fabric, forming a layer of color that is fixed using special polymer binders. Once dry, these binders bond the pigment to the material, creating a durable print.

In order to achieve optimal print quality and durability, the fabric can be properly prepared in advance. This process, known as sizing, involves applying a special substance that increases pigment adhesion and improves color intensity. After printing, the pattern is fixed at a high temperature—usually in a calender or heating oven—which activates the thermosetting binder and permanently bonds the pigment to the surface of the material. Generally, fixing takes place at 150 °C for 4 min (Rainert et al., 2024).

The main advantages of pigment printing include:

- no need to use water and chemicals in the fixing process,
- no evaporation and washing stages, typical for reactive printing,
- compatibility with print-on-demand and short runs,
- low energy and consumables consumption.

Thanks to these features, pigment technology is considered one of the most environmentally friendly methods of digital textile printing, meeting the needs of sustainable production.

The disadvantage of this technology is slightly lower color intensity compared to reactive or sublimation printing. Pigments, when deposited on the surface of the fabric, can give a more matte effect, and it can be more difficult to achieve deep blacks and bright colors (Fig. 1.14). Therefore, pigment printing works best in designs with pastel colors, non-uniform patterns, or artistic designs.

Fig. 1.14 Comparison of pigment and reactive printing (*Source* https://drukcyfrowynadzianinie.pl/pl/blog/poradnik-drukowania-i-wyboru-dzianin/druk-reaktywny-a-pigmentowy-kluczowe-roznice-w-technikach-druku-na-tkaninach)

Compared to other methods, pigment printing does not require complex infrastructure or large amounts of auxiliary media. The prints are resistant to abrasion after just one curing stage, without the need for additional drying or rinsing. This makes the technology particularly attractive to manufacturers looking for fast, clean, and economical solutions.

Interest in pigment printing is growing rapidly. Printing equipment manufacturers such as Epson and ink suppliers such as SunChemical are intensively developing this segment, offering increasingly advanced technological solutions (SunChemical, 2025). Industry data (Textileprint, 2025) indicates that, compared to reactive printing, pigment printing allows for:

- 50% lower water consumption,
- 50% lower energy consumption,
- 60% lower wastewater production,
- 60% higher color lightfastness (according to an 8-point scale, PN-EN ISO 105-B02:2014-11 standard).

Other sources (Epson, 2025), report that compared to traditional analog printing, water consumption can be reduced by up to 97%, making pigment printing one of the most promising technologies of the future in the textile industry.

1.5.2 Reactive Printing

Reactive printing is a method of digital fabric decoration that uses reactive dyes, which chemically bond with the fibers of the material, penetrating deep into its structure (Javoršek & Javoršek, 2011). Thanks to this, prints made with this technology are exceptionally durable—they are resistant to washing, abrasion, and UV radiation. Reactive dyes have a particular ability to react with natural fibers such as cotton, silk, linen, and viscose (Pasternak, 2024), making this technology ideal for materials that are difficult to print using the pigment method.

One of the greatest advantages of reactive printing is the intensity and saturation of colors. Thanks to the deep bonding of the dye with the fiber, it is possible to obtain vivid, deep colors and excellent color fastness. This technology is primarily used in the production of high-quality clothing, home textiles (e.g., bedding, curtains), and functional textiles, where the aesthetics and durability of the print are of key importance.

However, the reactive printing process is more complex than other digital methods. After printing, the material must undergo a steaming stage, which activates a chemical reaction between the dye and the fiber. This is followed by a finishing wash, which removes excess dye and auxiliary substances. Although the print dries fairly easily, full fixation requires specialized infrastructure. It is worth noting that the dried material can be stored before steaming, which allows for some flexibility in production planning.

The limitations of reactive technology include:

- higher production costs—resulting from the complexity of the process and the need for additional stages,
- longer lead times—printing houses often accept orders for entire rolls of material, and lead times can be several weeks (Ctnbee, 2025),
- less production flexibility—compared to pigment printing, which allows for short runs,
- greater environmental impact—due to the use of water, chemicals, and the generation of wastewater.

Despite these limitations, reactive printing remains irreplaceable in the production of textiles with the highest quality requirements. Its development is currently focused on optimizing processes and reducing environmental impact, including through the use of more efficient evaporation and wastewater treatment systems.

1.5.3 Acidic Printing

Acid printing is a digital inkjet printing method designed for printing on fabrics such as wool, silk, and polyamide (PA). This technology uses acid dyes which, like reactive dyes, chemically bond with the fibers, penetrating deep into the structure of the material (Babar et al., 2023). As a result, the prints are highly durable, resistant to washing, and have intense colors.

The acid printing process is similar to reactive printing. After applying the pattern to the fabric using a digital printer, the material can be dried or immediately subjected to steaming, which activates a chemical reaction between the dye and the fiber. Steaming can be performed immediately after printing, but dried fabrics can also be stored and fixed at a later time, which provides flexibility in production planning.

After the steaming process is complete, a finishing wash is necessary to remove excess dye and auxiliary substances. Only after this stage does the fabric achieve full durability and is ready for further use or packaging.

Acid printing is mainly used in the production of high-quality textiles, such as:

- luxury and elegant clothing (e.g., silk scarves, dresses, ties),
- sportswear made of polyamide (e.g., swimsuits, leggings),
- decorative and functional textiles made of protein fibers.

The main advantages of this technology include:

- excellent color saturation and the ability to obtain deep, vivid colors,
- high durability of the print thanks to the chemical bonding of the dye with the fiber,
- the ability to print on materials that are difficult to print on using other methods.

A limitation of acid printing is the need to use specialized finishing processes (steaming and washing), which increases the complexity and cost of production. In addition, due to the use of chemicals, this technology is less environmentally friendly than, for example, pigment or sublimation printing.

1.5.4 Direct Disperse Printing

Disperse printing is a digital inkjet printing technology designed for printing on synthetic fibers, in particular polyester (PES), but also nylon and acetate. This method uses disperse dyes, which are poorly soluble in water and occur in the form of small particles suspended in an aqueous carrier.

Dispersion printing can be carried out in two ways: direct or transfer (sublimation). In direct printing, the pattern is applied directly to the fabric and then heat-set, usually using steam or hot air. This process requires precise drying and specialized finishing washing to remove excess dye and auxiliary substances. In practice, however, the most commonly used form of dispersion printing is sublimation printing (see Sect. 1.5.5), in which the dispersion dye, under the influence of high temperature, turns into a gaseous state and penetrates the structure of the polyester fiber, creating a durable and resistant print.

Direct dispersion printing, despite its potential, has certain limitations. Prints made using this method may have lower color intensity and be prone to smudging. In response to these challenges, research is being conducted on the pre-treatment of fabrics to improve dye adhesion and print quality. Methods such as plasma treatment and the application of substances such as chitosan or acrylic resins are used to reduce the capillary effect and improve color intensity (Gao et al., 2020).Prints made with disperse dyes, after proper fixation, are characterized by:

- high resistance to washing, abrasion, and light,
- good mechanical resistance,
- intense, saturated colors.

This technology is used in the production of:

- carpet fabrics,
- flag and advertising materials,
- sails and technical textiles,
- sports and outdoor clothing.

1.5.5 Sublimation Printing

Sublimation printing is one of the most important technologies for digitally decorating synthetic fabrics, especially polyester and its blends, in which the polyester fiber content usually exceeds 60–80% (Rainert et al., 2024). This technology is not suitable for printing on natural fibers such as cotton, silk, or viscose, as sublimation dyes do not bind with their chemical structure.

The basis for this technology is the phenomenon of sublimation, i.e., the transition of a substance from a solid state directly into a gaseous state. In the context of printing, this means that the sublimation dye, under the influence of high temperature (usually 180–200 °C), turns into a gas and binds with polyester fibers, penetrating their structure. This makes the print an integral part of the material, ensuring its exceptional durability and resistance to abrasion, washing, and UV radiation.

Sublimation printing can be carried out in two ways: direct and indirect. In direct printing, the design is applied directly to the polyester fabric using a digital inkjet printer with sublimation inks. The material is then subjected to a heat treatment process, which fixes the print. In indirect (transfer) printing, the design is first printed on special transfer paper and then transferred to the fabric using a calender or flat press, which provide the appropriate temperature and pressure. This process can be carried out dry or with steam, depending on the type of equipment and application. Prints made with intermediate technology are characterized by lower print saturation, as some of the ink remains after printing on the pre-print paper (Khadzhynova et al., 2017).

Sublimation technology is widely used in many sectors of the textile industry. It is particularly valued in the production of:

- sports and technical clothing,
- advertising textiles (flags, banners, walls),
- interior decorations (curtains, cushions, tablecloths),
- utility textiles where high print durability is required.

The advantages of this technology include:

- very high quality of color and detail reproduction,
- imperceptibility of the print—no layer of ink on the surface,
- resistance to weather conditions and UV radiation,
- environmental friendliness—no need to use water or chemicals in the fixing process.

Thanks to its efficiency, quality, and durability, sublimation printing is currently one of the most popular technologies in digital printing on synthetic fabrics.

1.6 Advantages and Disadvantages of Digital Printing

Digital fabric printing, as one of the modern methods of applying patterns to textiles, has a number of significant advantages that make it an attractive solution for both industrial production and individual projects. First and foremost, it enables the production of short runs and any quantity—even single items of clothing or fabric fragments as short as one metre. Such flexibility opens the way to extensive product personalisation, which is particularly valued in the fashion and interior design industries.

One of the greatest advantages of this technology is its high print quality, which allows for excellent visual results. Digital printing is ideal for reproducing photographs, detailed patterns, subtle tonal transitions and complex graphic designs. Importantly, this process does not generate costs associated with the preparation of printing forms, as is the case with traditional techniques such as screen printing. The cost of printing remains constant regardless of the number of colours used in the design or the size of the production run—the price of printing the first metre is essentially the same as the hundredth, which is a significant advantage over analogue methods.

An additional advantage is the possibility of making a test print, which allows the final effect to be assessed before starting actual production. This technology does not impose any restrictions on the type of design—it is possible to reproduce any graphics, regardless of the number of colours or the complexity of tonal transitions. By using the basic CMYK palette and any additional colours, printing houses are able to reproduce almost any shade from a typical colour palette (Kašiković et al., 2016).

Furthermore, digital printing is in line with the trend towards more sustainable and environmentally friendly production, reducing the consumption of water, energy, and chemicals. Water and energy consumption can be reduced by 95% and 30%, respectively, compared to traditional fabric printing methods (screen printing) (Koseoglu et al., 2019).

Despite its numerous advantages, this technology is not without its limitations. One of the main drawbacks is the limited possibility of using special colors, such as metallic, fluorescent, or shades from the PANTONE palette. In addition, digital printing is most often done on white material, which poses certain difficulties for projects requiring the reproduction of light elements on a dark background. For example, to achieve the effect of white dots on a black background, it is necessary to print the entire surface in black and leave unprinted areas that will act as "white dots" (CTNBEE, 2025).

1.7 Materials, Process Efficiency and Costs

In digital printing on textiles, both fabrics and inks are of key importance, as they determine the quality of the print and the compatibility between the substrate and the dye. The range of materials used is very wide (cotton, viscose, silk, polyester, polyester blends, etc.)—both natural and synthetic fibers are printable, with weights ranging from 45 to 300 g/m^2. The most commonly printed material is cotton (48% of production), followed by cotton-polyester blends (19%), polyester (15%), and viscose (13%). Other materials, such as polyamide, polyacrylic, wool, and silk, account for a small percentage of global production (Promet, 2025).

There are many types of inks available on the market, which must be chemically and physically adapted to high printing speeds and resolutions. It is important that they do not damage piezoelectric heads or cause nozzle clogging.

Digital printing is particularly effective for personalized clothing and short production runs. It reduces material consumption, making it both an economical and environmentally friendly solution. However, in order for it to compete with traditional methods in terms of price, it is necessary to further reduce operating costs and increase printing speed (Itoh, 2016).

Although the differences in performance between digital printing and rotary screen printing are gradually decreasing, increased efficiency remains key to the further development of the digital printing market. Rotary machines can achieve productivity of over 1000 m^2/h, while commercial inkjet printers currently achieve between 2 and 150 m^2/h (Wang, 2017). Solutions such as reducing ink replacement time or using large cartridges that can be replaced during operation minimize downtime and increase productivity.

One of the main advantages of digital printing is that there is no need to retool the machine when changing designs and no costs associated with preparing printing forms. In traditional textile printing, the costs of preparing the machine and consuming materials (fabrics, inks) before the actual production begins are significant, which makes this method less cost-effective for small and medium runs. Digital printing eliminates this problem—each print is part of the final production.

For short runs, conventional textile printing is uneconomical due to the high cost of screen production. In addition, the inability to produce a proof means that any errors detected after production has started generate large losses—it is necessary to cancel the order, order new screens, and retool the machine.

In digital printing, errors can be detected and corrected almost immediately—all you need to do is correct the file and resume production.

The cost of digital inks is higher than that of screen printing inks (Wang, 2017). Printheads are also expensive and sensitive to improper use. Therefore, it is crucial to properly train operators and ensure the right environmental conditions (temperature, humidity, air cleanliness).

In summary, digital printing technology compared to traditional methods (e.g., screen printing) is: faster (3–14 days vs. 45–60 days), more environmentally friendly (lower water consumption and no ink waste), ideal for low-volume and diverse production (Kawamura et al., 2022).

1.8 Innovation and Environmental Aspect

Advanced digital printing technologies on fabrics enable the creation of high-quality and durable designs. In addition to innovations aimed at increasing the efficiency of printing processes, an important direction of development is activities aimed at reducing the negative impact on the environment and increasing the versatility of the technology.

Pro-environmental activities are supported, among other things, by the development of pigment printing, which significantly reduces water and energy consumption, eliminates the need for chemical preparation of fabrics before printing, and enables the use of concentrated inks and more environmentally friendly inks. The versatility of the technology can be achieved by developing new inks compatible with a wide range of materials.

Synthetic dyes currently dominate the market. However, in recent years there has been growing interest in natural dyes, known as "green dyes". Their advantage is that they are produced without the use of strong acids, bases, solvents, and heavy metals, and with reduced energy consumption. The limitations of natural dyes are their cost and limited durability. If they could be produced at a comparable price and had similar durability properties to synthetic dyes, they could become an alternative to synthetic dyes. This would reduce the use of toxic substances, as the raw materials are environmentally friendly, which has advantages in terms of waste disposal and occupational safety. Furthermore, production could be decentralized, which would reduce transportation problems and costs, and the waste after dye extraction in the form of biomass and nutrient medium could be appropriately used as biomass as an energy source and recyclable waste. In summary, the renewed interest in natural dyes in recent years stems from growing consumer awareness and concern for the environment and the need to introduce non-toxic, non-allergenic, environmentally friendly textiles to the market (Savvidis et al., 2012, 2014).

Further development of digital printing may also be stimulated by innovations in the clothing industry, such as smart clothing and computer simulations of prints. With the development of smart clothing, the importance of digital printing—which enables on-demand production—may increase (Papachristou & Bilalis, 2015). Due to the high costs of inks and pre- and post-treatment, the proper preparation of designs and print simulations becomes even more important (Javoršek & Javoršek, 2011).

The textile industry is one of the most polluting sectors in the world, both in terms of water and chemical consumption and greenhouse gas emissions (Hassan et al., 2024). Textile wastewater can contain dyes, detergents, pesticides, lubricants, solvents, heavy metals, inorganic salts, and fibers—its composition depends on the process used (Šostar-Turk et al., 2005).

It is estimated that the textile industry accounts for 10% of global greenhouse gas emissions. Fabric dyeing and finishing processes generate over 3% of global CO_2 emissions and over 20% of water pollution. If current practices remain unchanged, by 2050 this sector could account for 25% of the global carbon footprint (IMI, 2024).

In response to these challenges, the fashion industry has taken international action to protect the climate. The Fashion Industry Charter for Climate Action (FICCA), developed by the UN and signed by major fashion companies, aims to reduce greenhouse gas emissions by 30% by 2030 and achieve net zero emissions by 2050 (Hirsh, 2020). The document also emphasizes the importance of using energy from renewable sources and developing low-carbon transport.

The use of digital printing instead of traditional technologies also brings environmental benefits. A comparison of printing 1,000 m of fabric using a Monna Lisa digital machine and a rotary press showed a 27% reduction in water consumption (which translates into less wastewater and lower energy consumption for heating process water) and a 40% reduction in carbon footprint (85.66 kg CO_2eq for the Monna Lisa system vs. 139.56 kg CO_2eq for the rotary press) (API, 2025).

1.9 Color Reproduction and Print Quality

The quality of digital printing on fabrics and the colors obtained depend primarily on three main factors: the printing technology used, the type of ink, and the properties of the substrate (see Fig. 1.15) (Gazibarić et al., 2024). The quality of the image itself is also an important factor—the key parameters are resolution and sharpness. The quality of the print is also influenced by the structure of the fabric and the direction of printing (Gao et al., 2020).

The selection of the appropriate textile printing technique is important for the appearance, durability, and tactile properties of the material. The choice of technology should take into account the type of fabric and the size of the print run. The optimal choice of printing method (see Sect. 1.2) should meet the customer's expectations. In the case of pigment printing, the color durability and functional properties of the fabric depend on the stability of the binder layer (Sahin & Tufan, 2018).

In digital textile printing, inkjet technology is most commonly used, where print quality depends on parameters such as:

- drop shape and volume (2–32 pl)—smaller drops ensure higher precision,
- drop movement speed—faster movement reduces the risk of errors,
- parameter stability (volume, speed)—crucial for process control,
- straightness of the stream—affects the accuracy of ink application (Khadzynova & Jakucewicz, 2016).

Fig. 1.15 Comparison of print quality on different substrates (*Source* Own photo of the pattern template printed by Printella)

The type of ink also has a significant impact on print quality. The depth of black and color intensity vary depending on the type of dye—in pigment printing, they are lower than in acid, reactive, or sublimation printing (see Sects. 1.5.1–1.5.4).

The physical properties of inks, such as surface tension and viscosity, affect the repeatability and reliability of the process. The viscosity of dispersion inks is typically in the range of 1–25 mPa·s. The surface tension should be lower than the surface energy of the polyester fiber (43 mN/m) to ensure adequate wetting and high print quality. The optimal surface tension range for disperse inks is 21–48 mN/m. Increasing the viscosity and reducing the mobility of water in the ink improves print sharpness—a higher content of bound water (e.g., through the addition of PVP, DEG, EG) reduces ink spreading (Gao et al., 2020).

Another important aspect is the set of colors used in printing. The basic CMYK process colors (cyan, magenta, yellow, black) allow for the reproduction of approximately 400,000 colors. However, to reproduce corporate colors (e.g., logos), it is often necessary to use additional colors. Pantone offers over 1,000 such shades, of which only about 60% can be accurately reproduced using CMYK. Therefore, more and more digital printers are equipped with an extended color gamut—e.g., orange, green, purple—which allows for accurate reproduction of about 90% of Pantone colors. Additional inks in light cyan and light magenta shades improve tonal gradation, reduce graininess, and expand the color space (FESPA, 2025a, b; Hajipour & Shams-Nateri, 2019).

Due to the roughness of fabrics, the quality of inkjet printing on textiles is lower than on paper—the image may be less uniform (Hajipour & Shams-Nateri, 2019). The main problems are ink spreading and the influence of the weave structure. On paper, the drops retain their shape, while on fabric they penetrate the fiber structure, leading to the phenomenon of TVI (Tone Value Increase) (Mendizza & Urbas, 2024).

The fabric structure (weight, capillarity) and print direction have a significant impact on quality. Fabrics with higher weight and lower capillarity provide better sharpness. Printing in the weft direction gives better results than in the warp direction (Gao et al., 2020). Tests on different weaves (plain, twill, satin) showed high accuracy in reproducing details <1 mm. Better color saturation was achieved on pure fiber fabrics than on blends (Min & Hassan, 2024). The choice of material and number of ink layers is crucial for the quality and durability of the print—a greater number of layers improves color intensity (K/S), light and weather resistance, although it does not significantly affect press resistance (Kašiković et al., 2012).

The quality of the print is also affected by the preparation of the fabric before printing. Research on commercial pretreatment agents has shown that they improve the color quality of pigment printing on cotton and polyester, but reduce mechanical durability (abrasion resistance). The preparations increase color intensity and gamut range—the greatest on cotton, less on PET. The increase in gamut results from an increase in brightness (L*) and saturation (a*, b*). Unfortunately, pigments settle more superficially, which increases intensity but reduces durability (Ding et al., 2018).

Modification of the fabric surface by plasma or chemical methods can improve ink wettability and adhesion, which translates into better image reproduction and print durability (Vodyashkin et al., 2025). However, as shown by Gao et al. (2020), high-quality inkjet printing on polyester can be achieved without the need for material preparation, which further reduces resource consumption and pollutant emissions.

Print quality also depends on the image transfer method and ink fixation method, which is particularly important in DTF and screen printing thermal transfer technologies. It is worth noting that in these techniques, no speckle effect was observed, which can occur in direct printing and negatively affect color reproduction quality (Gazibarić et al., 2024).

Heat treatment causes structural changes in the material, which affects color perception (Kašiković et al., 2011). The properties of textile materials also change as a result of the printing process—the WVR (Wash Visual Rating) values of the tested knitted and woven fabrics confirm this relationship. Changes in process parameters, such as tonal value (TV) or the number of print head passes, combined with the appropriate choice of material, can be used to achieve the desired WVR values (Stančič et al., 2017).

Images intended for printing must have a sufficiently high resolution to allow them to be scaled without losing quality. According to the principle used in printing, the image resolution in pixels per inch (ppi) should be at least twice as high as the screen ruling. In the past, 150 lpi was the standard, but higher values are now commonly used. In practice, printing houses often require graphic files with a resolution of 300 ppi. For logos and other vector graphics that can be scaled without loss, it is essential to prepare them in vector format—preferably as spline curves.

1.10 Application of Digital Printing of Textile and the Market

Digital printing on textiles is used in many sectors, the most important of which are: fashion and clothing, home textiles, advertising and signage, and industrial textiles. The fashion and clothing industry has the largest market share, particularly the fast fashion segment, where rapidly changing trends require flexibility and short lead times. Reactive and pigment printing technologies dominate this area due to their compatibility with natural fibers and high print durability.

The second important area of application is home textiles, such as curtains, bedding, pillows, and upholstery fabrics. In this case, pigment printing is most commonly used, valued for its precision, versatility, and lower environmental impact. The third segment is advertising and signage, including banners, flags, and promotional products. Here, sublimation printing dominates, which ensures excellent color durability and is dedicated to synthetic fabrics. Industrial textiles, such as automotive, medical, and protective clothing materials, are also becoming increasingly important, where pigment and sublimation printing are mainly used due to their durability and versatility (Fespa, 2025a, b; Promet, 2025).

In recent years, there has been a growing role for fabrics and textile printing in the development strategies of companies involved in product personalization, advertising, and interior decoration. The global textile market reached a value of 49.7 billion square meters in 2022, and forecasts indicate growth to 53.4 billion m^2 by 2028 (CAGR 0.3%). During the same period, the digital textile printing market will grow from 3.3 to 5.5 billion m^2 (CAGR 8.3%) (Smithers, 2023). The value of the digital textile printing market was $2.86 billion in 2024 and is expected to reach $5.49 billion by 2033 (CAGR 7.13%) (Imarc, 2025).

Segments such as apparel, interior decoration, signage, and technical textiles remain key to the growth of inkjet printing. The outlook is very positive, with changes in supply chains, consumer habits, and growing environmental awareness driving the growth of this technology. As the digital printing market grows, so does the demand for printers and inks. In 2023, the value of the digital textile printer market was $2.58 billion, and by 2032 it is expected to grow to $4.19 billion. The market for digital printing inks will grow from $3.43 billion in 2023 to $5.54 billion in 2030 (CAGR 7.1%) (Fortune, 2025). In turn, the market for digital printing inks for textiles reached a value of USD 1.64 billion in 2023 and is expected to grow to USD 6.4 billion by 2032 (CAGR 16.37%) (Global Growth Insights, 2025).

The main factors driving market growth include advances in ink formulations, technological innovations, growing environmental awareness, and environmental regulations. In addition, growing demand for fast fashion, home textiles, and personalized clothing products is fueling further growth. In developing countries (e.g., China, India, Vietnam), growth is driven by a rising middle class and fashion awareness. In highly developed countries (Europe, North America), digital printing meets the needs of the premium segment (Global Growth Insights, 2025; Research & Markets, 2022).

The automation of fabric printing processes enables increased speed, precision, and production efficiency, which translates into shorter lead times, lower material consumption, and reduced waste—in line with the principles of the circular economy.

Current trends in the market for digital printing inks for textiles include: growing demand for eco-friendly and biodegradable water-based inks, the increasing importance of personalization, the development of sublimation and UV-curable inks, and integration with e-commerce and digital marketing. Brands are increasingly using digital printing to produce short runs that respond to the rapidly changing needs of the market.

Europe remains the market leader, with a share of over 37% in 2022 and 34.7% in 2024 (Imarc, 2025), mainly due to demand in fashion centers such as Italy and France. Italy has a particularly strong position thanks to the presence of numerous fashion brands in Milan, Rome, Venice, and Palermo. Asia–Pacific is the fastest-growing region (CAGR 15.1%), with China and India dominating and Bangladesh and Vietnam growing in importance (Research & Markets, 2022).

In 2021, the apparel segment accounted for more than half of the market and is expected to maintain its dominant position. Sublimation printing dominated the market in 2021 and is expected to reach over 53% share by 2030. It is expected to be the fastest growing type of ink. Cotton was the most commonly used material (over

52% share in 2021), but forecasts indicate that polyester will be the fastest growing segment (CAGR 12.9%).

However, it is worth noting that the introduction of new US tariffs on textile imports in 2025 may disrupt global supply chains and clothing prices (ITMF, 2025) and affect earlier forecasts for the digital printing market.

1.11 Conclusion

The prospects for the development of digital printing on textiles are very promising and result from both technological progress and changing market expectations. The dynamic development of this technology observed in recent years is leading to increasing automation, faster printing speeds, and greater precision, which translates into growing competitiveness compared to traditional methods. In the context of shorter production cycles, the need to quickly introduce new products to the market, and the growing importance of personalization, digital printing is becoming an increasingly attractive alternative. In addition, its potential in implementing the principles of the circular economy may contribute to its further growth in importance in the textile industry (Digital Textile Manufacturing Technologies Conference, 2024).

New directions for development include the integration of digital art with textile printing, which allows art to be disseminated in everyday life and increases the value of clothing as a form of artistic expression (Kawamura et al., 2022). At the same time, smart fabrics are being developed—lightweight, multifunctional and integrated with electronics—used, among other things, for temperature regulation, protection, communication, and health monitoring (Papachristou & Bilalis, 2015). Digital printing, which enables on-demand production, can play a key role in the development of this segment.

An important area for further research remains the development of pigment technology, including the modification of print nozzles, the improvement of pigment chemistry, and the development of more environmentally friendly inks (Koseoglu et al., 2019). At the same time, innovative hybrid solutions are being developed—printers that combine flat screen printing with digital printing—which increase the flexibility and production potential of companies (HPRT, 2025).

One of the technological challenges remains the development of a universal ink that would be compatible with a wide range of fabrics. Currently, most textile inks work on several types of materials, but the best results are achieved when the ink is matched to a specific substrate. Market pressure to simplify processes and increase production flexibility makes the development of such a solution increasingly desirable (FESPA, 2025a, b).

Declaration The authors declare that they have no known competing financial interests or personal relationships that could have appeared to influence the work reported in this chapter. Generative AI and AI-assisted technologies have been used in the writing process to improve the readability and language of the work only. After using this tool, the authors carefully reviewed and revised the content to ensure accuracy and quality. The authors take full responsibility for the final version of this publication.

References

API. (2025). Digital printing on textiles is environmentally friendly. Retrieved July 2, 2025, from https://druknabawelnie.com.pl/atramenty-apretury/.

Babar, A. A., Gianchandani, P. K., & Jhatial, A. K. (2023). Overview of different digital textile printing machines. In: Elsevier eBooks. https://doi.org/10.1016/b978-0-443-15414-0.00014-5.

Cahill, V. (2006). The evolution and progression of digital printing of textiles. In H. Ujiie (Ed.), *Digital printing of textiles* (pp. 1–15). Woodhead Publishing.

Carr, W. W., Sarma, D. S., Cook, L., Shi, S., Wang, L., & Pfromm, D. S. (2000). Xerographic printing of textiles: Polymeric toners and their performance. *Journal of Applied Polymer Science, 78*, 2425–2434.

Chakraborty, S., & Biswas, M. (2020). 3D printing technology of polymer-fiber composites in textile and fashion industry: A potential roadmap of concept to consumer. *Composite Structures, 248*, Article 112562. https://doi.org/10.1016/j.compstruct.2020.112562.

CTNBEE. (2025). Printing on materials–which printing technology to choose? Retrieved March 10, 2025, from https://ctnbee.com/blog/pl/druk-na-materialach-jaki-wybrac/.

Ding, Y., Shamey, R., Parillo Chapman, L., & Freeman, H. S. (2018). Pretreatment effects on pigment-based textile inkjet printing–colour gamut and crockfastness properties. *Coloration Technology, 135*, 77–86.

Epson. (2025). Epson at FESPA 2025: Personalization and more sustainable textile printing. Retrieved May 15, 2025, from https://oohmagazine.pl/2025/05/13/epson-na-fespa-2025-personalizacja/.

Fashandi, S. (2017). Digital printing of acrylic fabric with cationic dyes using conventional inkjet printer. *Color Research and Application, 42*, 244–249. https://doi.org/10.1002/col.22066.

FESPA. (2025). Colour management in digital textile printing. Retrieved June 6, 2025, from https://www.fespa.com/en/news-media/colour-management-in-digital-textile-printing.

FESPA. (2025). What is digital printing on textiles and what possibilities it offers. Retrieved May 5, 2025, from https://www.fespa.com/pl/media-informacyjne/czym-jest-druk-cyfrowy-na-tekstyliach-i-jakie-daje-mozliwosci.

Fortune. (2025). Digital printing ink market size, share & COVID-19 impact analysis. Retrieved June 30, 2025, from https://www.fortunebusinessinsights.com/digital-printing-ink-market-108924.

Gao, C., Xing, T., Hou, X., & Chen, G. (2020). The influence of ink viscosity, water and fabric construction on the quality of ink-jet printed polyester. *Coloration Technology, 136*, 45–59. https://doi.org/10.1111/cote.12439.

Gazibarić, Z., Stančić, M., & Ružičić, B. (2024). Evaluation of print mottle in textile printing: Impact of printing methods on macro non-uniformities. In: *GRID Symposium*. https://doi.org/10.24867/GRID-2024-p9.

Global Growth Insights. (2025). Digital textile printing ink market size. Retrieved June 20, 2025, from https://www.globalgrowthinsights.com/market-reports/digital-textile-printing-ink-market-100892.

Gminsights. (2023). Digital textile printer market analysis. Retrieved July 2, 2025, from https://www.gminsights.com/industry-analysis/digital-textile-printer-market.

Hajipour, A., & Shams-Nateri, A. (2019). Improve neural network-based color matching of inkjet textile printing by classification with competitive neural network. *Color Research and Application, 44*, 65–72.

Hassan, R., Acerbi, F., Rosa, P., & Terzi, S. (2024). The role of digital technologies in the circular transition of the textile sector. *Journal of the Textile Institute*. https://doi.org/10.1080/00405000.2024.2414162.

Hirsh, S. (2020). 100+ Fashion brands have signed the UN fashion industry charter for climate action. Retrieved November 25, 2025, from https://www.greenmatters.com/p/un-fashion-industry-charter-climate-action.

HPRT. (2025). Digital fabric printing technology development and prospects. Retrieved July 2, 2025, from https://pl.hprt.com/blog/2024-Digital-Fabric-Printing-Technology-Development-and-Prospects.html.

Imarc. (2025). Digital textile printing market size, share, trends and forecast 2025–2033. Retrieved July 2, 2025, from https://www.imarcgroup.com/digital-textile-printing-market.

IMI. (2024). Digital textile manufacturing technologies conference. Retrieved July 2, 2025, from https://www.imiconf.com/digitaltextile-2024.

ITMF. (2025). ITMF on US tariffs. Retrieved May 20, 2025, from https://www.itmf.org/images/dl/press-releases/2025/2025-04-04-ITMF-on-US-Tariffs.pdf.

Itoh, T. (2016). New trends of inkjet printing technology for digital textile. *J Imaging Soc Jpn, 55*, 718–722.

Javoršek, D., & Javoršek, A. (2011). Colour management in digital textile printing. *Coloration Technology, 127*, 235–239. https://doi.org/10.1111/j.1478-4408.2011.00304.x.

Kašiković, N., Novaković, D., Vladić, G., & Klančnik, M. (2011). Influence of heat treatment on characteristics of inkjet prints on textile material. *J Graph Eng des, 2*(1), 24–30.

Kašiković, N., Novaković, D., Karlović, I., & Vladić, G. (2012). Influence of ink layers on the quality of ink jet printed textile materials. *Tekstil Ve Konfeksiyon, 2*, 115–124.

Kašiković, N., Vladić, G., & Novaković, D. (2016). *Textile printing–past, present, future*. Glas Hem Tehnol Ekol RS 35.

Kawamura, H., Nemoto, Y., Maruyama, S., Kozai, A., Tosa, N., & Nakatsu, R. (2022). Development of art fashion by integrating digital art and digital textile printing. In: *Nicograph International Conference* (pp. 25–28). IEEE. https://doi.org/10.1109/NICOINT55861.2022.00012.

Khadzhynova, S., Jakucewicz, S., & Piłczyńska, K. (2017). *Drukowanie natryskowe (ink-jet)*. Łódź: Lodz University of Technology Monographs. (Inkjet Printing)

Khadzynova, S., & Jakucewicz, S. (2016). *Sposoby drukowania cyfrowego*. Łódź: Lodz University of Technology. (Methods of Digital Printing)

Koseoglu, A. U., Gungor, Y., Arık, Y. D. (2019). Innovations and analysis of textile digital printing technology. *International Journal of Science, Technology and Society, 7*(2), 38–43. https://doi.org/10.11648/j.ijsts.20190702.12

Marra, F., Minutillo, S., Tamburrano, A., & Sarto, M. S. (2021). Production and characterization of graphene nanoplatelet-based ink for smart textile strain sensors via screen printing technique. *Materials and Design, 198*, Article 109306. https://doi.org/10.1016/j.matdes.2020.109306.

Mendizza, A., & Urbas, R. (2024). Evaluation of textile prints printed with inkjet printer. In: *GRID Conference* (pp. 107–112). https://doi.org/10.24867/GRID-2024-p11.

Min, Y., & Hassan, H. (2024). Innovative application of digital printing technology in "China Chic" clothing. *International Journal of Engineering Science, 13*(12), 81–89.

Papachristou, E., & Bilalis, N. (2015). How to integrate recent development in technology with digital prototype textile and apparel applications. *Marmara Journal of Pure and Applied Sciences* (Special Issue-1), 32–39. https://doi.org/10.7240/MUFBED.96632.

Pasternak, J. (2024). Reactive versus pigment printing–key differences in textile printing techniques. Retrieved September 25, 2025, from https://drukcyfrowynadzianinie.pl/pl/blog/poradnik-drukowania-i-wyboru-dzianin/druk-reaktywny-a-pigmentowy-kluczowe-roznice-w-technikach-druku-na-tkaninach.

Promet. (2025). Nowa fala technologii cyfrowego drukowania tkanin. Retrieved July 2, 2025, from https://promet-alu.pl/nowa-fala-technologii-cyfrowego-drukowania-tkanin/.

Prybeha, D., Koshevko, Y., Smutko, S., Onofriichu, V., Skyba, M., Synyuk, O., & Zlotenko, B. (2021). Analysis of methods of printing images on textile materials and evaluation of their quality. *Vlakna a Textil 28*(2), 63–74. Retrieved September 2, 2025, from https://elar.khmnu.edu.ua/handle/123456789/11046.

Rainert, K. T., de Oliveira, C. R. S., Felipe, B. H. S., Herpich, H., de Cássia Siqueira Curto, V. R., & Valle, J. A. B. (2024). Textile printing: An integrated view of processes, properties, and future prospects. *Coloration Technology, 140*(4), 556–570. https://doi.org/10.1111/cote.12746.

Research and Markets. (2022). Digital textile printing market size, share & trends analysis report 2022–2030. Retrieved July 2, 2025, from https://www.researchandmarkets.com/reports/5644968/digital-textile-printing-market-size-share-and.

Sahin, U. K., & Tufan, H. A. (2018). Effects of binder solution on color fastness of digital printed cotton fabric. In: *18th World Textile Conference (AUTEX 2018). IOP Conference Series: Materials Science and Engineering* (Vol. 460). https://doi.org/10.1088/1757-899X/460/1/012038.

Savvidis, G., Zarkogianni, M., Karanikas, E., Lazaridis, N., Nikolaidis, N., & Tsatsaroni, E. (2012). Digital and conventional printing and dyeing with the natural dye annatto: Optimisation and standardisation processes to meet future demands. *Coloration Technology, 129*, 55–63.

Savvidis, G., Karanikas, E., Nikolaidis, N., Eleftheriadis, I., & Tsatsaroni, E. (2014). Ink-jet printing of cotton with natural dyes. *Coloration Technology, 130*, 200–204.

Smithers. (2023). The future of digital textile printing to 2028. Retrieved July 2, 2025, from https://www.smithers.com/services/market-reports/printing/the-future-of-digital-textile-printing-to-2028.

Šostar-Turk, S., Simonič, M., & Petrinić, I. (2005). Wastewater treatment after reactive printing. *Dyes Pigments, 64*(2), 147–152. https://doi.org/10.1016/j.dyepig.2004.04.001.

Stančič, M., Kašiković, N., Grujić, D., Novaković, D., Milošević, R., Ružičić, B., & Geršak, J. (2017). Mathematical models for water vapour resistance prediction of printed garments. *Coloration Technology, 134*, 82–88.

Stawski, R., Kozłowski, Ł., & Kuchta, B. (2023). Digital revolution in textile printing. *Fashion in Poland, 2*, 32–34.

SunChemical. (2025). Modern solutions, sustainability and market education. Retrieved May 20, 2025, from https://swiatdruku.eu/Czytelnia/Z-ostatniej-chwili/Mix-News/Sun-Chemical-na-targach-Fespa-20253.

Textileprint. (2025). Pigment printing. Retrieved July 10, 2025, from https://textileprint.pl/druk-pigmentowy/.

Tippett, B. G. (2002). The evolution and progression of digital textile printing. Retrieved July 2, 2025, from https://brookstippett.com/docs/Print2002-BGT.pdf.

Tyler, D. J. (2005). Textile digital printing technologies. *Textile Progress, 37*(4), 1–65. https://doi.org/10.1533/tepr.2005.0004.

Vodyashkin, A. A., Makeev, M. O., & Mikhalev, P. A. (2025). Inkjet printing is a promising method of dyeing polymer textile materials. *Polymers, 17*, 756. https://doi.org/10.3390/polym17060756.

Wang, M. (2017). *Digital inkjet textile printing*. Bachelor's thesis, Savonia University of Applied Sciences.

Xiao, Y. Q., & Kan, C. W. (2022). Review on development and application of 3D-printing technology in textile and fashion design. *Coatings, 12*, 267. https://doi.org/10.3390/coatings12020267.

Chapter 2
Sustainability Aspects in Digital Textile Printing

Aadhar A. Mandot and Sanjay Bambhaniya

Abstract If we are concerned about nature, environment, and climate change, then we are heading towards the dream of a Sustainable World. In order to achieve this dream, sustainable development is required, which simply means "Fulfilment of present needs without compromising the ability to fulfil future demands." This chapter deals with the various ways of achieving environmental sustainability through the 3R concept (i.e., Reduce, Reuse and Recycle) of textile waste, specifically in the Garment & Fashion Industry as well as the products which are not used up to their average limit. The 3R concept in textiles can serve as a means of providing solutions to many economic, environmental and social issues. The tempting arrival of the latest trend in the market makes it difficult for one to be satisfied with what they already have, which leads to over-purchase of new products and less usage of existing ones, which is not used even up to the average limit. Though textile recycling has a past history, in recent years it has gained prime importance due to the fast-changing fashion culture in the Western world which has resulted in overconsumption of textiles with less usage and corresponding waste generation. Innovations are made in terms of methods of applying this concept in every textile good possible, development of recycling machinery, design inputs, and innovative products with value addition to make recycling a profitable business. The various efforts made by industries and small recycling units aim to achieve a higher recycling rate in the textile industry to achieve sustainability. There are a number of textile industries that offer recycling of post-consumer waste. The brands offer innovative and sustainable textile solutions from 100% recycled materials. The ultimate goal is to keep textiles away from landfills by repurposing them into new apparel, closing the loop in the fashion industry. This chapter will highlight prevailing recycling practices, which are mechanical as well as chemical, and provide their efficacy in recycling textile waste into reusable products. Also, the idea of upcycling will be included as a novel measure for recycling waste material into high-value-added products, which will decrease virgin resource requirements and reduce waste generation. The chapter will also highlight the need for cooperation between stakeholders, such as designers,

A. A. Mandot · S. Bambhaniya (✉)
Department of Textile Engineering, The Maharaja Sayajirao University of Baroda, Vadodara, India
e-mail: bambhaniya.b-ted@msubaroda.ac.in

A. Athalye and H. Patil (eds.), *Digital Printing*,
https://doi.org/10.1007/978-981-95-2593-5_2

manufacturers, retailers, and consumers, to bring about sustainable change. Also, the latest development in fashion sustainability, for example, the emergence of circular fashion platforms and sustainable business models, will offer useful insights on the future of green practices in the fashion sector.

Keywords Sustainability · Fashion · Digital printing · Garments · Recycle

2.1 Introduction

Sustainability has evolved from a specialist issue to an international imperative that influences every facet of modern life. From how we generate energy and food to how we construct buildings and make products, efforts toward more sustainable methods indicate increased awareness of our planet's limited resources and the long-term implications of human activities. Among the many industries affected, fashion is particularly noteworthy due to its worldwide reach, cultural impact, and significant environmental impact. The fashion world is one based on creativity, expression and the pursuit of innovation. However, behind the glamour of the runways and the endless cycle of trends is a complex network of production, supply chains, and consumption patterns that have important consequences on the environment and society. The increasing trend of 'Sustainability in Fashion' attempts to address these problems by reshaping the designing, manufacturing, distributing and discarding the clothes.

One of the most widely used definitions, provided by the Brundtland Commission (1987), states sustainability as "The development that meets the needs of the present without compromising the ability of future generations to meet their own needs." This definition emphasises intergenerational equity and the balance between economic growth, environmental management, and social inclusion (Martinell, 2024).

John Elkington, a pioneer in corporate responsibility and sustainable development, introduced the concept of the Triple Bottom Line: People, Planet, and Profit as a roadmap for sustainability. John states that "True sustainability occurs when an organisation or society considers the environmental, social, and economic impacts of its actions and decisions." According to this view, industries are urged to extend beyond profitability to include their ecological and social responsibilities (Russo, 2008).

Herman Daly, a prominent ecological economist, emphasises the significance of preserving natural capital, stating that "Sustainability means that human activity has to be carried out in a way that enables ecosystems to sustain their functioning, biodiversity, and regenerative capacity." Daly's perspective highlights the Earth's biophysical constraints and the necessity for economic processes to function within these limits (Daly, 2007).

An environmentalist and author, Paul Hawken presents a systematic perspective and defines sustainability as "Stabilising the currently disruptive relationship between Earth's two most complicated systems, i.e., human culture and the living world."

This integrated view emphasises the interdependence of cultural, economic, and ecological systems (Birnbaum & Fox, 2014).

Although these definitions differ in their vision but they have in common: a long-term perspective, ecological balance, social responsibility, and breaking with unsustainable modes of production and consumption.

Sustainable development is not merely a theory; it is a necessity. As the population of the world continues to grow and consumption levels increase, pressure on natural resources is also increasing. Climate change, loss of forests, loss of biodiversity, pollution, and inequality are only a few of the challenges that sustainable development aims to tackle. Sustainable development encourages the responsible use of nature, with an assurance that the ecosystems are preserved and functioning properly. It helps to reduce harm to the environment through the support of renewable energy, waste elimination, water management, and clean production technologies (Leddin, 2024). In fashion, this means that the utilisation of harmful dyes is minimised, textile waste is reduced, and the carbon footprint of the supply chain is minimised. Sustainability also includes social aspects such as ethical labour practices, equitable resource access, and empowering communities. The clothing industry, which is mainly dependent on global supply chains, has to think about the welfare of garment workers, who work at low wages under deplorable conditions. Sustainable development promotes ethical labour practices and community involvement (Leal Filho et al., 2024).

Though sustainable development is often regarded as limiting or expensive efforts, it has the potential to stimulate long-term economic resilience. Companies adopting sustainable strategies can cut on operational costs, improve brand reputation, and gain conscious customers. New opportunities for economic growth, based on innovation in materials, circular business models, and technology, are accessible and in line with a sustainable approach. Sustainable development is in line with international contexts like the United Nations Sustainable Development Goals (SDGs), which present a vision for peace, prosperity, and protection of the environment by 2030 (Opoku, 2016). Several of these goals, such as responsible consumption and production, gender equality, and climate action, are immediately applicable to the fashion industry. As consumers get more educated, they are more and more insistent on transparency and sustainability in the products they purchase. This change in consumer culture is not only influencing the market but also holding brands responsible for their social and environmental footprint. In essence, sustainable development is essential to the preservation of a habitable world, an equitable society, and a prosperous economy. It is a blueprint for advancement that happens without sacrificing individuals or the planet.

As per reports from organisations such as the Ellen MacArthur Foundation and the United Nations Environment Programme (UNEP), the fashion sector generates about 10% of global carbon emissions and is the second-largest water consumer globally. The manufacturing of textiles, especially cotton and synthetic fabrics like polyester, consumes vast amounts of water, energy, and chemicals (Leal Filho et al., 2024; Dolzhenko & Churakova, 2022). For example, the textile dyeing and finishing industry is a significant cause of water pollution, with millions of tons of untreated or inadequately treated wastewater released into oceans and rivers. Microplastics from man-made fibres also present an increasing threat to marine life since they find their way into water bodies. The emergence of fast fashion has led to quick production, cheap products, and fleeting trend cycles, resulting in a staggering amount of textile waste. Clothing is used only a few times before being discarded in much of the globe. Globally, an estimated 92 million tons of textile waste is produced every year, much of which will end up as landfill or be burned (Chen et al., 2021). The use of non-renewable resources is another issue affecting the industry. Fashion supply chains are notoriously multifaceted, with many suppliers, manufacturers, and distributors located all over the world. This multifaceted nature usually allows transparency to be hidden and traceability to be hampered for the environmental and social impact of a product. Brands are being pushed increasingly to map the supply chain and reveal sourcing and labour practices.

Despite these challenges, the fashion sector also has great potential for positive transformation. Innovations in sustainable materials (e.g., organic cotton, bamboo, and lab-grown leather), circular business models (such as clothing rental, resale, and recycling), and digital technology (for transparency and supply chain traceability) are transforming the sector. Those brands that embrace sustainability can minimise their environmental impact, attract eco-aware consumers, and create more sustainable business models. Sustainability is not a passing trend, it is a call to action that encompasses all facets of the way we live, work, and shop. For the fashion industry, where culture, commerce, and ecology intersect, going green is not only an ethical imperative but a business imperative. Knowing what sustainability is, why sustainable development is important, and the specific environmental issues related to fashion, we set ourselves up for impactful change. Going forward in this chapter, we will see how the fashion sector can shift from being part of the problem to being a compelling force for sustainable innovation and transformation.

In this context, digitally printed textiles have rapidly gained traction in the global fashion market due to their versatility, design flexibility, and potential to reduce waste. According to recent market research, the global digitally printed textile market was valued at approximately USD 5.80 billion in 2024 and is projected to grow at a CAGR of over 12.7% during 2025–2030 (Grand View Research, 2024). Trends in digitally printed textiles highlight the rise of customised prints and rapid design-to-market cycles. Digital printing enables brands to produce limited-edition collections and react quickly to market trends, reducing waste by avoiding large minimum order quantities typical in traditional printing methods. This aligns with the broader goals of sustainability in the fashion sector by minimising overproduction, enhancing resource efficiency, and meeting consumer demand for personalized, eco-friendly designs.

2.2 The Concept of 3R

The fashion industry is one of the most significant contributors to global pollution, resource depletion, and environmental degradation. With the rise of fast fashion, the industry is experiencing increasing challenges due to excessive generation of waste, water and energy usage, and greenhouse gas emissions. As mentioned earlier, an estimated 92 million tonnes of textiles waste is set to rise to 134 million tonnes by 2030 if trends continue. This increase owes much to the rapid fashion industry, which churns out more than 100 billion clothes per annum, with much of it finding its way into waste sites shortly after being bought. The ecological consequences go beyond waste generation as the fashion industry contributes to almost 10% of the world's carbon footprint. Additionally, textile manufacturing depends significantly on water-based processes as it uses about 93 billion cubic meters of water every year and harmful chemicals, further adding to environmental issues (Chen et al., 2021; Igini, 2023; Business Waste, 2025).

In the context of Digital Textile Printing (DTP), however, the 3R principle: Reduce, Reuse, and Recycle, emerges as a powerful solution. Unlike conventional printing processes (e.g., rotary screen printing), DTP offers on-demand, precise printing with minimal wastage of materials, water, and chemicals. This technology enables the customisation and small-batch production of textiles, which aligns seamlessly with circular economy principles. DTP significantly reduces the need for large inventories and overproduction, supporting sustainable business models. Applying the 3R concept to the digital printing segment of the fashion industry can therefore substantially mitigate environmental degradation, cut waste, and ensure resource efficiency.

The application of 3R practices in the fashion sector is influenced by numerous inherent causes and challenges. Table 2.1 below highlights these vital factors that can lead to the application of 3R principles in the fashion sector. The following section will discuss practical methods of implementing the Reduce, Reuse, and Recycle approach in the fashion industry, ensuring that the sustainable model is adopted at each phase of the production and consumption cycle.

2.2.1 *Reduce: Minimising Waste Generation*

The rising level of population and increasing consumer demand have led the textile and fashion sector under tremendous pressure to minimise the quantity of waste it produces and decrease its carbon footprint. Waste reduction is a progressive approach that targets waste at the source rather than coping with it once it has been produced. It is worth noting that Americans alone generate around 11 billion pounds of apparel waste every year, and only 15% of that is recycled. This urges the urgent need for sustainable options that promote waste minimisation strategies (Mandot et al., 2023).

Table 2.1 Factors of 3R (Reduce, Reuse, and Recycle) Application in the Fashion Industry

Sr. No.	Reduce factors	Reuse factors	Recycle factors
1	Excessive fabric waste in cutting	Fast fashion trends leading to short product lifespans	Limited infrastructure for textile recycling
2	Overproduction due to inaccurate demand prediction	Lack of consumer awareness of reuse options	High cost of recycling processes
3	Inefficient pattern-making techniques	Poor quality materials reducing the reuse potential	Limited market for recycled textile products
4	Excessive use of water and chemicals in textile processing	Limited second-hand markets in some regions	Difficulty in separating blended fibres for recycling
5	Lack of sustainable design practices	Consumer reluctance to buy pre-owned clothing	Low investment in textile recycling technologies
6	Inefficient inventory management leading to deadstock	Lack of incentives for clothing repair and upcycling	Inconsistent textile recycling policies
7	Poor material selection leading to high waste generation	Changing fashion trends discouraging prolonged use	Difficulty in collecting post-consumer textile waste
8	Lack of awareness and training on waste minimisation	Absence of take-back programs by brands	Limited sorting facilities for textile recycling
9	High energy consumption in manufacturing	Inadequate infrastructure for clothing donation	Contamination issues in recycled textiles
10	Lack of policy enforcement for waste reduction	Limited corporate initiatives for reuse strategies	Weak consumer participation in textile recycling programs

Digital Textile Printing plays a critical role in the reduce paradigm by eliminating the need for screens, reducing sample requirements, and significantly lowering water and energy consumption compared to traditional printing methods. Unlike screen printing, DTP enables precise color application and design placement, reducing errors and overruns that often contribute to pre-consumer waste. Moreover, DTP supports "just-in-time" production, helping brands produce textiles closer to actual demand and avoid overstocking, which otherwise leads to deadstock waste (Ayyoob & Khan, 2023).

Organizing clothing swaps among communities and friends is an effective approach to extending the life of clothes. Swapping clothes at gatherings allows individuals to refresh their wardrobes without purchasing new clothing, thereby avoiding landfill waste. DIY fashion concepts have also gained popularity, shaping creative ways of upcycling second-hand clothing. There are numerous websites with several concepts regarding how to transform old clothes into stylish items of upcycled clothing. Selling second-hand clothing online or through consignment stores provides a platform where clothes end up with new consumers rather than in the trash. Reselling promotes second-hand economies in addition to creating revenue streams for vendors. Apart from that, donating used clothing to charities and non-profit organisations benefits the most people who need it and keeps textiles out of landfills. Most donation programs are also tax-deductible, and it's a socially and economically responsible choice. Old clothes can also be recycled as useful items within the home. Used t-shirts, for example, can be turned into rags for use when cleaning, and outdated skirts can be transformed into aprons or other functional accessories (Mandot et al., 2023).

In addition to waste minimisation, sustainable fashion also involves the reduction in the utilisation of non-renewable raw materials. Product, process, and business model changes can substitute synthetic and non-biodegradable materials with biodegradable, renewable, and recyclable materials. This reduces the utilisation of natural resources and facilitates circularity in the production of fashion. One of the most inspiring examples of the reduce paradigm applied is Haathi Chaap, a firm that produces paper from the raw material of elephant dung. They recycle animal waste, which not only reduces their environmental footprint but also generates green employment opportunities for rural society. Its distinctive business strategy consists of raw material collection, processing into paper form, and distribution of green products through networks of online and retail outlets. The firm's low-cost manufacturing process significantly saves on resource usage, which serves as an exemplar of how sustainable innovation is effective. In addition, wastewater from manufacturing is utilised as organic manure, thus enhancing sustainability in agriculture. Such processes demonstrate the potential of circular economy practices in the minimisation of waste and effective usage of natural resources (Goyal et al., 2018).

Reducing the generation of waste in the fashion sector involves a transition towards sustainable consumption patterns and environmentally friendly production methods. Adhering to the reduce principle, all producers, vendors, and buyers can work towards a greener future as one. Adopting waste-reducing processes as part of regular fashion choices has the ability to make a notable contribution to the conservation of the environment, in line with general sustainability.

2.2.2 *Reuse: Extending the Product Life Cycle*

'Everything Has a Second Life Except Your Old Clothes'

The average American disposed off around 82 pounds of textiles yearly, with people underestimating the ecological effects. What seems to be a harmless process of getting rid of old clothes eventually ends up piling in the landfills, where they break down and release toxic greenhouse gases into the environment that lead to climate change. Therefore, it is necessary to explore methods of prolonging the life of apparel (Mandot et al., 2023).

Digital textile printing enhances reuse by enabling customisation and personalisation, making garments more valuable to consumers and prolonging their use. For instance, DTP technologies allow brands and individuals to refresh, overprint, or embellish existing garments with new designs, logos, or artwork. This not only gives a second life to clothing but also supports circular business models (Ayyoob & Khan, 2023).

Thrift shopping is key to establishing a circular textile economy. Organisations such as Planet Aid gather and resell donated apparel overseas to finance sustainable development projects globally. Consumers can save money by buying second-hand apparel while minimising waste. Numerous charitable organisations receive, sort, and redistribute quality used clothing to those in need. But with such a large amount of donated clothing, much of it is sent overseas to developing nations, where it contributes to local economies and offers inexpensive apparel choices. Recycling old clothes into new items can go a long way in minimising textile waste. Domestic use items like handbags, slippers at home, curtains, pillow covers, and floor mats can be produced using discarded cloth. Programs such as 'Eco India' emphasise innovative methods of re-designing heirloom sarees into trendy clothing, making fashion sustainable (Mandot et al., 2023).

The reuse model is focused on maximising the functional life of products and steering discarded materials toward productive ends. This encourages a parallel economy where waste is remade into usable resources, prevents landfills from filling up, and creates new streams of income. An ideal case study for the paradigm of reuse in operation is 'Goonj', a social enterprise in India that converts waste city clothes into resource material for deprived communities. When Goonj saw that cloth was an elemental need among the rural people, it formulated progressive programs for harnessing excess resources towards development and earning projects. Used clothing is traded for community development activities, including constructing schools, roads, and sanitation facilities in rural areas. Waste fabric is converted into low-cost sanitary napkins, fulfilling an essential hygiene requirement of rural women and supporting decent living standards. Also, rural women are trained to make mattresses out of garbage clothing, averting landfill build-up while encouraging local enterprise (Goyal et al., 2018).

Adopting second-hand fashion and upcycling materials is crucial in reducing the environmental burden of clothing waste. Donating, reselling, or upcycling out-of-date garments can help promote a circular fashion economy with a focus on resource preservation and sustainability. Initiatives such as those by Goonj and Planet Aid prove that unwanted textiles can bring social good with a reduction in landfill waste. The solution to sustainable fashion is to change our mindset, each piece of clothing has a second life yet to be fulfilled. With mindful consumption and environmentally friendly disposal, the fashion world can head toward a greener and more sustainable future.

2.2.3 Recycling: Converting Waste into Usable Products

Recycling is crucial for minimising textile waste and promoting sustainability. Most recycled textiles are organised and reused in several countries, mainly in Eastern Europe, the Middle East, and Africa. Textile recycling involves gathering used garments and other fabrics for material repurposing or recovery. This process significantly decreases landfill waste and conserves essential resources (Mandot et al., 2023). Material for recycling originates from two major sources:

(1) Post-consumer waste: The waste includes worn-out clothes, home textiles, and automotive upholstery material.
(2) Pre-consumer waste: This involves leftover material from yarn and fabric manufacturing and post-industrial textile waste from other industries.

Unwearable fabrics, frequently labelled as being damaged, get converted into rags and delivered to industries to be used in wiping and flocking. Reclamation of other fibres is taken through and recast into fresh clothing, upholstery, or insulation. Unwinding fibres for re-weaving, reducing textiles into yarns (shoddy), mixing with different fibres, and then spinning them before weaving or knitting are operations done in recycling them. Fibres from plants as well as synthetic ones get recycled using the above process (Mandot et al., 2023).

The recycling paradigm closes waste loops by converting abandoned materials into productive resources. This paradigm achieves a balance between consumption and production, prioritising recovering and reusing solid waste to produce new products and achieve maximum economic value (Goyal et al., 2018). Recycling can be divided into two broad types:

(1) Closed-loop recycling: Waste materials are recycled to produce new products without changing their intrinsic properties,
(2) Open-loop recycling: Materials that are recovered are transformed into lower-value products, prolonging their use before they are disposed of.

Traditional recycling methods usually involve incineration, disassembly, and shredding in uncontrolled environments, typically conducted by unskilled workers. Recycling based on the circular economy involves an end-to-end system ensuring efficient collection, sorting, and segregation. The model follows international quality standards and employs trained staff with advanced technology to ensure maximum material recovery. One of the most prominent examples of an integrated recycling network is Attero, India's largest electronic asset management firm. Founded in 2007, Attero initiated an environmentally friendly recycling centre based on international standards. Though primarily e-waste-oriented initially, its sustainability strategy can be applied to the textile industry as well (Goyal et al., 2018). The five essential parameters of success for the company were:

(1) Personalised recycling solutions that include the adoption of standardised procedures for the secure disposal of assets.

(2) National network of collections that reverse logistics and IT-supported tracking to manage waste.
(3) Sustainable recovery of resources, which offers new approaches to recovering valuable materials.
(4) Reuse and refurbishment, which lengthen material life by reconditioning it.
(5) Integration with e-commerce, which allows direct online sales of refurbished goods.

This model demonstrates the possibility of an organised and sustainable textile recycling business. Through the incorporation of circular economy thinking, textile recycling can become a more effective, environmentally friendly process with reduced waste and increased recovery of resources.

2.2.4 The 3R Approach in Fashion: Economic, Environmental, and Social Impact

The 3R (Reduce, Reuse, Recycle) strategy is the crucial solution to drive the sustainability of fashion and textile firms. This yields economic, environmental, and social benefits by minimising waste, extending product life, and encouraging mind consumption. Economically, companies can save on the usage of new materials and the amount of waste created. Second-hand, rental, and resale markets bring in entirely fresh income, and circular economy processes generate textile recycling and repair jobs. It also boosts the brand image and consumer trust in sustainability and creates an avenue for profitability in the long run (Romero-Hernandez & Romero, 2018). Environmentally, the 3R does that by minimising greenhouse gas emissions by consuming fewer resources. The lesser the emissions associated with virgin fibre extraction, for example, cotton and synthetics, the less pollution is created by keeping textile waste from landfills. The use of biodegradable materials and environmentally viable dyes acts as a further deterrent to adverse environmental impacts. Socially, awareness on the part of the consumer is raised, thereby incentivising responsible fashion purchases through the concept of 3R. Sustainable fashion brands engage in ethical labour practices, promote fair trade, and offer decent working conditions. The fall of fast fashion does not provide the impetus for exploitative work but instead augments small businesses through local recycling endeavours, making them sustain cheap clothes for impoverished groups (Manickam & Duraisamy, 2019).

The implementation of the 3R strategy in fashion and textiles on a universal basis poses far-reaching effects for the industry. This would allow for sustainability to capitalise through the establishment of novel mechanisms like clothing rental, refurbishment programs, and resale platforms. This will promote a more sustainable trend in fashion among the public: lengthen the lifecycle of the clothing, decelerate excessive consumption, etc. The 3R values are used for implementing environmental sustainability, and they will help reduce the ecological footprint of the fashion industry, in turn (Manickam & Duraisamy, 2019).

Every piece of clothing has the potential for a second life. Globally, 70% of second-hand clothing is recycled as second-hand clothing, 6% are made into dust bags and zips, 8% are recovered for fibre recycling to produce new goods, 7% are used as wiping material, and 9% become stuffing materials via a process called shredding. Interestingly, over 70% of the world relies on second-hand clothing, hence the significance of recycling for waste reduction. Recycled products will also be cost-effective and environmentally friendly for manufacturers, hence making the green approach to production more viable (Mandot et al., 2023). The highest reuse value will be obtained through the separate collection of clothing, distinguishing between those containing natural fibres, synthetic fibres, and mixes thereof. Sustainability in the fashion industry entails waste reduction, recycling, and reuse. Since it comes from nature, the ultimate end of textiles would be to return to nature without creating damage to it, supporting the principles of circular and sustainable fashion. 3R strategies allow the fashion industry to advance towards a sustainable and circular economy, ensuring economic sustainability, conservation of the environment, and social fairness in generations to come.

2.3 Overconsumption and Waste Generation in the Fashion Industry

The fashion industry has designed and built the platform for rapid change, based on standardised production and overconsumption. This growth has made fashion accessible to everyone, yet it has also generated a series of environmental and social issues. 'Overconsumption and Waste Generation' is also rapidly becoming mainstream in modern discourses. This section describes the problems of misuse and wasted resources, and promotes sustainable behaviour for mitigating their ill effects.

2.3.1 Fast Fashion and Its Impact

Fast fashion is a business model characterised by rapid production cycles, low-cost manufacturing, and the constant renewal of fashionable clothing. Zara, H&M, Shein, and others are all renowned brands that have frequently employed this model to offer trendy products at affordable prices. Their popularity has led to enormous environmental damage, workers in developing nations working under unsafe conditions, and so much waste. The production of fast fashion products consumes massive amounts of natural resources such as water, energy, and raw materials (Degli Esposti et al., 2024).

Textile production consumes the most water of any industry globally. To produce just one cotton shirt, thousands of litres of water are required (Devi & Devi, 2024). Polyester in synthetic clothes contributes to clumping in seas and rivers. In the processes of dyeing and finishing, waste disposal results in water pollution. It causes the destruction of sea life and mankind (Muthu, 2020). Low-cost production of clothes happens quite easily at the expense of the rights of labour. Some fast fashion companies were accused of establishing factories in developing nations and thus, pursuing a higher profit margin. Many production hubs have seen such exploitative activities as child labour and unsafe working environments that indicate the ethical dilemma involved in this sector. Fast fashion makes clothes disposable. Much of the abandoned clothing goes into landfills or is burned in incinerators, contributing to greenhouse gas emissions and soil contamination. Because fast fashion items have such a limited lifespan, much clothing is discarded after only a few wears (Kalk & Tulder, 2004).

2.3.2 Consumer Behaviour and Over-Purchasing

Consumer behaviour plays a pivotal role in the overconsumption crisis. The promotional or marketing tactic of constantly buying, combined with the allure of fast fashion, has created a never-ending cycle of purchasing apparel and discarding it. This is because of several factors. Psychological and social factors, fashion trends shift quickly, influenced by social media, celebrity endorsements, and influencer marketing. Being aware of the current fashion trends is necessary, as it provides the impression of not being inexpensive. The "buy now, pay later" culture makes individuals shop excessively since they do not experience the immediate effects of their excessive spending. Excessive availability of online shopping e-stores has revolutionised the manner in which consumers shop (Lia & Natswa, 2021).

Purchasing clothes is now just a click away. Online shopping has become so convenient that individuals prefer to buy clothes through online websites. Due to all the discounts and offers, individuals usually end up buying more than they require. In an act of shopping in excess, most consumers tend to buy things in bulk, only to return them after using them very sparingly. This adds to soiling the air since the entire process of bringing back is emissions-intensive because of transport. Individuals overlook the environmental and societal effects of their dress code. Consumers are unaware of the dangers of the dress they are wearing. Fashion brands are ambiguous regarding their fashion cycle and environmental sustainability practices, further contributing to increased possibilities of making the wrong decision. Awareness is increasing via educational campaigns and slow fashion, however, there is a long way to go (Gabler & Reynolds, 2013; Lia & Natswa, 2021).

In contrast to impulsive buying associated with fast fashion, digitally printed textiles are gaining traction with consumers looking for personalisation or originality. Digital printing has enabled a type of short run production with customisation opportunities to attract young environmental/digital consumers looking for expressiveness through made-to-order items. Numerous consumers desire the stunning visuals and originality of digitally printed clothes as they tend to choose items based on individual identity rather than mass-market or trend indicators. Despite a demonstrated interest in digitally printed clothing, challenges remain. A large segment of consumers still prefers cost and speed over sustainability. The average consumer remains unaware of some of the environmental benefits of digital textile printing (e.g., less water consumption, less waste). Also, consumers' willingness to purchase sustainable, digitally printed fashion remains capped due to fiscal constraints and a lack of transparency from brand designs. Therefore, there is a large gap between consumer interest and responsible purchasing that needs to be closed, and it starts with education, clearer product labels and accountability from brands.

2.3.3 Underutilization of Garments and Textile Goods

One of the places where waste is incurred is from the underutilization of clothing. Most clothing is discarded too soon, thus unnecessarily wasting resources. With the shorter lifespan of clothing, it is being discarded more. Studies estimate that a typical consumer will use an item seven to ten times before discarding it. Profit comes ahead of durability for most fashion companies in order to produce low-quality clothing that wears off fast and thus promotes replacement. With decreasing prices of clothes, people are stockpiling them. Many closets hold numerous items that people never or hardly use at all. This is wasteful and does not allow people to utilise the use of clothing to its full extent. Recycling schemes exist, but due to the fibre blends, dyes, and treatments applied to make the fabric these days, recycling is problematic. Only a small percentage of apparel is recycled back into fabric. Most will be "downcycled" into other, lesser-value items, such as insulation or cleaning (Klepp et al., 2020).

Fast fashion, shifting consumer behaviour and under-use of apparel are at the centre of the overconsumption issue in the fashion industry. Although fashion is an influential tool of expression and culture, presently it is moving in the wrong direction, which is unsustainable. Fast fashion provides unhealthy production cycles and social injustice. As the industry develops, the demand for sustainability and ethics is even more essential. To decelerate the fashion industry is to shift towards slow fashion, that is, quality, durable, and ethically produced clothes. Raising awareness among people about conscious consumerism can go a long way in reducing the industry's carbon footprint. They need to make informed decisions on buying sustainably produced clothing and activities such as up-cycling, repairing, and re-wearing rather than throwing them away after wearing them once. Fashion companies are also expected to incorporate sustainability in their business, from the use of green materials alongside labour processes to simple supply chains. The implementation of efficient recycling systems is also essential in lessening textile waste (Krishnakumar & Venkatesh, 2020).

Investment in new recycling technologies, such as fabric-to-fabric recycling and biodegradable fabrics, will reduce the industry's role in landfills. Sustainable production and proper waste disposal regulations should assist businessmen and policymakers. In addition, companies can be motivated to support circular economy practices where clothes are designed for recyclability, reuse, and durability. In order to build a more responsible fashion system, we require the active involvement of people, companies, and politicians (Krishnakumar & Venkatesh, 2020). If we all do everything together to encourage sustainable fashion habits, cut back on overconsumption, and drive better innovation in greener production models, we could possibly build an industry where fashion no longer comes at the expense of the planet, nor people, in the future. Everyone must make a conscious effort if we are to shift to sustainability, but together, we can revolutionise fashion for the better so that we leave behind a world where fashion is sustainable.

2.4 Textile Recycling: Past, Present, and Future

The textile industry, despite its historically resource-conscious practices, continues to generate substantial waste each year. Waste generation in this sector is closely linked to production levels; higher output typically results in increased waste. A significant portion of this waste originates from clothing manufacturing, much of which ends up in landfills. Among the various operations within the textile sector, chemical processing is particularly harmful, contributing nearly 70% of the industry's total pollution. Additionally, the escalating costs related to energy consumption, raw material usage, and waste disposal are intensifying the financial burden on textile businesses worldwide. Notably, the industry is responsible for roughly 10% of global carbon emissions, making it the fifth-largest source of such emissions (Athalye, 2012; Muthu, 2015).

In addition, the rapid turnover of fashion trends and the widespread failure to recycle textiles have led to considerable economic and environmental consequences. Although numerous global initiatives have been proposed and executed by experts, their success remains uncertain without the support of enforceable regulations. For example, the Textile Exchange has introduced several initiatives aimed at lowering greenhouse gas (GHG) emissions associated with fibre and raw material production. One of their key targets is to achieve a 45% reduction in GHG emissions by the year 2030 (Textile Exchange, 2025). While the objectives outlined in future-focused programs are well-defined, reaching these targets demands coordinated industrial efforts, the adoption of advanced technologies, and the implementation of wide-ranging strategies. As per these goals, greenhouse gas (GHG) emissions are expected to decrease from 336 billion kilograms in 2019 to 203 billion kilograms by 2030. However, instead of declining, emissions rose to 340 billion kilograms by 2021, emphasising the urgent need for more effective action (Textile Exchange, 2025). This data indicates that, in the initial two years of the program, the trend in greenhouse gas (GHG) emissions moved in the opposite direction of the intended targets. Historically, natural fibres served as the primary raw materials for the textile industry before the advent of synthetic alternatives. The introduction of synthetic fibres in the late 1930s marked a turning point, leading to a significant shift in material usage. Today, synthetic fibres, such as polyester, nylon, and others, account for approximately 70% of global fibre consumption (Echeverria et al., 2019; Gounni et al., 2019). Hence, textile reuse and recycling are vital in promoting this innovative act. This section look into the history of textile recycling, determines the current trends of existing textile waste along with the innovation in reuse and recycling technologies.

2.4.1 History of Textile Recycling

Textile recycling is not new; its history goes back centuries. Prior to industrialisation, clothing was a precious item. Old clothes were repaired, reused, or passed down from generation to generation. In agricultural communities, old clothes were turned into rags, quilts, or padding. In these eras, recycling clothes was more about necessity and thrift than about the environment. During the 18th and 19th centuries, recycling of textiles occurred on a more systematic basis. Wool was most widely recycled in industrialised countries (Wang, 2006). In the United Kingdom, West Yorkshire developed a reputation for the production of "shoddy" and "mungo," both reclaimed woollen products. They were produced through grinding up used woollen clothing and re-spun into new yarns. The material found application in low-grade woollen items, such as blankets and uniforms (Leather, 1917).

The Industrial Revolution was a turning point in textile production and consumption. With mass production, there came greater availability and affordability of clothing. Subsequently, the recycling of textiles also became more organised. The introduction of mechanical means of recycling made large quantities of wool and cotton available for reuse. Clothing collection was also made formal in certain places, with merchants and rag collectors having a crucial role in collecting old clothes and fabric waste. They were then sold to mills that were experts at reprocessing the clothing into new material. However, with the introduction of synthetic fibres such as polyester in the mid-20th century, recycling became difficult. Synthetic fabrics had different chemical and physical challenges compared to natural fibres, and the industry took its time in developing recycling processes that can deal with these fibres (Haque & Naebe, 2024).

2.4.2 Current Trends and Importance of Recycling in Textiles

Like all waste, textile waste is produced through multiple channels, including fibre production, textile and garment manufacturing, consumer usage, and commercial or service sectors. A significant portion of this waste originates from households. On average, clothing items have a lifespan of approximately three years before being discarded. In many cases, garments that are still in good condition are thrown away simply because they are no longer fashionable or desirable. Within the apparel industry, textile waste is typically categorised into three main types:

(i) Pre-consumer waste
(ii) Post-consumer waste and
(iii) Industrial waste.

Pre-consumer textile waste refers to the material discarded during the manufacturing process, which occurs when natural, synthetic, or blended fibres are converted into yarns, fabrics, and final products such as garments, footwear, home textiles, or technical textiles. This category includes items like off-cuts, selvedge waste, rejected batches, and nonwoven scraps. Since pre-consumer waste is typically clean, uniform, and free from contaminants, it is generally easier to recycle compared to post-consumer waste. Additionally, unsold inventory and returned goods from both physical retail and e-commerce platforms also fall under this category.

Post-consumer textile waste refers to garments or fabrics that have completed their useful lifespan and are discarded by consumers due to wear and tear, damage, size changes, or loss of relevance in fashion or style (McCauley & Jestratijevic, 2023; Stanescu, 2021). This category often includes garments of fair to good quality, which can be recovered and reused as second-hand clothing. A significant portion of this recovered apparel is exported and sold in developing countries.

Industrial textile waste arises from commercial and industrial applications, including items such as carpets, curtains, hospital textiles, filtration materials, and conveyor belts. This type of waste is often classified as "dirty waste" due to contamination or complex material composition. A large portion of these end-of-life products is either incinerated or disposed of in landfills. Because of the challenges associated with collection and the intricate chemical makeup of these materials, recycling this category of waste is particularly difficult. As a result, incineration and landfill disposal remain the most common practices (Chavan, 2014). Some industry experts classify unsold stock and returned items as a form of industrial waste, as the fashion industry is responsible for their disposal. According to the European Union (EU) Waste Framework Directive, certain types of waste, specifically 24 identified items, should be minimised and are only permitted for landfill disposal when no further use is possible. Garments often feature various embellishments such as buttons, zippers, and decorative elements, and are frequently made from blended materials. These factors significantly complicate the recycling process (Chavan, 2014; Eppinger, 2022).

2.4.3 Innovations in Textile Recycling Technologies

Textile recycling encounters several challenges in terms of cost, time, and technological limitations. However, with the growing emphasis on sustainability, ready-to-wear manufacturers and other industry stakeholders have stepped up their efforts to enhance recycling processes. Consequently, initiatives aimed at recycling and recovering textile waste have expanded significantly in recent years. Today, a variety of technologies are available to support textile waste recycling, including biotechnological methods such as anaerobic digestion, fermentation, and composting.

Anaerobic digestion (AD) is a commonly employed method for processing biodegradable organic waste, primarily to generate biogas. Its growing adoption is attributed to the dual benefits of organic waste recycling and renewable energy generation. Cotton, which contains over 50% cellulose, serves as a suitable material for biological conversion. Several investigations have explored the use of cotton waste in AD processes to produce biogas with a high methane content (Saravanan et al., 2009). Anaerobic digestion (AD) of waste from the medical cotton industry under thermophilic conditions, utilising cattle manure as an inoculum, has shown a significant enhancement in biogas production, with yields reaching around 92% (Ismail & Talib, 2016). The application of pre-treatment techniques has been found to substantially improve both the quantity and quality of biogas, as well as the quality of the resulting digestate. These pre-treatments enhance the overall efficiency of the digestion process, leading to greater recovery of valuable end-products (Moorthy et al., 2022).

Research into the use of cotton waste as a potential feedstock for ethanol production dates back to 1979. Cotton gin waste, often referred to as cotton gin trash, is a by-product generated during the cotton ginning process. This lignocellulosic material can serve as a raw substrate for ethanol production, a renewable fuel that, when blended with gasoline, helps reduce greenhouse gas emissions and reliance on fossil fuels. The production process involves the breakdown of cellulose into fermentable sugars, which are subsequently converted into ethanol through microbial fermentation. This approach offers a sustainable method of energy generation by utilising agricultural residues. Additionally, the impact of alkali pre-treatment on ethanol yield was investigated using polyester/cotton (polycotton) blends. Among the tested conditions, pre-treatment with a NaOH/urea solution at 20 °C led to the highest ethanol yield, achieving 70% through simultaneous saccharification and fermentation, and was identified as the most effective strategy (Gholamzad et al., 2014).

Composting is a natural biodegradation process through which organic materials, such as cotton waste, are transformed into nutrient-rich soil amendments. As a low-tech, bio-oxidative method, composting can reduce the volume of organic waste by up to 50% during its active phase (Hamawand et al., 2016). This aerobic process relies on microbial communities, including bacteria and fungi, to break down complex organic compounds into simpler, stable forms. The resulting compost returns essential nutrients to the soil, aids in the rehabilitation of nutrient-depleted lands, and serves as a beneficial additive for garden and agricultural soils. It also contributes to improved soil structure, reduced erosion, and enhanced water retention. Vermicomposting, a biotechnological variant of composting, employs earthworms to accelerate decomposition and enhance compost quality. When cotton waste is used as the substrate, both compost and vermicompost exhibit similar bacterial diversity. However, vermicompost typically shows a higher density of bacterial isolates, resulting in superior humus formation and greater improvements in soil fertility (Mahitha et al., 2016).

Textile waste can be processed into fibres and repurposed as reinforcement material in cement-based composites, leading to the development of lightweight, strong, and durable bricks and concrete. Advancements in this area are driving the potential replacement of conventional construction materials with more eco-friendly and sustainable alternatives. Such recycled textile-based materials can be effectively utilized in various construction applications, including thermal and acoustic insulation, as well as fillers or reinforcement agents in concrete structures (Woodings, 1995).

Textile Reinforced Concrete (TRC) is an advanced composite material that incorporates textile fibres as reinforcement within fine-grained concrete. It is increasingly applied in precast construction, structural rehabilitation, repair works, and the strengthening of existing infrastructure. Originating as a sustainable innovation within the construction industry, TRC offers an effective way to repurpose textile waste, aligning with environmentally responsible building practices. The integration of multi-axially oriented textile reinforcements imparts several advantages, including reduced structural thickness, high load-bearing capacity, corrosion resistance, enhanced ductility, non-magnetic properties, and overall lightweight performance (Williams Portal et al., 2015). In addition, textile waste has been effectively repurposed into durable ropes used for slope stabilisation, helping to prevent slippage and soil erosion (Hegyi et al., 2022). Recycled textile-derived materials such as cellulose and wool offer a sustainable alternative to conventional fibreglass insulation, providing effective thermal and acoustic performance while reducing environmental impact.

Fibre regeneration through the recycling of cotton waste garments represents a closed-loop upcycling approach that enables the sustainable recovery of textile fibres. This process involves converting discarded cotton fabrics into cellulose pulp, which is then dissolved using a solvent and re-spun into new fibres. One such solvent, N-methylmorpholine N-oxide (NMMO), is capable of dissolving cellulose entirely without causing degradation, making it an environmentally benign option. The regenerated pulp from cotton waste can also be blended with wood-derived pulp to produce fibres with properties comparable to Lyocell, thereby supporting circular textile manufacturing (Alexander & Shashikala, 2020).

2.5 Recycling Practices in the Fashion Industry

The ideal vision of a fully circular fashion industry, where all textiles are recycled, eliminating the need for new raw materials and minimising environmental harm, remains largely aspirational. Despite growing awareness and technological advancements, the current reality tells a different story. Alarming images of discarded garments accumulating in landfills or being incinerated highlight the pressing waste problem in the fashion sector. This raises a critical question: What proportion of our clothing is genuinely recycled? According to studies very small percentage of the clothing is recycled, and a majority of it is downcycled.

Recycling textiles and clothing allows materials such as old garments, linens, and other fabrics to be repurposed, extending their lifecycle. This recovery process offers significant environmental and economic advantages by avoiding many of the resource-intensive and polluting steps involved in producing textiles from virgin materials. It also helps decrease landfill usage, particularly since synthetic fibres do not readily biodegrade. Furthermore, recycling alleviates the demand for virgin resources like cotton and wool, conserving natural materials. Recycled raw materials typically incur lower production costs, making them attractive to manufacturers and encouraging the growth of markets for reclaimed fibres. Additionally, recycling reduces energy consumption and pollution, while diminishing the need for dyes and fixing agents, thereby lessening the environmental impact associated with their production and use.

2.5.1 Types of Recycling

Recycling is a key concept of modern waste management. Recycling is the reprocessing of waste materials into new or reusable products. Ninety-nine percentage of used textiles are recyclable. There is an array of different technologies that can be employed to promote and facilitate textile waste recycling and recovery. These technologies have the potential to provide an effective solution for reducing environmental impact and minimising landfill waste. Recycling plays a critical role in waste management by transforming discarded materials into new or reusable products. Remarkably, up to 99% of used textiles are recyclable. A variety of technologies exist to support and enhance the recycling and recovery of textile waste, offering promising avenues to mitigate environmental damage and reduce landfill accumulation. Beyond environmental benefits, textile recycling also addresses numerous economic and social challenges. Although recycling textiles has a long history, its importance has surged recently due to the rise of fast fashion in Western countries, which has driven excessive textile consumption and increased waste production. The most environmentally and economically favourable recycling method is 'closed-loop' recycling, where materials are reused to produce the same product. The next best option involves repurposing textiles into different products with less stringent quality requirements, for instance, recycling car seat fabric into backing materials. The following sections will explore key mechanical, chemical, and biochemical technologies utilised in textile recycling.

(i) Physical recycling
(ii) Chemical recycling
(iii) Biochemical recycling.

Physical Recycling

Mechanical recycling is widely regarded as the simplest and most effective technique for textile recycling. This approach involves two main processes. The first process consists of cutting, shredding, and carding or garnetting the textile waste, which is applicable to a broad range of fabric types. However, the fibers obtained through shredding are shorter than those from virgin materials, resulting in reduced fiber strength and lower-quality yarns. Despite this, the recycled fibers can be spun into yarn for use in woven or knitted fabrics or can be directly incorporated into non-woven textile products (Dissanayake et al., 2018). Furthermore, the fibers produced through shredding, often called "shoddy" yarns or fibers, are commonly used in the manufacture of products such as rags, carpets, insulation fillings, and materials for industrial and construction applications. A notable advantage of mechanical recycling is the reduced need for additional chemical treatments like dyeing, especially when textiles are sorted and processed in batches by colour, as the fibres retain their original dye. However, a primary limitation of this method is that the quality of recycled fibres is generally inferior to that of virgin fibres, which can affect the performance of the final products (Dissanayake et al., 2018).

Thermo-mechanical processing is another widely used technique, particularly in the open-loop recycling of plastic bottles into recycled polyester or nylon fibres. This method shows significant promise for recycling thermoplastic textiles such as polyester and nylon. The process begins with mechanically shredding the waste material, followed by heating, agitation, and filtration steps that break down the fibres and eliminate impurities. The outcome is a high-quality recycled material suitable for diverse applications (Williams Portal et al., 2015). Thermo-mechanical recycling is generally faster and more cost-effective than chemical recycling, as it avoids the use of chemical solvents. It also consumes less energy and tends to cause less degradation in fibre quality compared to many traditional mechanical recycling methods. This approach offers multiple advantages, including reducing landfill waste, conserving natural resources, and lowering greenhouse gas emissions linked to the manufacture of new textiles (Alexander & Shashikala, 2020).

Physical recycling involves reprocessing manufacturing waste and post-consumer products into new materials through mechanical means or commingled plastics waste processing. Because it is simpler, more cost-effective, and environmentally friendly compared to chemical recycling, physical recycling is generally preferred in many applications.

Chemical Recycling

Chemical recycling refers to breaking down high molecular weight polymers into smaller molecules. This process typically involves depolymerising textile waste into oligomers and monomers, which can then be repolymerised into new polymers. Textile waste, including materials like nylon and polyester, can be converted into valuable resin precursors through chemical recycling. However, the specific techniques and processes depend heavily on the chemical composition and polymer structure of the materials. Compared to mechanical recycling and producing virgin fibres, chemical recycling tends to be more expensive. It also requires the separation of components such as blended fibres, dyes, finishes, and processing additives to isolate usable feedstock. The recovered substances can serve as raw materials for synthesising other chemicals and polymers.

Biochemical technologies

Biochemical recycling is an environmentally friendly approach for processing cellulosic textile waste, utilising enzymes to break down polymer chains into monomer units. Enzymes act as biocatalysts that enhance the efficiency and speed of chemical and biochemical reactions. Typically, biochemical recycling begins with acidic or alkaline pre-treatments that disrupt the fiber structure, separating it into amorphous and crystalline regions. The choice of pre-treatment, such as acidic, alkaline, or specific cellulase enzymes, depends on the type and origin of the textile fibers being processed.

2.5.2 Recycling of Digitally Printed Materials

Digitally printed textiles offer various challenges and opportunities. Digital printing is advantageous over conventional dyeing as it consumes less water and less dye. On the other hand, the inclusivity of digital printing means that there is a risk of complicated recycling processes and products that incorporate a variety of inks, pre- and post-treatment chemicals, and combination fabrics (Ayyoob & Khan, 2023).

One approach that may help mitigate these challenges is by utilising design-for-recyclability methods, where digitally printed textiles are essentially designed for recycling at the end of their life cycle. For example, by ensuring they are either single-material substrates or that the inks and finishes will allow for recycling. Digitally printed textiles can be mechanically recycled as long as sorting by material type and colour occurs, as further sorting reduces contamination and helps maintain the integrity of the fibre. For example, polyester-based digitally printed textiles can ultimately be shredded and repurposed into insulation, carpet underlay, or nonwoven products (Glogar et al., 2025). Chemical recycling provides a promising opportunity, particularly with synthetic fibres like polyester, where depolymerisation can pull the fibre away from inks and coatings, creating quality monomers for new fibre production (Darai et al., 2024). Pilot projects also exist that are developing closed-loop recycling systems that combine digital printing and recycling infrastructure through the use of recyclable ink formulations and modularity to help disassembly and recycling of objects at end-of-life (Harkema et al., 2024).

As digital print technology continues to expand into the fabric and textile industries, it will be important to integrate recyclability into the supply chain. Working together, designers, printers, recyclers, and policymakers, can begin innovations in recycling methods and develop the necessary infrastructure to make sure digitally printed items represent another step toward a circular textile economy.

2.6 Upcycling as a Sustainable Approach

The idea of "built-in obsolescence" assures that consumers will continually buy replacements, where products are specifically designed for short-term use. For instance, in the case of nylon stockings, the manufacturers chose to forgo any improvements to prevent laddering. The result of obsolescence is millions of tons of textiles being discarded into landfills each year, creating a significant ecological risk (Reynolds, 2011).

Recycling is frequently flaunted as part of the solution to waste, however, there are many limitations to recycling clothing. The recycling process typically degrades the quality of the materials, and as a consequence of downcycling, the fibres lose strength and durability as a result of reprocessing. Mechanical recycling weakens the length of the textile fibres, while chemical recycling is more effective but expensive and energy-intensive. Moreover, both methods of recycling require extensive water and chemical treatment, which can cause further environmental damage in some circumstances. These issues urge a necessity and desire for more innovative methods, ones that not only repurpose waste but also elevate its value (Czekała, 2023).

Upcycling has become an effective and socially conscious alternative to traditional recycling. In contrast to recycling, which involves breaking materials down and remanufacturing them, upcycling maintains the sanctity of the current materials and transforms them into an elevated quality of a higher-value item. This concept ties back to the cradle-to-cradle and circular economy models, where materials would stay in use continuously and are kept out of the waste cycle after their lifecycle. Upcycling is particularly relevant in the textile and fashion sectors, where unused or non-desired garments can be re-imagined and redesigned into high-quality, fashionable, tailored items (Czekała, 2023).

This section discusses the concept, the benefits, and the increasing significance of upcycling in the fashion industry. Understanding how upcycling addresses issues of waste management and creates the foundation for sustainable business practices allows us to better understand upcycling's potential possibilities for changing the future of the industry. As fashion moves toward a future where sustainability becomes a must and not an option, upcycling can become a legitimate solution to down the remaining wasted resources while redefining consumer perceptions and valuation of materials.

2.6.1 Benefits of Upcycling

Upcycling refers to the repurposing of used materials into a product that is of better quality or functionality. The term upcycling became widely used in the 1990s and is now recognised as one of the keys to sustainable development within most industries, including fashion, engineering, and product design (Braungart & Mcdonough, 2010). The following are various benefits of implementing upcycling in the industries.

Reducing Environmental Impact

One of the major benefits of upcycling includes reducing environmental pollution by keeping waste out of landfills and incinerators. Upcycling extends the life of materials, which reduces the need for virgin raw materials and lessens the use of resources needed, such as water, energy, and raw fibres to produce a new garment. For example, it takes almost 2,700 L of water to make one cotton T-shirt, and upcycling old garments avoids using all of that excess resource. New textiles require large amounts of energy, transportation, and chemical treatments, which all create greenhouse gas emissions (Fletcher, 2013). Industries can lower their overall carbon footprint and meet sustainability requirements by moving towards a more resource-efficient means of producing textiles.

Promoting Circular Economy

Upcycling is an essential aspect of the circular economy concept, which aims to prioritise the continuous use and regeneration of resources. Linear economies function in a "take-make-dispose" system, where products are produced, used, and thrown away without consideration for environmental issues. Circular economies facilitate the reuse, repair, and repurposing of materials and resources to minimise waste. The implementation of upcycling into business practices can contribute to a sustainable economic system. Industries can also gain notoriety as sustainability leaders. Additionally, upcycling enables a relationship between the shift to a circular economy, which creates resource efficiency, spurs innovation, and promotes environmental responsibility (Bocken et al., 2016).

Economic Advantages

The economic advantages of upcycling are substantial for small enterprises, independent designers, and artisans. Due to the availability of surplus materials, entrepreneurs can offer their goods at lower labour and material costs, while still providing one-of-a-kind, distinctive products that serve a niche market. Many consumers are willing to pay extra for unique and sustainably produced fashion items, furniture, and home decor, making upcycling an appealing business opportunity (Gwilt, 2014). Upcycling also creates jobs while growing the local economy. Small and medium-sized seed businesses (SMEs) focused on upcycled products will sometimes hire skilled labour (designers, craftspeople, and artisans) to convert waste into marketable products. This may also help preserve and promote traditional craftsmanship and the creative industries more broadly. Businesses within the corporate sector can also benefit from upcycling as part of their sustainability strategies by saving costs, having a better reputation, and customer loyalty. Many organisations, like Patagonia and Eileen Fisher, have implemented some form of upcycling program within their organisation, and have demonstrated that sustainable businesses can be environmentally responsible and profitable (Niinimaki & Hassi, 2011; Patagonia, 2025).

Promoting Creativity and Innovation

Upcycling promotes creativity and innovation by pushing designers, engineers, and artists to repurpose unwanted goods and materials. Upcycled products will often display the uniqueness of the discarded materials, some with interesting patterns, textures, and styles, thus having creative and cultural significance. The creative act of upcycling has penetrated the fashion business in shoes and garments, in furniture design, and is even entering the architectural industry as waste materials are employed to produce aesthetically pleasing and functional designs. Upcycling fashions include turning worn-out denim into jackets, upcycling vintage materials into fashionable style garments, and upcycling outdated or broken accessories into jewellery (Niinimaki & Hassi, 2011).

Consumer Awareness and Changing Perceptions

The active involvement of consumers is essential to their participation in upcycling. Informing consumers about the environmental harm caused by waste and the benefits of upcycled products is a strong motivator for purchasing sustainable alternatives. A brand may involve consumers through awareness campaigns, DIY upcycling workshops, and storytelling to share the efforts they make towards sustainability. Further, the invention of social media and other digital platforms has resulted in a powerful promotion tool for upcycling. Many independent designers/brands have used these platforms to showcase their upcycled product offerings, share videos and tutorials, and connect with the eco-friendly consumer. As consumers become increasingly motivated to purchase products that are sustainably sourced, brands that embrace upcycling can stand out in a competitive market (United Nations, 2025).

Contribution to the Sustainable Development Goals (SDGs)

Upcycling contributes to multiple United Nations Sustainable Development Goals (SDGs), especially SDG 12 (Responsible Consumption and Production). Upcycling facilitates both organisational and self-alignment with global sustainability goals by reducing waste generation and using materials again. Upcycling also contributes to SDG 8 (Decent Work and Economic Growth) by generating green jobs and sustainable business processes and ultimately encouraging a more resilient and just economy (Degli Esposti et al., 2024).

Despite having numerous advantages of upcycling, there are challenges that limit its widespread applicability. Scalability is one of the significant challenges; while upcycling can operate in small productions and individual makers, it is challenging to execute in a mass-produced way. Availability and consistency of waste material are another challenge, as upcycling methods rely on the predictable prevalence of used products in recycled resources. Also, consumer perceptions affect the adoption of upcycled products. While more people are thinking in terms of sustainability, there are still consumers who are strongly connected to inferior quality or used/second-hand products. Changing these attitudes and perceptions requires successful marketing efforts and thrust education focused on the visual and environmental value of upcycled products. Technologies in material processing and digital making could better realise the efficiencies of upcycling approaches, providing better applicability for mass-industry systems. Innovative technologies such as 3D printing, artificial intelligence-based designs, or intelligent sorting machines allow for scaled applications in upcycling without compromising the original product quality.

2.6.2 *Innovative Upcycled Products: Case Studies*

Innovative upcycling in the fashion industry has led to the development of exciting and sustainable products from waste materials, creating desirable fashion products from previously discarded materials. Waste fabrics are donated, recycled, and upcycled by brands, designers, and artisans, resulting in inclusive practices that address waste and showcase the sustainability aesthetic. The following case studies of upcycling illustrate how design and craft are leading a revolution in fashion.

The Textile Recycling Association, India, has been pivotal in uniting second-hand clothing recyclers and distributors working in Kandla, Gujarat, which employs approximately 3,000 people per year and plays a vital role in the circular economy through the repurposing of old textiles. Similarly, Panipat, India (the "cast-off capital"), houses around 150–200 mills that import used clothing from Western countries and convert this into recycled textiles. According to a report by the All India Woollen and Shoddy Mills Association in September 2017, these mills employ a total of 20,000 individuals and have revenues of $62 million per year. The industry remains largely unregulated and uncoordinated, but the mills' evidence of producing up to 10,000 kg of yarn each day from 20 tonnes of used clothing demonstrates its possibilities for upcycled textile production. All yarn produced is used to make various products, such as blankets, school blazers, and even the red and black check fabric worn by the Masai in Tanzania and Kenya (Pandit et al., 2019).

Ksenia Schnaider, a Kyiv-based fashion label founded in 2011 by Ksenia and Anton Schnaider, is notably recognised around the world for its creative repurposing of second-hand denim. One of the brand's signature designs is the 'demi-denims', a combination of culottes and skinny jeans that creates a new fashion silhouette. Each year, the brand reuses approximately 5.3 tonnes of denim jeans, or 7,000 pairs, to make around 5,000 new pieces. The company's sustainability practices therefore reduce textile waste and create unique limited-edition garments, with customers including Celine Dion and Bella Hadid (Kseniaschnaider, 2025).

Looptworks, a Portland-based company, specialises in upcycling abandoned material and designing limited-edition products such as bags and accessories. Scott Hamlin and Gary Peck founded the organisation in 2009 as a way to "break the cycle of waste," rethinking resources as products, pre-consumer and post-consumer waste, to promote sustainable consumption. For example, Looptworks gathered scrap neoprene, which is left over from making wetsuits, and upcycled it for functional laptop sleeves. They further reduce resource needs while diverting a large volume of textile waste from landfills to help minimise the environmental impact of the fashion industry's impact planetary crisis (Looptworks, 2024).

In Mumbai, "I Was A Sari" aims to repurpose traditional saris into contemporary garments such as jackets, vests and bags. Established in 2014, this initiative was founded by Stefano Funari and partners with local NGOs to empower women by providing them vocational training, skills and employment opportunities. They collect second-hand saris from Chor Bazaar, Mumbai, and process approximately 15–20 tonnes of garments every day. They transform the garments into high-quality products to be exported to 30 countries. This initiative aims not only to upcycle textiles but also to contribute to their zero-waste policy and has saved approximately 1.7 million square meters of fabric from landfills (Khanna, 2016).

British designer Christopher Raeburn is famous for using surplus military materials for innovative fashion. His "Remade" collection takes apart old military parachutes, uniforms, and flight suits to create some high-end jackets, parkas, and other accessories. The fabric and materials are from surplus military stockpiles to ensure some durable, high-quality fabric has a second life. The project saves millions of square meters of military fabric from becoming landfill while reducing the need for virgin synthetic materials. It also advocates transparency in the supply chain and support for environmentally conscious fashion production. The Raeburn Parachute Parka is made from upcycled military-grade nylon parachute, giving it a lightweight, windproof, and water-resistant appeal in a unique urban style (Raeburn, 2025).

Elvis and Kresse, a UK-based company, takes decommissioned fire hoses, military parachutes, and leather scraps from luxury car seats, and makes them into high-quality bags, belts, and wallets. In the process, they clean, cut, and then handcraft them to make stylish accessories with a vintage look. The company recaptures over 300 tons of waste every year as a result, rather than accumulating in landfills. They also donate 50% of their profits to firefighters' charities in the UK. The Elvis and Kresse Weekend Bag is made from reclaimed fire hose and features a water-resistant and scratch-proof surface while maintaining a stylish and handmade design (Elvis & Kresse, 2025).

Switzerland's Freitag and Finland's Globe Hope recycle aeroplane waste products that have been rejected, like seat belts, life jackets, and retired aeroplane seat covers, to make them functional travel items in the form of backpacks, wallets, and laptop cases. These organisations scavenge overstocked aviation fabrics, cleanse and process them, and recreate them as usable, water-repellent commodities. The upcycled materials, particularly seat belts and life jackets, are exceptionally waterproof and durable and hence perfectly suitable for functional travelling equipment. Freitag's "F707 Stratos" travel backpack, consisting of recycled aeroplane seat belts and truck tarpaulin material, is an example of the ultimate sustainable travelling fashion statement piece (Globe Hope, 2025).

In addition to these cases, digitally printed fabrics are also increasingly being upcycled in innovative ways. Digitally printed fabrics, with their vibrant colors, unique patterns, and custom designs, present excellent opportunities for creative reuse. Brands and small designers have experimented with patchwork techniques, combining digitally printed offcuts with plain or solid-colored fabrics to create one-of-a-kind garments. This approach not only reduces textile waste but also showcases distinctive design aesthetics. Some designers are also engaging in remanufacturing;

deconstructing surplus or second-hand digitally printed garments and reassembling them into new, fashion-forward pieces that celebrate print diversity.

These case studies illustrate the vast potential of upcycling as a revolutionary solution in the fashion industry. The recycling of waste materials into functional, beautiful, and valuable products, by designers and brands, is not just a way to minimise the environmental footprint but also challenge new standards for sustainable innovation. This convergence of creativity and mindful design opens the door to a more conscious future in fashion and beyond, leading to further investigation and use of upcycled solutions throughout the industry.

2.6.3 *Business Potential of Upcycling*

Textile upcycling presents an intriguing opportunity for sustainable development in the fashion and textile industry. As environmental awareness, consumer consciousness, and corporate responsibility increase, the market prospects of upcycling of textiles are more viable than it has ever. This potential lies in the dynamic yet sophisticated system that plays on a multidisciplinary axis comprising multiple parties, such as manufacturers, retailers, customers, collectors, processors, and recycling end-use markets. As the planet transitions from a linear to a circular economy, upcycling represents a singular convergence of profitability and environmental care.

Over the last few years, consumers and brands have come to appreciate the value of circular fashion systems. This change is gradually decreasing the dependence on virgin material as more companies adopt sourcing approaches that prefer upcycled and recycled materials. Brands not only contribute to sustainable development goals but also establish an increasing market for upcycled goods. The more universal the use of upcycled material is, the more it will be within reach and at a lower cost, further sustaining the demand–supply cycle. This shift to a circular textiles economy means that materials are cycled back into the system once used. Brands need to take responsibility for the entire lifecycle of their products, from collecting and processing to reusing as much as they create. This reduces waste but also creates new business models, including take-back schemes and product-as-a-service models, increasing potential revenue streams.

One of the major stimulants of business prospects in upcycling is innovation. Traditional recyclers are now using high-end technologies to produce appropriate feedstock for fresh textile manufacturing. Fibre-to-fibre recycling, chemical recycling, and mechanical recycling are all being enhanced to be able to reclaim material previously thought to be non-recyclable. Such innovations can enable high-quality textiles to be remade and reused, many times with lower environmental effects than virgin materials. Apart from lowering the long-term operational expenses, firms practising upcycling also acquire market segmentation. Through the provision of unique, limited collection pieces created using upcycled products, just like in the case of Loop Works, firms can attract aware consumers. These consumers will not only pay extra for environmental reasons, but they are increasingly becoming drivers of trends.

The network of infrastructure backing textile collection and sorting is enormous and diversified. Donation centres, consignment stores and thrift, curbside programs, and retail take-back programs all form part of post-consumer textile collection. When collected, these textiles are carefully sorted into several grades, like wearable, reusable, recyclable, or unusable. Such a system is responsible for making textiles retain their highest environmental and economic value. As per SMART (Secondary Materials and Recycled Textiles Association), about 45% of recovered textiles are resold as usable clothing, 30% are converted into wiping cloths, and 20% are torn into fibre materials. The other 5% is discarded. This material segmentation creates different streams of revenue, from direct resale in second-hand outlets to providing material for industrial and automotive uses. The open-loop textile recycling schemes, which reuse fabrics in products such as insulation, mattress filler, or horticultural padding, highlight the versatility and scalability of textile upcycling.

There are places like East Africa and China that are limiting the importation of second-hand clothes to safeguard local industries, however, others like Pakistan, Guatemala, and Honduras remain thriving with dual economies of textile production and second-hand markets. Opportunities for upcycling businesses are in grasping these dynamics and finding opportunities in nations where sustainable fashion and second-hand clothing are accepted. Moreover, policy-influenced incentives in nations such as the UK, where post-Brexit currency devaluation resulted in heightened affordability of British second-hand apparel overseas which imply that geopolitical and economic pressures can shape the success of upcycled items. Companies that remain abreast and responsive to these changes can access growing markets and realign their strategies as a result.

Despite the potential being great, the upcycling industry is hindered by challenges. They encompass inefficient recycling markets, hefty search and transaction costs, non-transparency, and consumer distrust. Sorting, grading, and quality assurance demand considerable investment in terms of labour and technology. Further, a shortage of standardisation and market regulation can complicate the organisation of consistent pricing models and scaling up. Government action can be instrumental in alleviating these issues. Public collection programs, recycling content requirements, and subsidies for recycling technology can reduce operational barriers. Markets for recycled products, facilitated by certification schemes and market intelligence systems, can decrease transaction costs and establish trust between buyers and sellers. For instance, public authorities may create platforms that link recyclers with producers and offer real-time information on material availability, price, and quality.

The upcycling of textiles can not only decrease environmental damage but also create substantial social value. In poorer countries, the availability of affordable, quality clothing raises standards of living. Current systems generally benefit intermediaries at the expense of end users. Upcycling projects that value open book accounting, fair labour, and community development can lead to more balanced results. Brands that make investments in local communities, endorse fair trade practices, and engage consumers in their sustainability initiatives can establish stronger customer connections and brand loyalty. Additionally, informing consumers of the environmental advantages of upcycling as well as encouraging circular consumption habits can also enhance demand.

The business opportunity of upcycling textiles is that it can convert trash into cash in a way to solve the critical global issues at hand. Economic, social, and environmental thriving of the industry of upcycling can become a reality, provided there's the right balance of innovation, cooperation, and policy backing. Innovative brands, governments, and entrepreneurs who embrace circular systems today will be among the leaders for the next phase of sustainable textiles and fashion consumption.

2.7 Role of Stakeholders in Achieving Fashion Sustainability

The pursuit of a sustainable fashion industry is not the responsibility of a single player. It requires collective and coordinated action across the entire fashion ecosystem. This ecosystem is composed of a diverse group of stakeholders, each with a unique role and set of responsibilities. Fashion brands and designers influence sustainability through product design, material choices, and business models. Manufacturers and suppliers are critical in implementing cleaner technologies, ensuring fair labour practices, and maintaining transparency. Consumers, through their purchasing decisions and awareness, can drive demand for ethical products and hold brands accountable. Governments and policymakers play a pivotal role by setting regulations, offering incentives for sustainable practices, and enforcing labour and environmental laws.

Additionally, non-governmental organisations (NGOs) act as watchdogs, advocates, and facilitators of change, bridging the gap between industry and civil society. The media and fashion influencers shape public narratives and consumer attitudes, which is important in normalising sustainable choices. Academic and research institutions contribute by developing new materials and sustainable production techniques, and educating future industry leaders. Lastly, financial institutions and investors can direct capital toward sustainable ventures and pressure companies to adopt ethical, environmentally conscious practices. Together, these stakeholders form the backbone of the sustainability movement in fashion. Only through their collaborative engagement, shared accountability, and commitment to systemic change can the fashion industry be transformed into a model that is not only creative and profitable but also responsible, regenerative, and resilient.

2.7.1 Role of Fashion Brands and Designers

Designers have the unique ability to initiate sustainability from the earliest stages of product development. Through intentional design choices, they can greatly reduce the environmental footprint of clothing before it even reaches production. A growing number of designers are also embracing circular design principles, where garments are designed with their entire lifecycle in mind. This includes ensuring durability, ease of repair, recyclability, and biodegradability. For example, some brands are incorporating modular designs that allow consumers to replace parts of a garment (such as collars or sleeves), thereby extending its usable life. Others are exploring cradle-to-cradle models, in which every component of a product can be safely returned to the environment or repurposed into new items. These approaches challenge the traditional linear "take-make-dispose" model that has dominated fashion for decades.

Sustainable fashion brands are redefining success by embracing the principles of slow fashion, which emphasise quality, skilled craftsmanship, and mindful consumption. Brands like Patagonia, Stella McCartney, Eileen Fisher, Reformation, and FabIndia have demonstrated that it is possible to operate profitably while maintaining a strong commitment to environmental conservation and social responsibility.

One of the most critical areas where brands can lead change is transparency. Consumers today are increasingly demanding to know where, how, and by whom their clothes are made. Brands that proactively disclose their supply chain practices, labour standards, material sources, and environmental impact earn the trust of ethically minded consumers. Initiatives such as publishing sustainability reports, obtaining third-party certifications and using blockchain for supply chain tracking are helping to build a culture of accountability in the fashion world.

Material selection plays a foundational role in sustainably. Eco-friendly, renewable, and responsibly sourced materials can significantly reduce their environmental footprint. This includes minimising water and energy usage, decreasing reliance on harmful chemicals, and lowering greenhouse gas emissions (di Poratelli, 2023). Natural fibres such as organic cotton, hemp, bamboo, and linen are preferred for their biodegradability and lower resource demands compared to conventional materials. These fibres do not require excessive use of pesticides or synthetic fertilisers, making them gentler on the environment. Moreover, innovative textiles like Tencel, which is made from sustainably sourced wood pulp, and Pinatex, derived from pineapple leaves, offer creative and sustainable alternatives. Recycled materials such as Econyl, produced from industrial and ocean waste, also contribute to reducing the growing problem of textile waste (Damioli, 2022).

In addition to environmental concerns, material choices have a direct influence on social sustainability. Choosing fabrics that are certified by organisations such as the Global Organic Textile Standard (GOTS) or OEKO-TEX ensures that production processes are free from harmful substances and follow ethical labour standards. This includes providing fair wages, maintaining safe working conditions, and upholding workers' rights. Ethical sourcing contributes to a more transparent and accountable fashion supply chain (Good On You, 2025). Another important aspect of sustainable material selection is its role in promoting circular fashion. Using durable, long-lasting materials allows garments to be worn longer, reused, or eventually recycled. High-quality fabrics that resist wear and retain their shape reduce the frequency of disposal and replacement. This shift towards longevity and circularity encourages consumers to engage in more mindful consumption and reduces the environmental burden associated with short-lived fashion (Mazzitelli et al., 2024).

2.7.2 Role of Manufacturers and Suppliers

Manufacturers and raw material suppliers serve as the backbone of the global fashion supply chain. Their responsibilities span a wide range, from sourcing and cultivating raw fibres to processing textiles and assembling finished garments. Because these operations form the foundation of fashion production, they bear a significant share of the industry's environmental and social impact. According to the study findings in the International Business Education Journal (2021) (Shafie et al., 2021), sustainable manufacturing practices have the potential to reduce pollution, lower costs, and contribute to community development—all while addressing global environmental challenges.

Raw material suppliers play a pivotal role in supporting sustainable fashion by selecting environmentally friendly sources such as organic cotton, natural fibres, and recycled textiles. Experts interviewed in the study emphasised that using natural, biodegradable materials helps limit soil and water pollution and avoids the toxic residues associated with synthetic fibres and chemical-based processing. On the manufacturing side, industrial processes consume vast amounts of energy, water, and chemicals. Sustainable production demands a shift toward cleaner technologies. Manufacturers can reduce their environmental footprint through innovations such as solar-powered operations, energy-efficient machinery, and water-recycling systems. Techniques like waterless dyeing, laser cutting, and 3D knitting were highlighted as emerging methods that conserve resources while maintaining production quality.

In addition to environmental sustainability, the human element of fashion production must not be ignored. Social sustainability hinges on ethical labour practices, which include providing fair wages, ensuring safe working conditions, and respecting workers' rights. Initiatives that educate and empower workers, especially marginalised groups such as single mothers, as seen in the case of Malaysian community programs, can turn fashion manufacturing into a platform for social upliftment. Manufacturers must adopt circular strategies to manage waste. Upcycling and recycling unused or post-consumer textiles reduce the volume of fabric discarded in landfills and open avenues for new product innovation.

2.7.3 Role of Consumers

Consumers play a pivotal role in driving fashion sustainability, as their choices directly influence demand, production strategies, and market behaviour. Opting for durable, timeless, and ethically produced clothing over trendy, disposable items can significantly reduce the frequency of purchases and the demand for fast fashion. Sustainable alternatives such as second-hand shopping through thrift stores, vintage boutiques, or digital resale platforms like Poshmark, ThredUp, and Depop offer eco-conscious options while promoting reuse and extending product life cycles. However, sustainable fashion does not end at the point of purchase. The way consumers care for their clothing also contributes greatly to sustainability outcomes. Simple habits such as washing garments in cold water, air drying instead of machine drying, and washing less frequently can conserve water and energy while prolonging the life of textiles. When items are damaged, consumers can choose to repair, alter, or upcycle them creatively instead of discarding them. These practices not only reduce waste but also encourage a deeper appreciation of craftsmanship and personal style. Consumers can engage in responsible disposal by donating unwanted garments, participating in clothing swaps, reselling lightly used items, or recycling through community programs and brand take-back schemes. Each of these actions helps reduce the volume of waste and supports the circular economy.

Perhaps most powerfully, consumers can use their collective voice to influence the industry at large. Through social media platforms, product reviews, petitions, and public feedback, consumers can hold brands accountable for unethical practices and push for greater transparency and sustainability. As the demand for responsible fashion continues to grow, brands are increasingly responding to these pressures by improving their sourcing, production, and communication strategies. In essence, consumers are not passive participants in the fashion system they are active agents of change. Through conscious purchasing, sustainable habits, and advocacy, they have the power to reshape the industry and promote a culture of fashion that respects both people and the planet.

2.7.4 Role of Governments and Policy Makers

Governments and policymakers act as regulators, enablers, and educators in the sustainable fashion movement. Through legislation and policy enforcement, they can establish minimum environmental and labour standards across the textile and apparel industry. Environmental regulations are crucial in limiting the use of harmful chemicals, managing waste disposal, reducing emissions, and regulating water consumption during textile production. Additionally, labour laws, especially in regions with weak oversight, ensure fair wages, decent working conditions, and protection against exploitation.

Enforcement mechanisms such as fines, import bans, and mandatory disclosures help deter non-compliance and promote ethical operations. In tandem with regulation, economic incentives such as tax benefits, grants, and low-interest financing for sustainable fashion startups and innovations encourage brands and manufacturers to adopt greener practices. Public investment in infrastructure, including textile recycling plants, renewable energy systems, and waste collection networks, further supports the industry's transition to sustainability. Governments also have a key role in public education. Through national awareness campaigns, integration of sustainability topics in school curricula, and community-based programs, policymakers can encourage a culture of responsible consumption and educate future generations about the environmental and social costs of fast fashion. Progressive initiatives like

Extended Producer Responsibility, which hold brands accountable for the end-of-life impact of their products, are also gaining global momentum and may prove transformative in reducing textile waste.

2.7.5 *Role of Academia and Research Institutions*

One of the most impactful ways academia contributes to fashion sustainability is through education and curriculum development. Many universities and design schools have begun integrating sustainability into their fashion and textile programs, offering specialised courses that address topics such as sustainable design, ethical sourcing, life cycle assessment, and circular economy models. These programs equip students, not only future designers but also merchandisers, supply chain managers, and entrepreneurs, with the knowledge and skills to make responsible decisions in their careers. Students learn to assess the ecological and social implications of their design and business choices, preparing them to become informed agents of change within the industry.

In addition to teaching, research institutions drive innovation by developing innovative technologies and alternative materials aimed at reducing the environmental footprint of fashion. Collaboration between academic institutions and the fashion industry is another crucial aspect of their role. Through partnerships with manufacturers, designers, and sustainability-focused organisations, universities act as incubators for experimentation and testing. For example, textile departments may work with dye manufacturers to develop low-impact dyeing technologies or partner with recycling firms to improve the efficiency of textile recovery systems. These collaborations enable research findings to be translated into scalable, real-world applications and commercial solutions.

2.8 Fashion as an Ecology

The concept of fashion as an ecology shifts the perspective of industry from isolated production and consumption to a living, interdependent system comprising various components, including designers, producers, retailers, end-users, and waste management. In ecological terms, fashion is not an isolated activity; instead, it is a network of interactions that are influenced by and impact the local and global environment, economy, and social structures. Recognising fashion as an ecology means considering the life cycle of garments, from raw material extraction to disposing of used clothing, and vice versa, and considering the implications of each stage on the entire ecology (Whitty, 2021).

The approach of considering fashion as ecology emphasises balance and symbiosis as opposed to depletion and exploitation. For instance, organic cotton farming without synthetic pesticides that do not harm biodiversity is more ecologically sustainable than conventional farming practices (Letourneau & Bothwell, 2008). Likewise, manufacturing that is local and at small scales supports local economies and infrastructure while mitigating carbon emissions created by transportation. Fashion ecology recognises the social dimension of practice with an awareness of the benefits of fair labour practices and inclusive design (Heinze, 2020). Stakeholders can create systemic interventions with resilience building, harm reduction, and new regenerative practices in mind by framing the conversation with the ecology of fashion. The fashion industry's place in natural ecosystems is improved by switching from a linear "take-make-dispose" process to a more agitated and cyclical.

2.8.1 Circular Fashion and Closed-Loop Systems

A circular economy operates on the principles of eliminating waste, keeping products and materials in continuous use, and regenerating natural systems. In the textile industry, adopting a circular model involves integrating recycling into the supply chain to extend the lifecycle of materials, reduce dependency on virgin resources, and minimise environmental impacts (Chang, 2020). Transitioning from a traditional linear economy (take, make, dispose) to a circular economy in the textile sector requires significant innovation in product design, manufacturing processes, and waste management. Researchers and companies are actively working on technological advancements, such as fibre-to-fibre recycling, which aims to improve the quality and longevity of recycled fibres (Panda & Sangle, 2019). There is also a growing interest in the development of biodegradable materials that can safely reintegrate into natural ecosystems after their life cycle ends.

Closed-loop recycling is a key objective of the circular economy in textiles. In this system, textile waste is recycled back into new textile products without a loss in material quality, thus avoiding downcycling (Bocken et al., 2016). However, achieving this model is challenging due to technical hurdles, including the separation of blended fibres, maintaining the strength and quality of recycled fibres, and scaling up the recycling processes to meet global demands.

Several fashion brands are leading the way in closed-loop systems by establishing take-back programs. For example, Patagonia offers initiatives where consumers can return their used garments, which are then recycled into new products (Salfen, 2019). This approach not only reduces waste but also promotes sustainability by closing the loop within the production-consumption cycle. However, closed-loop recycling also faces challenges as many garments are made of a mix of natural and synthetic fibres, making it difficult to fully recycle the material back into new fibres. Ensuring that recycled fibres retain the strength and durability of virgin fibres is critical for achieving high-quality recycled products. Developing recycling processes that are both environmentally and economically sustainable on a large scale is necessary for widespread adoption.

Eco-design is the practice of designing products with their entire lifecycle in mind, from production to disposal, with a focus on reducing environmental impact. In the context of textiles, this could involve creating fabrics that are easier to recycle, using fewer blended materials, or choosing biodegradable fibres that naturally decompose after use. Eco-design also encourages the use of recycled fibres in the creation of new garments, reducing the reliance on virgin materials. One of the most critical elements in fostering a circular textile economy is the adoption of Extended Producer Responsibility (EPR). EPR shifts the responsibility for the entire lifecycle of a product, including disposal, back to the manufacturer (Braun et al., 2021). This encourages brands to design products that are more sustainable and invest in recycling infrastructure. EPR policies push companies to:

- Use more sustainable materials.
- Create products that are easier to recycle or biodegrade.
- Take ownership of the disposal and recycling of products after consumer use.

Countries and regions that have implemented EPR schemes have seen significant improvements in waste management and resource efficiency (Braun et al., 2021). For instance, EPR policies incentivise brands to reduce packaging waste and ensure proper disposal of products, which ultimately supports circular practices within the textile industry.

The implementation of circular economy principles, including closed-loop recycling and eco-design, offers promising pathways for reducing waste and promoting sustainability in the textile industry. As recycling technologies advance and more brands adopt sustainable practices, the textile sector can move closer to a truly circular model where resources are continuously reused and environmental impacts are minimised. However, scaling these solutions requires ongoing innovation, regulatory support, and consumer participation (Braun et al., 2021).

2.8.2 Sustainable Business Models and Digital Platforms

The transformation of the fashion industry requires not only new materials and technologies but also innovative business models that prioritise sustainability. Sustainable models prioritise long-term value rather than short-term profit. They integrate environmental and social considerations in the form of core strategies, operations, and value propositions.

One model that has emerged as a popular way to incorporate sustainability into a business is through renting or subscribing to a Product-as-a-Service (PaaS) model. PaaS models reduce overproduction and overflow by decoupling ownership from use (Crespo et al., 2024). Brands like Rent the Runway and Le Tote have created new business models based on PaaS and have been successful in demonstrating their feasibility in the fashion industry. Another model in progress is the made-to-order (or on-demand) model. This model reduces unsold inventory and conserves resources. It allows consumers to purchase products that are made for them specifically, under a direct relationship between the producer and consumers, and in conjunction with data-driven algorithms that predict what will be in demand rather than guessing. This benefits consumers by reducing waste and the risk of purchasing a product that is undesirable to others while creating a thoughtful and customised product experience (Davis, 2016).

Digital platforms are central to sustainable business activities. They enable transparency in supply chains to let brands and consumers track the production source and the environmental impact of their clothing. For example, blockchain can create unique and unbreakable digital records about products and manufacturing, adding an extra layer of accountability. Ethical brands using e-commerce platforms, virtual fitting rooms, and AI-driven design tools all help bring down the return costs, reduce the number of items produced, and potentially cut down on the amount of materials used in the production process. Likewise, social media and other digital marketing strategies can stimulate awareness and even changes in consumer behaviour, encouraging more sustainable consumption.

Sustainable design tools in digital fashion are revolutionising how designers conceptualise, develop, and produce garments, with significant implications for waste reduction and resource efficiency. Digital prototyping, 3D rendering, and virtual sampling tools allow designers to create and refine garments in a virtual environment, thereby minimising the need for physical samples and reducing material waste. These tools also facilitate faster iterations and enable designers to visualise fit, drape, and style on digital avatars before committing to production. Platforms such as CLO3D, Browzwear, and TUKA3D enable designers to develop detailed digital garments with accurate fabric simulations, improving design accuracy and reducing reliance on multiple prototypes (Browzwear, 2021). These tools support sustainable practices by enabling brands to test and visualise new collections virtually, thus avoiding the environmental impact of physical sample production and reducing carbon emissions associated with shipping and logistics. Virtual sampling also contributes to sustainable supply chains by allowing real-time collaboration between designers, manufacturers, and suppliers. Stakeholders can review and approve designs digitally, streamlining the production process and minimising the risk of errors and rework that often lead to waste.

2.8.3 Future Challenges and Opportunities in Eco-Friendly Fashion

The movement towards sustainability in the fashion industry presents challenges as much as it does opportunities. One of the challenges is the existing fashion industry's infrastructure, which relies on a fast-fashion ethos, meaning that quick and cheap don't equal sustainability. Creating a more holistic, sustainable approach will necessitate investment, policy changes, and a cultural shift among consumers. Limiting technology in textile recycling, non-standardisation in sustainability measures, and the expense of sustainable materials are other issues associated with sustainability. Additionally, the continued use of so-called greenwashing tactics when companies create opportunities for consumers to compare the sustainable benefits of a product limits the effectiveness of innovative efforts and cuts into the consumer's trust factor. But with the rise of conscious consumerism, there's an opportunity. When consumers demand more sustainable products, brands will be more likely to innovate and showcase their potential sustainability initiatives as a differentiation tool. Recent changes in legislation for extended producer responsibility (EPR) and environmental labelling will create more pressure for responsibility amongst the industry.

Innovation remains a key driver. Research and development of new biotechnology around lab-grown leather, biodegradable textiles, and next-generation alternatives are opening up possibilities for new sustainable textile materials. Digital tools are enabling more streamlined processes for designing and producing garments, and data analytics are being applied to speed up supply chains and minimise excess.

Digital printing represents both an opportunity and a challenge within this evolving landscape. For fashion aspirants, digital printing technology offers significant advantages, including small-batch production, entry-level costs, and unparalleled creative freedom. It allows independent designers and startups to experiment with new designs and produce limited quantities of garments without large capital investments, thus reducing overstock and textile waste. Digital printing empowers designers to respond quickly to trends and consumer preferences, supporting agile and sustainable business models.

For fast fashion, digital printing enables rapid prototyping and quick design changes which supports the industry's demand for constant newness and speed to market. While this flexibility allows brands to adapt swiftly to shifting trends and consumer demands, it also presents sustainability challenges by potentially reinforcing overconsumption and contributing to waste if not managed responsibly. Hence, balancing the benefits of digital printing, such as reduced water use, lower chemical consumption, and less fabric waste with the risks of overproduction and increased consumption is critical in aligning fast fashion with sustainability goals.

Collaboration is essential in addressing future challenges. Joint efforts between brands and industries with governments, NGOs, and academia will expand and speed up the learning, development, and implementation of sustainable practices. Supporting education and training opportunities focused on sustainable practices will allow designers, manufacturers, and consumers to make deliberate decisions. Overall, the pathway to a sustainable future in fashion is complicated, but is positively supported by a positive public narrative. By framing fashion as an ecology, working within circular systems, leveraging digital platforms, and considering the challenges of the future, the fashion sector has the opportunity to open a pathway to a more sustainable and equitable future.

2.8.4 Digital Printing and Sustainability Aspects

Digital printing technology has multiple routes towards sustainability in the textile and fashion industry. In comparison to traditional printing methods that are often characterised by high water use, high levels of chemicals, and heavy energy use, digital printing can make a real difference to the sustainability of textile production.

The ink itself has a significant impact on sustainability in digital textile printing. Water-based inks have become a preferred option for many digital textile printing applications, with manufacturers consistently introducing products that have less VOC emissions and subsequently pose a lower risk to the health of workers, as well as contribute to reduced air pollution. These inks are also easy to clean from printing equipment, which has led to less use of harsh cleaning products. Additionally, various manufacturers are reformulating their modern pigment inks and reactive dyes to improve wash-fastness and durability while minimising toxic by-products. Modern digital printing machines are also designed to be energy-efficient, incorporating advanced printheads and heating systems that consume less power per unit of textile printed. Some systems even feature intelligent software that optimises energy use depending on fabric type and design complexity, contributing further to sustainable manufacturing (Ayyoob & Khan, 2023).

Digital textile printing contributes meaningfully to the three pillars of sustainability; social, ecological, and economic, by enabling innovative, flexible, and responsible production practices. Digital printing encourages local production through small-batch and on-demand production, minimizing the need for inventories and long supply chains. This also supports fair labour practices, as production can happen in areas that have better working conditions and oversight. It puts the power back into the consumer's hand, as they can now customise and co-create their designs, increasing inclusivity and supporting consumer-led choices (Ayyoob & Khan, 2023).

From an ecological perspective, digital printing presents numerous advantages. It requires less water to produce (and less chemical waste). This means a safer environment for the worker and industry while reducing the environmental footprint. Water-based, low-VOC inks also produce less air pollution and harm to worker health. Energy-efficient equipment produces CO_2 emissions that are lower than conventional printing, which will help negate climate change issues. Economically, digital printing enables two important advantages over conventional printing—flexibility in production with rapid design changes and running small production runs to reflect market demands. This mitigates any overproduction and waste, which ultimately reduces expenses and increases profits. In addition, digital printing allows for innovation opportunities, and the ability to experiment with new materials, new processes, and new business models which enhances competitiveness and resilience (Ayyoob & Khan, 2023).

2.9 Conclusion

The path towards sustainability in fashion is not only a passing trend, it is an imperative necessity dictated by environmental, economic, and ethical requirements. An industry that has previously existed on the quick turnaround of seasons and the myth of perpetual freshness must now contend with its environmental impact. This chapter has engaged with the concept of sustainability in fashion not as a limitation, but as a creative, systematic rethinking of garment creation, usage, and eventual return to the material cycle. It is accepted that the current fashion system (one that relies on mass production and consumption) is fundamentally incompatible with sustainable practices. The momentum of fast fashion has raised a culture of disposability, while the linear model of extraction, production, consumption, and disposal suggests that the consequences of these issues are too significant to be ignored. However, this sense of accountability presents opportunities. The need to reverse this unsustainability trend is necessary and reversible with collaborative and value-driven efforts.

At the centre of this change is the idea of the 3R's: Reduce, Reuse, Recycle, which does not operate as isolated actions, but provides a way of thinking about fashion as it relates to the environment. Reduce provokes restraint at every level, from mindful purchasing to minimalist design. It requires that we slow down, and not just production timelines, but the pace at which we experience clothing and assign value to it. Reuse demonstrates the fact that the life cycle of a clothing product is not only based on its durability, but on sentimentality, versatility, and timeless design. Lastly, recycling, which was seen as a last option, is now one of the most exciting realms of innovation, where waste becomes a raw material and used fibres are revived with the aid of innovation. However, sustainability in fashion is more than just the material or reducing waste; it is rethinking systems that include the business models behind production cycles, technologies that can articulate closed-loop manufacturing, and cultural narratives around fashion. The chapter's focus on upcycling is especially significant in this regard. Upcycling pushes against the socially constructed understanding of value, implying that value can exist in old and used clothes, encouraging us to look at old and waste as places of possibility, manufacturing products that are both sustainable and meaningful with character and history.

This has to be emphasized that none of this is possible without the willingness of all stakeholders to engage in this fashion sustainability ecosystem collectively. Designers, manufacturers, retailers, and consumers are all interconnected in these systems. Therefore, the designer must look beyond aesthetics and think about materials' health, modularity, and end-of-life solutions in the design process. Manufacturers must align production operations with leaner, cleaner technologies. Retailers can no longer simply exist as sellers; they must become sustainable teaching storytellers and advocates. Consumers must awaken to their agency and realize that every purchase is not just a product request, but a signal of demand for value. The integrated approach of fashion sustainability becomes even more effective with the development of circular fashion. Designing systems similar to natural cyclical processes, where nothing is wasted, nor excluded from value, is one way the fashion industry aligns more closely with the patterns of ecological harmony. Digital platforms will be crucial to this transition: from the tracking of garment lifecycles to ensuring resale, rental, and repair ecosystems. Sustainability is also not going to remain a niche market; ultimately, it will be the force of innovation, differentiation, and resilience.

Yet transformation does not come without difficulty. Technology must scale at an affordable price. Regulations must support urgency for the environment. Supply chains must become transparent. Above all, cultural beliefs must shift away from a tendency of 'trendy' fashion that lasts weeks into long-lasting, tangible values. Fashion, with its far reach and global influence, may be uniquely poised to lead that change forward, not simply less harmful fashion, but actively helpful to environmental regeneration and social justice. Sustainability is not a linear journey, it is iterative, layered, and nuanced at times. However, the path is clear, as we explored this chapter, tools, ideas, and initiatives are already in existence; some systems are rooted in long-standing social norms, while others are rooted in innovation. It is now time for action in fashion, to not only be committed to just changing, but to commit to integrating fashion that is ethical, ecological, and enduring.

Fashion cannot continue to be an extractive industry; it must shift to a regenerative industry, one that is no longer taking from the future to serve the present, but that aligns style with stewardship. The world we dream of, where the word sustainable can really mean something, is possible. It starts with one decision, one garment, and one action. The fashion industry can reimagine itself through collaboration, creativity, and courage. This transformation of 'Sustainability in Fashion' is not just about survival, but about becoming a system that thrives in harmony with the planet.

Declaration The authors declare that they have no known competing financial interests or personal relationships that could have appeared to influence the work reported in this chapter. Generative AI and AI-assisted technologies have been used in the writing process to improve the readability and language of the work only. After using this tool, the authors carefully reviewed and revised the content to ensure accuracy and quality. The authors take full responsibility for the final version of this publication.

References

Alexander, A. E., & Shashikala, A. P. (2020). Sustainability of construction with textile reinforced concrete-A state of the art. *IOP Conference Series: Materials Science and Engineering, 936*(1), Article 012006. https://doi.org/10.1088/1757-899X/936/1/012006.

Athalye, A. (2012). *Carbon footprint in textile processing* (Vol. 59, pp. 45–57).

Ayyoob, M., & Khan, M. (2023). Western markets for digitally printed textiles. In H. Wang & H. Memon (Eds.), *Digital textile printing* (pp. 207–223). Woodhead Publishing. https://doi.org/10.1016/B978-0-443-15414-0.00004-2.

Birnbaum, J., & Fox, L. (2014). *Sustainable revolution: Permaculture in ecovillages, urban farms, and communities worldwide*. North Atlantic Books.

Bocken, N. M. P., de Pauw, I., Bakker, C., & van der Grinten, B. (2016). Product design and business model strategies for a circular economy. *Journal of Industrial and Production Engineering*. https://www.tandfonline.com/doi/abs/https://doi.org/10.1080/21681015.2016.1172124.

Braun, G., Som, C., Schmutz, M., & Hischier, R. (2021). Environmental consequences of closing the textile loop-life cycle assessment of a circular polyester jacket. *Applied Sciences, 11*(7), Article 7. https://doi.org/10.3390/app11072964.

Braungart, M., & Mcdonough, W. (2010). *Cradle to cradle: Remaking the way we make things.* North Point Press. https://biblio.com.au/book/cradle-cradle-remaking-way-we-make/d/1519975292.

Browzwear. (2021). Browzwear 3D apparel design solutions. https://browzwear.com.

Business Waste. (2025, April 26). Textile waste facts. Business waste. https://businesswaste.com/waste-types/textile-waste/textile-waste-facts/.

Chang, D. (2020). *Second: Redesigning the textile recycling experience using a circular economy.*

Chavan, R. (2014). Environmental sustainability through textile recycling. *Journal of Textile Science & Engineering, s2*(01). https://doi.org/10.4172/2165-8064.S2-007.

Chen, X., Memon, H. A., Wang, Y., Marriam, I., & Tebyetekerwa, M. (2021). Circular economy and sustainability of the clothing and textile industry. *Materials Circular Economy, 3*(1), 12. https://doi.org/10.1007/s42824-021-00026-2.

Crespo, J. H., Riel, A., Duberg, J. V., Bunodiere, A., & Dawson, P. G. (2024). An exploratory study for product-as-a-service (PaaS) offers development for electrical and electronic equipment. *Elsevier, 122*, 521–526. https://doi.org/10.1016/j.procir.2024.01.076.

Czekała, M. (2023). Recycling, upcycling and downcycling in aspect environmental safety. *Scientific Journal of the Military University of Land Forces, 209*(3), 169–181. https://doi.org/10.5604/01.3001.0053.8960.

Daly, H. E. (2007). Ecological economics and sustainable development, selected essays of Herman Daly. https://doi.org/10.4337/9781847206947.

Damioli, E. (2022). The top 8 sustainable fashion materials for clothing brands. Bamboo Rose. https://bamboorose.com/blog/top-8-sustainable-fashion-materials-for-clothing-brands/.

Darai, T. E., Ter-Halle, A., Blanzat, M., Despras, G., Sartor, V., Bordeau, G., et al. (2024). Chemical recycling of polyester textile wastes: Shifting towards sustainability. *Green Chemistry, 26*(12), 6857–6885. https://doi.org/10.1039/D4GC00911H.

Davis, R. A. (2016). *Demand-driven inventory optimization and replenishment: Creating a more efficient supply chain.* Wiley & Sons.

Degli Esposti, P., Mortara, A., & Roberti, G. (2024). Unsustainable fashion: SHEIN and the fast fashion consumption model. *Italian Sociological Review*, 929–951. https://doi.org/10.13136/ISR.V14I3.822.

Devi, O. R., & Devi, L. J. (2024). Water consumption and microfibers: The biggest threat. In *Climate action through eco-friendly textiles* (pp. 73–90). Singapore: Springer. https://doi.org/10.1007/978-981-99-9856-2_6.

di Poratelli, F. (2023, August 2). Sustainable materials in the fashion industry: New data from textile exchange. http://www.cikis.studio/en/article/sustainable-materials-in-the-fashion-industry-new-data-from-textile-exchange.

Dissanayake, D. G. K., Weerasinghe, D. U., Wijesinghe, K. A. P., & Kalpage, K. M. D. M. P. (2018). Developing a compression moulded thermal insulation panel using postindustrial textile waste. *Waste Management, 79*, 356–361. https://doi.org/10.1016/j.wasman.2018.08.001.

Dolzhenko, I. B., & Churakova, A. A. (2022). Environmental responsibility of fashion industry multinational corporations (MNCs) in the context of industry 4.0. In *Industry 4.0* (pp. 79–89). Cham: Palgrave Macmillan. https://doi.org/10.1007/978-3-030-79496-5_7.

Echeverria, C. A., Handoko, W., Pahlevani, F., & Sahajwalla, V. (2019). Cascading use of textile waste for the advancement of fibre reinforced composites for building applications. *Journal of Cleaner Production, 208*, 1524–1536. https://doi.org/10.1016/j.jclepro.2018.10.227.

Elvis, & Kresse. (2025). Elvis & Kresse: Sustainable luxury designer accessories. https://www.elvisandkresse.com/?srsltid=AfmBOoqNWvobN4714i6bWWcT2TeAVOpgM-woYQDNc1_2ObnI5Y7Gmhlj.

Eppinger, E. (2022). *Recycling technologies for enabling sustainability transitions of the fashion industry: Status quo and avenues for increasing post-consumer waste recycling*. Sustainability: Science, Practice and Policy. https://www.tandfonline.com/doi/abs/, https://doi.org/10.1080/15487733.2022.2027122%4010.1080/tfocoll.2024.0.issue-sustainable-redesign-of-the-global-fashion-system.

Fletcher, K. (2013). Sustainable fashion and textiles: Design journeys. *Routledge*. https://doi.org/10.4324/9781315857930.

Gabler, C. B., & Reynolds, K. E. (2013). Buy now or buy later: The effects of scarcity and discounts on purchase decisions. *Journal of Marketing Theory and Practice*. https://doi.org/10.2753/MTP1069-6679210407

Gholamzad, E., Karimi, K., & Masoomi. (2014). Effective conversion of waste polyester–cotton textile to ethanol and recovery of polyester by alkaline pretreatment. *Chemical Engineering Journal, 253*, 40–45. https://doi.org/10.1016/j.cej.2014.04.109.

Globe Hope. (2025). We are globe hope. Forerunners in uncompromisingly sustainable design. Globe Hope. https://globehope.com/.

Glogar, M., Petrak, S., & Mahnic Naglic, M. (2025). Digital technologies in the sustainable design and development of textiles and clothing-a literature review. *Sustainability, 17*(4), Article 4. https://doi.org/10.3390/su17041371.

Good On You. (2025, March 27). What are the best lower-impact fabrics and fibres on the market right now? Good On You. https://goodonyou.eco/most-sustainable-fabrics/.

Gounni, A., Mabrouk, M. T., El Wazna, M., Kheiri, A., El Alami, M., El Bouari, A., & Cherkaoui, O. (2019). Thermal and economic evaluation of new insulation materials for building envelope based on textile waste. *Applied Thermal Engineering, 149*, 475–483. https://doi.org/10.1016/j.applthermaleng.2018.12.057.

Goyal, S., Esposito, M., & Kapoor, A. (2018). Circular economy business models in developing economies: Lessons from india on reduce, recycle, and reuse paradigms. *Thunderbird International Business Review, 60*(5), 729–740. https://doi.org/10.1002/tie.21883.

Grand View Research. (2024). *Digital textile printing market (2025–2030): Size, share & trends analysis report by printing process, operation, material, ink type, application, region, and segment forecasts, grand view research*. GVR-4-68039-957-5.

Gwilt, A. (2014). *Fashion design for living*. Routledge.

Hamawand, I., Sandell, G., Pittaway, P., Chakrabarty, S., Yusaf, T., Chen, G., Seneweera, S., Al-Lwayzy, S., Bennett, J., & Hopf, J. (2016). Bioenergy from cotton industry wastes: A review and potential. *Renewable and Sustainable Energy Reviews, 66*, 435–448. https://doi.org/10.1016/j.rser.2016.08.033.

Haque, A. N. M. A., & Naebe, M. (2024). *Future textiles*. Deakin University. https://dro.deakin.edu.au/articles/book/Future_Textiles/27221277/1.

Harkema, S., Rensing, P. A., Domensino, S. M. D. C., Vermeijlen, J. M., Godoi Bizarro, D. E., & van Schaik, A. (2024). Disassembly of in-plastic embedded printed electronics. *Journal of Cleaner Production, 450*, Article 141837. https://doi.org/10.1016/j.jclepro.2024.141837.

Hegyi, A., Vermeşan, H., Lazarescu, A. -V., Petcu, C., & Bulacu, C. (2022). Thermal insulation mattresses based on textile waste and recycled plastic waste fibres, integrating natural fibres of vegetable or animal origin. *Materials, 15*(4), Article 4. https://doi.org/10.3390/ma15041348.

Heinze, L. (2020). *Fashion with heart: Sustainable fashion entrepreneurs, emotional labour and implications for a sustainable fashion system*. https://doi.org/10.1002/sd.2104.

Igini, M. (2023, August). *10 Concerning fast fashion waste statistics*. Earth.Org. https://earth.org/statistics-about-fast-fashion-waste/.

Ismail, Z. Z., & Talib. (2016). Recycled medical cotton industry waste as a source of biogas recovery. *Journal of Cleaner Production, 112*, 4413–4418. https://doi.org/10.1016/j.jclepro.2015.06.069.

Kalk, A., & Tulder, R. V. (2004). Ethics in international business: Multinational approaches to child labor. *Journal of World Business, 39*(1), 49–60. https://doi.org/10.1016/j.jwb.2003.08.014.

Khanna, J. M. (2016, September 26). *These Indian labels are upcycling saris to make cool garments and products*. Vogue India. https://www.vogue.in/fashion/content/indian-labels-upcycling-sarees-into-outfits-products-i-was-a-sari-bodements.

Klepp, I. G., Laitala, K., & Wiedemann, S. (2020). Clothing lifespans: What should be measured and how. *Sustainability, 12*(15), 6219. https://doi.org/10.3390/su12156219.

Krishnakumar, R., & Venkatesh, S. P. (2020). *Stabilising the oscillations between consumption practices and circular economy business models (Masters)*. UPPSALA University.

Kseniaschnaider. (2025). Kseniaschnaider. Kseniaschnaider. https://www.kseniaschnaider.com/pages/about.

Leal Filho, W., Dinis, M. A. P., Liakh, O., Paco, A., Dennis, K., Shollo, F., & Sidsaph, H. (2024). Reducing the carbon footprint of the textile sector: An overview of impacts and solutions. *Textile Research Journal, 94*(15–16), 1798–1814. https://doi.org/10.1177/00405175241236971.

Leather, C. W. (1917). The Yorkshire section. Utilisation of waste products. *Journal of the Textile Institute*. https://doi.org/10.1080/00405001708630754.

Leddin, D. (2024). The impact of climate change, pollution, and biodiversity loss on digestive health and disease. *Gastro Hep Advances, 3*(4), 519–534. https://doi.org/10.1016/j.gastha.2024.01.018.

Letourneau, D. K., & Bothwell, S. G. (2008). *Comparison of organic and conventional farms: Challenging ecologists to make biodiversity functional*. https://doi.org/10.1890/070081.

Lia, D. A. Z., & Natswa, S. L. (2021). Buy-Now-Pay-Later (BNPL): Generation Z's dilemma on impulsive buying and overconsumption intention. *Advances in Economics, Business and Management Research, 193*, 130–137.

Looptworks. (2024). *Looptworks home*. Looptworks. https://looptworks.com/.

Mahitha, U., Devi, G. D., Sabeena, M. A., Shankar, C., & Kirubakaran, V. (2016). Fast biodegradation of waste cotton fibres from yarn industry using microbes. *Procedia Environmental Sciences, 35*, 925–929. https://doi.org/10.1016/j.proenv.2016.07.070.

Mandot, A. A., Lavate, S., & Kushwaha, R. (2023). Concept of 3R and sustainability for textiles-a review. *Eastland Publications Private Limited, 8*(11), 35–37.

Manickam, P., & Duraisamy, G. (2019). 3Rs and circular economy. In *Circular economy in textiles and apparel* (pp. 77–93). Elsevier. https://doi.org/10.1016/B978-0-08-102630-4.00004-2.

Martinell, D. (2024). What is sustainability? In *An introduction to sustainable aquaculture* (1st ed., pp. 24–36). Routledge. https://doi.org/10.4324/9781003174271-4.

Mazzitelli, M., Papile, F., & Del Curto, B. (2024). Materials selection and fashion design: Strengthening reflections on fibre's nature in fibres and textiles selection. *Discover Sustainability, 5*(1), 1–17. https://doi.org/10.1007/s43621-024-00294-3.

McCauley, E., & Jestratijevic, I. (2023). Exploring the business case for textile-to-textile recycling using post-consumer waste in the US: Challenges and opportunities. *Sustainability, 15*(2), Article 2. https://doi.org/10.3390/su15021473.

Moorthy, R., Rajarathinam, R., Kumar, P. S , Rangasamy, G., Gurunathan, B., Ethiraj, B., & Thanabal, V. (2022). Insight into the effective utilization of cotton spinning wastes from textile mills for the production of bioethanol. *Sustainable Energy Technologies and Assessments, 53*, Article 102770. https://doi.org/10.1016/j.seta.2022.102770.

Muthu, S. S. (2015). *Handbook of sustainable apparel production*. CRC Press.

Muthu, S. S. (2020). *Assessing the environmental impact of textiles and the clothing supply chain*. Woodhead Publishing.

Niinimaki, K., & Hassi. (2011). Emerging design strategies in sustainable production and consumption of textiles and clothing. *Journal of Cleaner Production, 19*(16), 1876–1883. https://doi.org/10.1016/j.jclepro.2011.04.020.

Opoku, A. (2016). *A sustainable built environment's role in achieving the post-2015 United Nations sustainable development goals* (Vol. 2, pp. 1149–1158).

Panda, S. S., & Sangle, S. (2019). *An exploratory study to investigate the relationship between social license to operate and sustainable development strategies*. https://doi.org/10.1002/sd.1957.

Pandit, P., Nadathur, G. T., & Jose, S. (2019). Upcycled and low-cost sustainable business for value-added textiles and fashion. In *Circular economy in textiles and apparel* (pp. 95–122). Elsevier. https://doi.org/10.1016/B978-0-08-102630-4.00005-4.

Patagonia. (2025). Used Patagonia® clothing & gear. Worn Wear. Patagonia Worn Wear®. https://wornwear.patagonia.com/.

Raeburn, C. (2025). Christopher ræburn-redress design award. https://www.redressdesignaward.com/academy/resources/case-study/christopher-raeburn.

Reynolds, N. Y. (2011). National socks and the "nylon woman": Materiality, gender, and nationalism in textile marketing in semicolonial Egypt, 1930–56. *International Journal of Middle East Studies, 43*(1), 49–74. https://doi.org/10.1017/S0020743810001194.

Romero-Hernandez, O., & Romero, S. (2018). Maximizing the value of waste: From waste management to the circular economy. *Thunderbird International Business Review, 60*(5), 757–764. https://doi.org/10.1002/tie.21968.

Russo, M. V. (2008). *Environmental management: Readings and cases.* SAGE.

Salfen, C. (2019, August 14). Recycling is broken. Now what? Patagonia Stories. https://www.patagonia.com/stories/recycling-is-broken-now-what/story-73479.html.

Saravanan, C. G., Sendilvelan, S., Arul, S., & Raj, C. S. (2009). Bio gas from textile cotton waste-an alternate fuel for diesel engines. *The Open Waste Management Journal, 2*(1), 1–5. https://doi.org/10.2174/1876400200902010001.

Shafie, S., Kamis, A., Ramli, M. F., Bedor, S. A., & Ahmad Paud, F. N. (2021). Fashion sustainability: Benefits of using sustainable practices in producing sustainable fashion designs. *ResearchGate, 14*(1), 103–111.

Stanescu, M. D. (2021). State of the art of post-consumer textile waste upcycling to reach the zero waste milestone. *Environmental Science and Pollution Research, 28*(12), 14253–14270. https://doi.org/10.1007/s11356-021-12416-9.

Textile Exchange. (2025). Textile exchange. Textile Exchange. https://textileexchange.org/climate-dashboard/.

United Nations. (2025). Transforming our world: The 2030 Agenda for sustainable development. Department of Economic and Social Affairs. https://sdgs.un.org/2030agenda.

Wang, Y. (2006). *Recycling in textiles.* Woodhead Publishing.

Whitty, J. (2021). Fashion design for holistic systems. In *Sustainable design in textiles and fashion* (pp. 1–22). Singapore: Springer. https://doi.org/10.1007/978-981-16-2466-7_1.

Williams Portal, N., Lundgren, K., Wallbaum, H., & Malaga, K. (2015). Sustainable potential of textile-reinforced concrete. *Journal of Materials in Civil Engineering, 27*(7), 04014207. https://doi.org/10.1061/(ASCE)MT.1943-5533.0001160.

Woodings, C. R. (1995). The development of advanced cellulosic fibres. *International Journal of Biological Macromolecules, 17*(6), 305–309. https://doi.org/10.1016/0141-8130(96)81836-8.

Chapter 3
Digital Textile Printing: Opportunities and Challenges for Sustainable Localised Production

Becky Gooby

Abstract This chapter examines how digital textile printing (DTP) has the potential to offer scalable and sustainable alternatives to conventional fashion and textile production by addressing two key challenges: inefficient overproduction, and consumer behaviour. As mass-produced, low-cost textiles dominate the market, consumers struggle to make sustainable choices. DTP enables on-demand production, facilitating localised supply chains and supporting innovative business models that engage consumers and reduce the risks of mass production. Drawing on a narrative review of sustainable fashion ecosystems and two commercial case studies the chapter seeks to address the role that DTP can play, the barriers to achieving, whilst also considering design-led approaches to drive change. It proposes repositioning DTP as a co-creation tool for *customise–print–make–wear*. By situating within micro factories and maker spaces this approach offers potential for community regeneration and small-scale economic growth, to support efficient production and behaviour change. However, barriers remain, including limited access to sustainable substrates and colourants, profitability challenges, a lack of knowledge-sharing and lack of proven environmental data. By balancing DTP's economic and creative potential with a realistic assessment of its limitations, this chapter contributes to the broader discussion on digital manufacturing's role in ethical and sustainable fashion.

Keywords Digital textile printing · Localised production · Design-led approach

3.1 Introduction

Digital textile printing has been heralded as a technologically advanced alternative to traditional fabric printing methods (Ujiie, 2021). It is often promoted as a more sustainable solution within the fashion and textile industries (Ujiie, 2006a, b). As the sector grapples with environmental and ethical issues across complex supply

B. Gooby (✉)
UWE, Bristol, UK
e-mail: becky@beckygooby.co.uk

A. Athalye and H. Patil (eds.), *Digital Printing*,
https://doi.org/10.1007/978-981-95-2593-5_3

chains (Macchion et al, 2018; Ujiie, 2006a, b), digital printing has the potential to offer opportunities to reduce waste, water consumption, and chemical use in the print process (Gooby, 2020), while also promoting safer and more sustainable working conditions. Yet, its widespread adoption and dominance in the industry have yet to materialise.

This chapter explores whether the unique selling proposition of digital textile printing lies in its capacity to facilitate localised production and manufacture-on-demand, through a workflow that integrates the wearer, via a *customise-print-make-wear* approach and opportunities for digital repair. The qualitative research seeks to propose a design-based solution to the ecological and globalisation-related problems of the fashion and textile industry by considering a micro factory model in which digital textile printing could be used as a co-creative tool within a collaborative community. It offers the context of two case studies: one exploring the feasibility of digitally printed denim, and the other examining a linen mill's expansion into digitally printed textiles. These examples, both from the UK, highlight the potential while also indicating the need for further refinement. Further, rather than assessing digital textile printing in isolation, this chapter aims to evaluate its sustainability within a broader context and necessary areas for improvement. The intention, to serve as a foundation for future recommendations and highlight the need for continued research and industry transformation.

It is written from the standpoint of the design community, as a designer who has been working with this technology since 2008 in the UK. Building on an initial focus on colour variables in digital textile printing (Gooby, 2020; 2023a), the research has expanded to explore the broader significance of the technology and consider how creatives can engage with it as a tool and study digital textile printing's potential value to the creative industries. As early-stage research, it does not aim to be comprehensive, but rather to contribute a design-led perspective alongside the work of more established scholars in textiles and sustainability. The chapter will ask: *What role can digital textile printing play in fostering more local and sustainable manufacturing? What are the material, logistical, and behavioural barriers to achieving a sustainable digital textile workflow? How might design-led approaches leverage this technology to drive systemic change in fashion production?*

To answer the question, *What role can digital textile printing play in fostering more local and sustainable manufacturing?* it is first necessary to consider the broader context and establish what the sustainability issues are in fashion and textiles, as well as survey current research paradigms in this area.

3.2 Contextualising Fashion's Sustainability Crisis

In the 21st Century, it has become standard practice for fashion and textile companies operating in the Global North to outsource manufacturing to lower-cost regions abroad. Grose and Mansfield (2023) highlight that the labour-intensive nature of textile production, coupled with narrow profit margins and the phasing out of the

Agreement on Textiles and Clothing by the World Trade Organisation, have significantly driven this shift. However, Fast Fashion and now Ultra-Fast Fashion business models have further increased demand, with mass-produced low-cost textile production predominantly occurring in countries in the Global South, taking advantage of minimal environmental regulations, low standards for working conditions, and cheap labour, causing environmental and ethical concerns (Grose & Mansfield, 2023).

Consequently, supply chains have become increasingly convoluted and complex, characterised by extensive outsourcing in pursuit of profit maximisation (Grose & Mansfield, 2023). This impedes transparency, making traceability and accountability for harm difficult. Furthermore, it heightens vulnerability to risk (Göçer Akder & Bahçecik, 2023). Lengthy sub-production lines and extended shipping times reduce flexibility and responsiveness, even as real-time sales data increasingly informs manufacturing decisions (Grose & Mansfield, 2023). Short-term contracts may be cancelled when styles fail to meet commercial expectations (Martino et al., 2017). This, coupled with systemic overproduction, results in surplus deadstock, which is often destroyed to prevent industrial espionage, landfilled due to the complexities of textile recycling, or cast off to swamp other clothing markets, such as Kantamanto Market in Ghana (Manieson & Ferrero-Regis, 2023).

It is incredibly challenging for consumers to make sustainable purchasing choices as it is unclear where, how and by whom their clothes are made (Street, 2024). Current production is linear with a *take-make-dispose* model (Goldsworthy, 2017). Thus, as textile production has increased, clothing utilisation, the number of times a garment is worn, has decreased. Fashion is perceived as throwaway and single-use due to changes in consumers' emotional connections to clothing in an abundant low-cost market, alongside a reduction in quality where clothes quickly fall apart, and it is cheaper and easier to discard and replace than repair (Goldsworthy et al., 2023).

Awareness of the negative impact of fashion and textile production and the untenability of existing practices, especially the heavy usage of finite resources, has increased with consumers leading to reputational risks. Many companies now have corporate responsibility programs and put out sustainability resorts but as these do not cover the entire supply chain accusation of greenwashing are rife (Goldsworthy, 2014).

As recognition of these issues grows, there is an urgent need for adaptive, scalable solutions, involving the consumer as a key mechanism to encourage longevity and responsible clothing selection and care.

To achieve sustainability, and moreover circularity, complex system transformations are needed, some of which are long term and will be slow to implement requiring international coordination and compliance (Goldsworthy, 2017; Goldsworthy et al., 2023). Thus, solutions which offer a *fast forward* approach (Goldsworthy, 2017; Goldsworthy et al., 2023) such as using more efficient production techniques have the potential to resolve many issues immediately whilst wider systematic change is implemented.

3.3 Current Research

Sustainability is a key and pertinent theme in textile research incorporating all aspects of design and manufacture; from fibre to fabric, critiquing and measuring the impact of existing practice (Tomaney, 2015), and implementation into industry (Goldsworthy, 2017; Goldsworthy et al., 2023). Many fashion and textile conferences and trade shows now have a focus on or are wholly devoted to sustainability. Areas of research include new systems thinking, for example, looking at redesigning production models (Stewart & Tooze, 2015), identifying and measuring the impact of industry problems (Miller & Woodward, 2011) and solutions offered by industry and fashion brands themselves. There is also concentrated research in specific areas such as dyeing and textile chemistry (Deraniyagala, 2017).

UK research from The Business of Fashion, Textiles and Technology Cluster (BFTT, 2025) which includes the Centre for Sustainable Fashion (CSF, 2016) scopes out fashion ecosystems, remodelling and transition. The Centre for Circular Design (CCD, 2025) focuses on advancing and supporting circular economies and design-led communities, as exemplified by the pioneering work of researchers such as Earley (2022) and Goldsworthy (2014, 2017, 2022). The Centre for Industrial Sustainability (CIS, 2025) develops knowledge and tools to bring about sustainable industrial change.

Other people of note include; De Castro who proposes practices such as upcycling, and the care and repair of garments, to highlight their role in extending product life and promoting sustainable fashion behaviours (De Castro, 2021); and Thomas, a prominent fashion journalist, who offers critical insight into sustainable fashion practices and new innovations (Thomas, 2019). Harrison, whose work on laundering, particularly her research into how travellers and van dwellers wash clothing in spaces not typically equipped with washing machines, offers valuable insight into the environmental impact of garment care and proposes alternative, lower-impact cleaning solutions (Whitson-Smith et al., 2019). Grose and Mansfield who address sustainability within supply chains, evaluate existing practice (Grose & Mansfield, 2023); and Sinclair, whose research focuses on the intersection of technology and textiles, pinpoints the role of the consumer and challenges faced, to provide an overarching view of fashion and textiles production and supply (Tomaney, 2015).

Sustainability roadmaps for the fashion, textiles, and British manufacturing sectors have been published by the UK Government (Policy Connect, 2025a), focusing on collaborative approaches; Centres of Excellence (CSF, 2025; CUSP, 2025), which assess the UK's distinctive designer fashion profile and promote collaborative business networks and community-based culture; and the Institute of Positive Fashion (IPF, 2025), which outlines reshoring strategies for print and cut, make, trim production, highlighting the role of automation (FESPA, 2021) and AI in improving efficiency.

The Ellen McArthur Foundation further support the dual approach of improved efficiency coupled with changing consumer behaviour (EMF, 2013). They call for academia and industry to develop demonstrator products which are 'conducted collaboratively' to test new models and provide evidence for their success. This is echoed by Kate Fletcher who cites consumer behaviour and changing production as crucial (Fletcher, 2013). Fletcher describes best practice for textile sustainability as; minimising numbers of processing stages, clean production techniques, reducing energy and water consumption, reducing waste production, and managing waste streams, whilst choosing clean or minimising the use of chemicals.

Research is also being conducted by non-governmental organisations such as Greenpeace (2011), Fashion Revolution (2022), Ellen McArthur Foundation (EMF, 2013) and The OR Foundation (OR, 2025) as well as the European Environment Agency (EEA, 2020) into the impact of the fashion and textile industry. As a result of this regulations are changing to reduce the impact of textile reproduction and chemical usage, embed circular design within industry and combat textile waste. The UN's sustainable Development Goals for 2030 include directives around clean water and responsible consumption and production which will all clearly impact fashion and textile production (UN, 2025). The UK Waste and Resources Action Programme estimate that 1.8 M tonnes of textile waste is generated by the UK alone (WRAP, 2019). The UN Environmental Programme recently launched with the governments of Indonesia, Bangladesh, Pakistan, and Vietnam provides a billion-dollar plan to tackle textile chemical waste (UN, 2022). In China, the largest producer of textiles, there has been a clear drive in the 13 and 14th Five-year plans towards green production as well as promoting digital textile printing in textile production (Hongsam, 2024). Net Zero Emission targets include those set by the UK for 2050 (UN, 2019) will require changes in fashion and textile production.

Digital textile printing aligns closely with these priorities, with its capacity for on-demand, localised production and reduced resource use within production. Given these urgent environmental and regulatory pressures, there is a clear imperative to accelerate the development and adoption of key technology, such as digital textile printing, for a more sustainable future in fashion and textiles.

However, technology alone is insufficient; it must be integrated with thoughtful design strategies and driven by industry commitment, to realise its full potential and importantly create behaviour change. Developments in software and technology such as those utilised by UnMade (2025) are enabling designers to connect with customers and offer accessible mass-customisation. Direct to consumer production and micro factories for localised production is explored by Weffan (2025), Assyst Bullmer (2025) and Future Maker Spaces (Stewart & Tooze, 2015). In London, Fashion Enter, a social enterprise providing accessible garment manufacturing services across the UK, is seeking to increase and upskill the fashion and textile industry workforce through educational workshops and access to cutting edge 3D technology (Fashion Enter, 2025). Research from University of Sheffield looking at alternatives to existing practices calls for micro factories, re-distributed and localised production, and maker spaces with embedded repair services to drive small-scale economic change (University of Sheffield, 2025). Building upon this momentum National Endowment for Science, Technology and the Arts (NESTA), a UK charity supporting innovation, contributed to the study *Making Sense of the UK Collaborative Economy*, to map the UK's Digital Maker Ecosystem (University of Sheffield, 2014; NESTA, 2014).

The impact of which shaped national digital strategy, with the UK Government's Libraries Taskforce (UK GOV, 2019), supporting a range of users with different skill levels to gain access to digital technology and knowledge, developing new skills and creativity in community maker spaces. Case study research demonstrated that the spaces enabled people to collaborate, share resources and knowledge, contributing to corporate priorities.

An example of a commercial maker space is The Factory, part of Knowle West Media Centre, situated in Bristol, UK equipped with digital fabrication technology as an accessible option for small scale prototyping for businesses and industry (KWMC, 2025). The site hosts community-orientated workshops designed to foster both upskilling and skill sharing, while also facilitating knowledge exchange and community-building initiatives. This growing network of maker spaces across the UK, both community and commercial demonstrates a clear demand for access to such facilities from both public and small-scale industry. With targeted investment, they hold significant potential to be scaled into localised micro factories creating opportunities for employment, enterprise, and skills development.

A desire by industry to initiate change may also be evidenced by the range of UKRI Innovate awards seeking to use artificial intelligence (AI) (Intelistyle, 2025; UKRI, 2015), digital technology (UKRI, 2020a), and circularity practices (UKRI, 2020b; UKRI, 2020c) to evolve and generate growth or promote behavioural change in fashion and textiles, for example, clothing rental (UKRI, 2020d), scan-to-fit technology (UKRI, 2012) and automated manufacture (FFF, 2025).

3.4 Digital Textile Printing Research

Research into digital textile printing spans foundational knowledge and intersects multiple disciplines, including textile technology, industry practices, colour management, textile chemistry, digital creativity, design education, and digital printing across other mediums. Key researchers in the field include, but not limited to Tyler (2005), Ujiie (2006a, b), Ding (2016, 2021), Carden (2016), Treadaway (2007), and the author Gooby (2023b; 2020). In addition, McKeegan contributing knowledge from a journalistic perspective (McKeegan, 2022). Figure 3.1 provides a diagram of key researchers in the field, detailing the field that they work within.

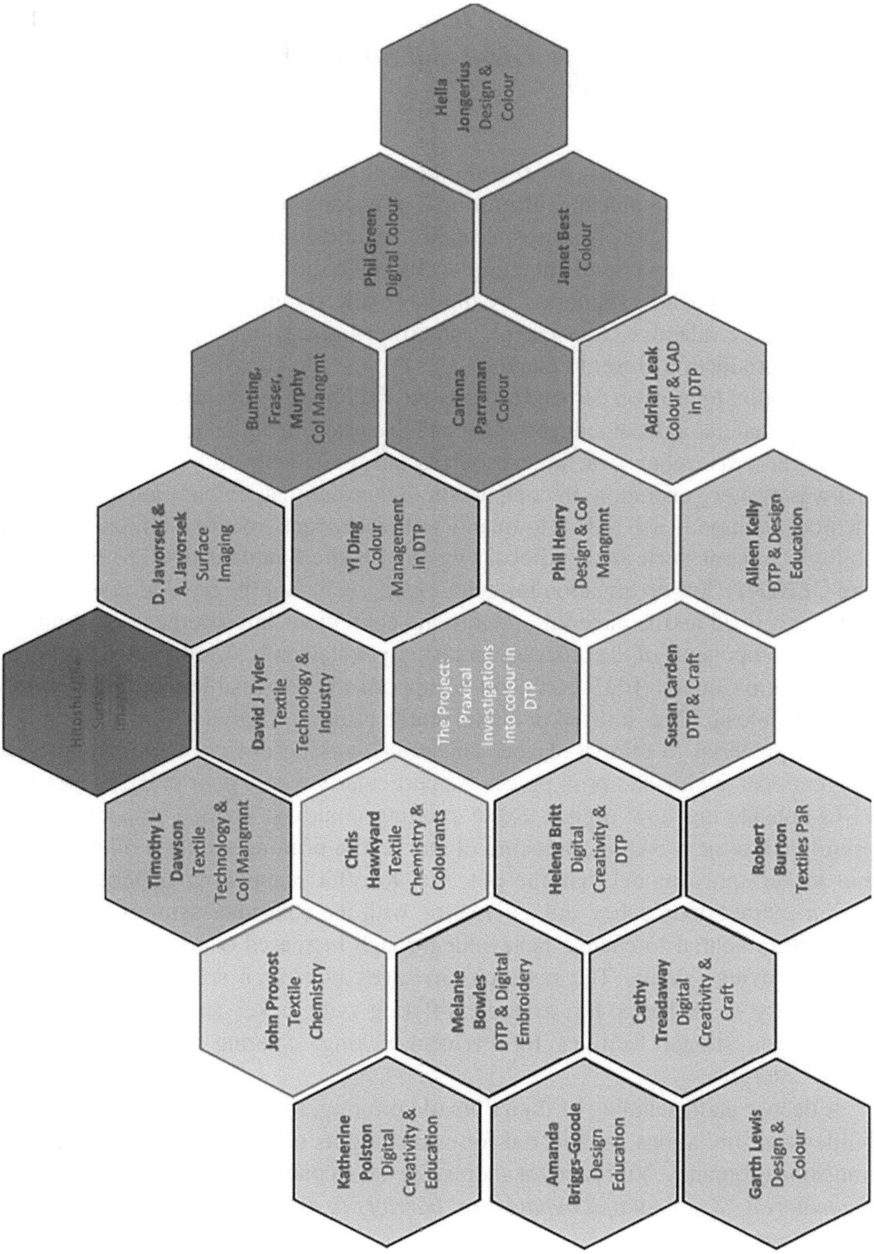

Fig. 3.1 Visualisation of key researchers in the field of digital textile printing based on a contextual review, Gooby (2023a)

3.4.1 *What Role Can Digital Textile Printing Play in Fostering More Local and Sustainable Manufacturing?*

This research proposes that the role digital textile printing could play in fostering more local and sustainable manufacturing is as a viable, immediate and complementary approach to support localised, scalable, manufacturing helping to address key industry challenges while maintaining the agility required for commercial success. Localised approaches, supported by digital textile printing, offer a potential shift towards more adaptive, transparent, and lower-impact manufacturing models by enabling production closer to market.

The ability to print on demand reduces the likelihood of overproduction. While it is widely understood that reduced water and dye usage in digital textile print production can decrease reliance on finite resources and dramatically minimise toxic effluent and wastewater generation, in comparison to conventional colouration and printing methods, comprehensive data to quantify these impacts remains limited. Sustainability standards such as the Global Organic Textile Standard (GOTS) for fabrics and OEKO-TEX® certification for colouration, including the regulation of harmful chemicals in all textile materials, are driving continuous improvements, making the various components of the process increasingly sustainable. Additionally, the Forest Stewardship Council (FSC) certification supports the responsible sourcing of wood-based cellulosic fibres utilised in digital textile printing. Digital textile printing also has the potential to offer a cleaner and safer working environment compared to conventional dyeing and print processes, particularly where poor practices persist.

In considering how digital textile printing could be applied to a customise-print-make-wear workflow, situating digital textile printing within a micro factory and maker space model, has the potential to offer compelling opportunities for design-centred approaches and co-creation with the consumer using mass production with potential to change behaviour through increased connection to clothing and improving quality. The research envisages that these spaces could become community workspaces for example offering cut, make, and trim techniques, and as educational facilities for garment-making, catering to individuals and micro-enterprises.

A deeper understanding of digital textile printing, including its processes, capabilities and limitations, is essential for evaluating its role in enabling a sustainable transition (Tomaney, 2015) and for addressing the second research question: *What are the material, logistical and behavioural barriers to achieving a sustainable digital textile workflow?*

3.5 Understanding Digital Textile Printing

The term digital textile printing encompasses a range of printing technologies including dye sublimation, solvent, latex, ultra-violet, and direct-to-garment printing, for example T-shirt printers (Romano, 2017). Unlike traditional textile printing methods, digital textile printing is a non-surface-impact print process. In inkjet printing, colourants, such as reactive dyes or pigment inks, are jetted in a dithered matrix of dots onto specially treated fabric to create the appearance of blended tonal colour (Carden, 2016). The colourant is deposited onto the fabric. Secondary processes allow dye molecules to permeate the fibres (Freire, 2017). Solvent, latex, ultra-violet, and direct-to-garment printing are all variations of inkjet technology. In contrast, dye sublimation printing involves heating solid disperse dyes into a gaseous state, which then bond with the substrate, either directly with polyester-based fibres or indirectly through transfer paper which requires an additional heat pressing, called calendaring, to fix the image and produce continuous-tone images (Romano, 2017).

Digital printing for textiles is a relatively emerging technology despite being in use since the 1970s in carpet printing (Cahill, 2006). The first recognisable digital textile printer was produced in 1991 by Stork. The Stork TruColour TCP2500 was designed for sampling (Kelly, 2014) and superseded by the Fashion Jet Textile Printer in 1995 for small print runs known as strike offs (Cahill, 2006; Dawson, 2003). Mimaki brought out their first TX series in 1998. It wasn't until 2003, at trade show ITM Birmingham, that digital textile printing became an established technology with the exhibition of machines by Du Pont, Mimaki, Reggiani (Moser & Wilson, 2000). Many specialist textile printer companies have since been bought by larger reprographic printing companies establishing digital textile printing as of viable economic interest (Belsham, 2017).

According to the Ellen MacArthur Foundation, the global textile industry is valued at approximately 1.3 trillion USD (EMF, 2019). Recent data from the World Textile Information Network indicates that digital textile printing now accounts for 6 percent of this market (WTIN, 2025). This reflects a gradual increase from earlier estimates of 3–5.5% reported by Hayward (2017), Scrimshaw (2017), and Ross (2018).

China is the biggest produce of digitally printed textile with figures showing 15% of production in 2016 (Scrimshaw, 2017). Before the technology took off in China, Como in Italy was the leader and retains many specialist companies. Both Western Europe and Asia are growth markets for production due to urban production and low costs (Hayward, 2017). In the UK government grants have prompted a return to inhouse printing and rapid prototyping providing some initial enhances to the UK digital textile printing industry (Ross, 2018). Acquisition of specialist companies by larger established first, for example Epson bought Robustelli and EFI bought Reggiani (Hayward, 2017) may have increased the promotion of pigment and dye sublimation printing over better quality colour, handle and print options ink-jet and reactive dyes due to existing reprographics expertise and practices from the parent companies.

Digital textile printing is seen as an instantaneous option achieving quick results. The industry has been driven by Just in Time and Concept to Consumer production and social media trends (King, 2006; Polston et al., 2015; Ujiie, 2006a, b). This has attracted entrepreneurial business models (Kelly, 2014). For example, drop shipping where items are commissioned after the order has bene placed, reduces the need for stock to be held. Digital textile printing has the potential to support both push and pull supply chain models, responding either to consumer-driven demand (pull) or to production based on market forecasts (push) (Grose & Mansfield, 2023).

3.6 Key Components of Digital Textile Printing

In inkjet printing, dots are emitted on demand, or continuously, which uses more ink and is therefore expensive (Parraman, 2017). The smaller the dots the paler the colour (Feisner & Reed, 2014). Dot gain describes the issue of dots altering in shape or size when they hit the surface of the substrate which impacts the colour perception (Fraser et al., 2004). This is a challenge in textiles substrates as they have textured rather than smooth surfaces so the dots spread differently to paper (Dawson, 2006). Textile substrates are usually pre-coated, mimicking the smoother and glossy surfaces applied to printing paper which can reduce dot gain (Tyler, 2005). Colour management processes which work with pixels rather than dots do not rectify this issue (Ding, 2016).

In dye sublimation printing, thermal activity exponentially heats disperse dyes into gas form to bond and form produce continuous-tone images on transfer paper or direct-to-fabric (Romano, 2017).

The jetting method and the print head determine the printer characteristics. Both inkjet and dye sublimation print heads suffer from clogging and misfiring and so machine maintenance and ensuring that an optimum print environment is in place, affect colour and print quality outcome. Colourants must have an incredibly pure and specific viscosity, less than 5mPas and tolerate 10 mPas (Tyler, 2005) for inkjet and 3–7 mPas for dye sublimation (Romano, 2017). Other variables are the make of the machine, whether the machine uses continuous or drop on demand print technology, the drop formation (Carr et al., 2006) and the speed positioning, make up of and type of print head (Freire, 2017). Bubble jet and drop on demand printing are favoured because of suitability as the alternative method, thermal jet, is a fire risk for textile substrates. Cost is also a consideration as drop-on-demand technology typically uses less ink than continuous tone systems, making it a more economical choice (Dawson, 2006).

Colourants can be classified as dyes, organic compound molecules dissolved in water or an organic solvent, or inks (pigments), non-organic particles in a suspension (Carden, 2016; Hoskins & Craine, 2025). The purity of colourants is determined by cost, stability (Polston et al., 2015) and multiple end-use requisites meaning quality may be compromised (Fraser et al., 2004) impacting colour and quality outcome.

Dyes have a superior colour range whilst pigments produce more permanent colour and do not require additional processing (Carden, 2016), for example steaming and washing for fixation. Dyes are soluble and more suitable for inkjet printing because they have low viscosity and can pass through the print head (Miles, 2010). Pigments can be ground small enough for inkjet printing but frequently clog print heads with small particles (Carden, 2016). Dyes and pigments have limited shelf life. Colour stability is therefore not guaranteed. A colourant's viscosity and consistency are key in managing colour outcome (Ingamells, 1993).

Similarly, the fibres used to make substrates are also variables (Carden, 2016). Colourant developers must consider how a colourant will react with different fibres, and any pre and post treatments they undergo (Tyler, 2005). The colour and quality outcome will also be affected by the amount of moisture in a colourant or substrate (ASBCI, 2014). Colouration is a complex and variable process and the method of transferring the colour and binding it to the fibre is not exact (Ingamells, 1993). An image reprinted under the same conditions can therefore still appear different because the chemistry is imprecise (Carden, 2016).

The construction yarn and weave of a fabric all affect the achievable colour gamut (Carden, 2016) and print outcome. The end product will ultimately define the fabric choice (McKeegan, 2018). Cotton polyester and hybrid fabrics are the most common fabrics in digital inkjet textile printing, but silk, wool, cellulosics and linens are all compatible. Dye sublimation is suitable for with polyester or poly mix substrates. Each has its own absorbency surface tension and potential for wicking when the fabric is over saturated which impacts the optimum amount of colourant that can be used (Campbell, 2006; Carden, 2016; Ingamells, 1993).

Fabric construction is complex and there are several processes to prepare for print (Provost, 2017) for example singeing, scouring, and bleaching (ASBCI, 2014). Fabrics are primarily three-dimensional constructions woven or knitted (Kim, 2006) and therefore the space between the fibres require consideration too, as this will impact colour perception (Miles, 2010). Protein fibres, made by animals, have good absorbency and colourant retention as does cotton and bamboo cellulose fibres, Other cellulose fibres such as linen and hemp which are oilier, have less absorbency and prints tend to fade. Synthetic fibres which are derived from petrochemicals, have poor to little absorbency and so colourants sit on the surface, reducing rub fastness (Ingamells, 1993).

Digital printed fabrics require additional pre-treatments (Hawkyard, 2006) to prepare them for printing. Pre coatings are adapted from knowledge of the binders and thickeners used in traditional dye and print pastes (Hawkyard, 2006). They also help reduce drop gain where an undesirable colourant spread occurs when it hits the textured surface. Pre coatings are also included to enhance print colour quality and provide properties for textile end uses. Once printed, substrates may undergo steaming and washing to allow the binders and thickener used with inkjet printed dyes to react, enhancing the achievable colour gamut and quality (Provost, 2017) or simply be washed to remove surface pre coatings. Coatings use conventional textile chemistry for example acid and alkaline treatments (Dawson, 2006) to aid fixation print and colour quality and end use (Provost, 2017). They have high viscosity levels so they cannot be jetted using typical print head and there are concerns about damage to the machinery if they were (Provost, 2017). A bespoke coating for the print environment, fibre and colourant combination, would give the best outcome but many fabrics are supplied with a generic pre coating (Tyler, 2005). Coatings must be consistent and stable (Provost, 2017) to achieve the chemical reactions necessary for colour yield penetration (Miles, 2010). Likewise, the steaming process must be uniform to achieve the desired chemical reactions (Carden, 2016). Kelly (2014) notes that adjusting steaming times improves colour outcome and advises lightning colours in design software to counter.

3.7 Engagement with Digital Textile Printing

Digital printing technology has had a major impact for the textile designer. Their design and process now enables them to create complicated multi-layered digital designs incorporating photographic images and intricate detail. Artwork is no longer restricted to several colours or a repeat pattern. Leading in new entrepreneurial business models in the textile industry, it has become possible to print a metre, or hundreds of metres, at the click of a button, without high set-up costs. In particular, digital textile printing provides unprecedented opportunity for designers to offer bespoke and customised designs (Polston et al., 2015). Production models (King, 2006; ASBCI, 2014) benefit from short print runs and expedited strike-offs, product sampling and testing made possible by digital textile printing (Dawson, 2006). With low set up costs and the ability to print small batches a designer can be free to customise and be innovative (Tyler, 2005). Downtime is minimal as designs can be printed to demand, responding to changes in trends and markets and wastage is low (Dawson, 2003; Scrimshaw, 2017). Digital textile printing can be used as the basis of further textile work such as embellishments (Britt, 2012). Digital textile printing can reduce pattern waste by printing direct-to-pattern outlines, although allowances must be made for shrinkage.

Digital textile printing has been used in costume design to print the illusion of heavy cloths such as tweed or wool onto light weight, flexible fabrics allowing freedom of movement for the performer or recreate delicate, rare or expensive fabrics. For example, the work of Hatley Print (Hatley Print, 2025), a studio specialising in printing for costume and theatre. Similarly tromp l'oeil effects have been used to print images of jewellery, sequins or embroidery onto fabrics used in fashion and soft furnishings. For example, the work of Hil Dreissen Vandeijne (2025), Katranzou (2009) and IA London (2025).

The designer will not commonly oversee any of the production process and prints are made in their absence (Carden, 2016). Commissions may even be conducted entirely online with fabric delivered, negating any need for contact. Designers need to learn how to communicate with a digital print bureau and request practical support from a digital print bureau on how to prepare the artwork, to which colour management processes they use, and information about their print environment (McKeegan, 2017, 2018) to ensure a satisfactory outcome and avoid reprinting.

Early industry research at InkDrop boutique found mass customisation difficult to manage, citing challenges with colour accuracy and image quality. To address these issues, the company focused on '*managing expectations*' (King, 2006, p. 333), offering user guidance, a downloadable customisable design library, and colour management tutorials. While these efforts highlight initial limitations, subsequent technological advancements and improved workflows may have alleviated some of these early barriers.

3.8 Digital Print Bureaus and UK Textile and Digital Textile Printing Industry

Digital print bureaus offer a digital textile printing production service to designers and SMEs. They are primarily located in urban environments speeding up textile production to the retailer (Ujiie, 2006a, b). There are international, commercial digital print bureaus such as Spoonflower and Contrado and smaller independent business. A survey of digital print bureaus in the UK revealed several established companies for example CAT Digital, Forest Digital, RA Smart, and PrinFab. However, operating a digital print bureau is challenging with many companies closing including Cameron Gilmartin, Be Fab, Surface Pattern Print and Silk Bureau. Some businesses have maintained visibility by specialising in niche areas. For example, Hatley, focuses on costume and theatre fabrics, or CAT Digital, affiliated with Glasgow School of Art, specialises in printing a range of archive designs.

Research has found that the poor adoption of good professional working practices, lack of institutional guidance and a wealth of specialised technology has led to concerns about quality and colour issues (Ding, 2016; Ross, 2018; Gooby, 2023b) potentially undermining business viability. Rising energy costs and narrow profit margins present significant challenges. In addition, the cost of reprinting, due to the myriads of variables that can impact quality and colour outcomes, can make it difficult to achieve profitability. The UK market is now predominantly focused on inkjet pigment and dye sublimation printing, with very few companies currently offering reactive dye inkjet printing. This is despite its superior colour quality, fastness, and fabric handle, features typically preferred by fashion brands, as the additional processing steps of steaming and washing remain a deterrent (Gooby, 2020; 2023a). Following the recent closure of the Silk Bureau, brands are increasingly having to seek overseas print bureaus for reactive dye inkjet printing on GOTS-certified fabrics, due to the now limited availability of such services in the UK (Esce-Tex, 2025).

An interview with a digital print bureau based in the Southeast of England attributed their successful and longstanding business to the integration of automation into their systems which reduced labour costs (Gooby, 2025a). This efficiency is supported by the team's background in software design, enabling a technologically streamlined approach to production. In addition, they streamlined their operations by reducing their range of printing machines to focus primarily on pigment inkjet technology. Combined with automation, this approach has minimised the need for reprints, which, when necessary, are typically occur in their business as the result of substrate-related issues, particularly with pre-coatings. They identified challenges in fabric supply as a significant area in need of improvement, noting that the current market is dominated by a limited number of suppliers. Despite operating with pigment inkjet technology, their processes still rely on substrates that have been pre-coated.

Another digital print bureau, based in Scotland, described how naive they felt entering the industry (Gooby, 2023b). They felt frustrated by lack of shared information which they felt were down to commercial sensitivity. Initially it had been advised to alter artwork to try and match the colours their customers expected but found that this caused problems when repeat batches were ordered. They upgraded their colour management processes to improve colour consistency and found that with increased experience they could achieve better colour accuracy.

With its industry expertise, technological capacity, and collaborative research base, the UK is well-placed to position digital textile printing as a driver of sustainable innovation in fashion manufacturing (UK GOV, 2025b; BFTT, 2025) for localised production. Real impact however, depends on changes in consumption and consumer behaviour as well as production, where design-led and co-creation approaches already embedded in UK's strong design culture and knowledge base, could be key. UKFT measures the contribution of the textile industry contributes £62 billion to the national economy (UKFT, 2023) highlighting a thriving sector with strong potential for growth through investment in innovations such as digital textile printing. To fully realise this potential, the UK fashion and textile industry, which employs approximately 800,000 people (ONS, 2025), will require significant upskilling in digital and technical skills in creative technologies.

Having reviewed the components and engagement with digital textile printing to contextualise the material, logistical, and behavioural factors that may present barriers to a sustainable workflow, the research next identifies three key areas of sustainability that show both positive developments and opportunities for improvement; reduced waste and water use, energy and chemical efficiencies, and integration into the sustainable supply chain.

3.9 Evaluating the Sustainability of Digital Textile Printing

3.9.1 Reduced Waste and Water Use

One of the most frequently cited sustainability benefits of digital textile printing is its ability to minimise resource consumption. Compared to screen printing or rotary methods, digital printing requires significantly less water and fewer dyes in production (Tomaney, 2015). The process applies colour directly to fabric without large-volume dye baths or extensive rinsing, which traditionally generate harmful wastewater. This makes it an important innovation in addressing one of the textile industry's most damaging phases: the dyeing and finishing processes, which have long been associated with toxic effluent discharge and high-water use (Greenpeace, 2011).

3.9.2 Energy and Chemical Efficiency

Digital textile printing also allows for greater control over chemical inputs, often using pretreated fabrics and reactive or pigment inks with more targeted application (Ujiie, 2006a, b). While energy is still required, particularly for fabric preparation and post-print curing, this is generally lower than for conventional methods that demand heating large dye baths or drying thick layers of ink (Ingamells, 1993). Furthermore, because printing is done on demand and at smaller scales, digital methods reduce overproduction, a key sustainability issue in fast fashion cycles that rely on volume and speed.

3.9.3 Integration into a Sustainable Supply Chain

Within the proposed micro factory maker space model, digital textile printing has the potential to offer complementary and scalable localised manufacturing. This approach could help address challenges in the current fragmented supply chain by enabling greater transparency and traceability. Consumers could either participate directly in the making process or gain a clearer understanding of who made their clothes, with opportunities to engage in mass customisation or bespoke design with the potential for behaviour change.

In traditional textile printing there may be an extensive supply chain which can include a textile mill, colouration laboratory, dye factory, cutting and sewing production, and the retailer. Many of these may also be overseas (Lawn, 2006). Digital print bureaus are likely to be located in an urban environment, local to their customers, with reduced transport costs, and occupying less space than a traditional textile mill or factory (Ujiie, 2006a, b). Short print runs and the ability to react quickly to market trends to allow businesses to carry less stock, reducing storage and costs, and the likelihood of stock wastage as fabric is printed to demand (Ujiie, 2006a, b). These changes offer exciting possibilities for designers and small and medium sized enterprises such as design studios working in the textile industry.

Data on the environmental impact of digital textile printing remains limited, and further research is needed to generate and make publicly accessible robust, transparent data that can inform sustainability assessments and best practice across the industry (Gooby, 2023a). It is important to note that water use and waste still occur throughout other stages related to digital textile printing. This includes the manufacturing of printing machines, the production and construction of textile substrates, the development and application of chemical pre coatings, and the manufacture of colourants. Additionally, digital textile printing itself generates waste, such as used printer cartridges and discarded materials from reprints. Post processing such as steaming and washing will also use water and produce waste (Gooby, 2025b). Further, the energy required to manufacture digital printing machines, along with the sourcing and production of their components such as metal and plastic parts must all be considered. However, novel technologies may offer solutions for example blockchain, automation and AI approaches (FESPA, 2021) are being adopted to enhance traceability, forecast demand, and cut down on surplus inventory and waste (Grose & Mansfield, 2023).

To accurately assess the environmental impact of digital textile printing, it is essential to use Life Cycle Assessment tools. These should capture the cumulative effects of energy use, emissions, water, and chemical inputs across the entire product journey, from fibre cultivation to final disposal. While digital printing may score well in several areas, its true sustainability depends on thoughtful integration into the design, production, and post-consumer phases. As the industry moves toward more transparent and circular models, digital printing must be paired with responsible sourcing, ethical labour practices, and consumer education to fulfil its sustainable promise. Future research should therefore prioritise the development of comprehensive, standardised life cycle assessments that encompass every stage of the digital textile workflow.

Despite the efficiencies outlined, digital printing alone cannot resolve the structural challenges of sustainability in textiles, but it offers some immediate solutions to reduce harm, generate efficiencies in production with potential to also improve consumer behaviour.

3.9.3.1 What Are the Material, Logistical, and Behavioural Barriers to Achieving a Sustainable Digital Textile Workflow?

Material challenges primarily concern the limited availability of truly sustainable fabrics and environmentally responsible colourants, which constrain the ecological benefits of the printing process and hinders circularity. Access to sustainable supply chains can be challenging due to the monopolisation of substrate supplies and the requirement for printing systems to use specific colourant setups. The origins and processing of materials to include fibres, fabric construction, colourants and chemical processes as well as exploitative labour conditions all require broader system redesign (EMF, 2019; UKFT, 2023). Positive developments come from GOTS certified fabrics and OEKO-TEX colourants but further improvements are still needed.

The logistical, practical and operational barriers concern operating space, waste management and business planning. Whilst set up costs are low, a digital textile printer will still require a temperature and humidity controlled operating space, and dependent on the type of print environment, post processing machinery. Further recycling of textiles is a challenge particularly of blended fibres so waste management may be complex. Disposal of print cartridges including residue liquor will require hazardous waste removal in addition to disposal of plastic and metal casings. Furthermore, as production is typically small-scale and labour-intensive, when implemented without automation, maintaining profit margins and engaging in effective forward planning can be difficult. Finally, and specifically addressing the proposed model for digital textile printing, the establishment of a micro factory within a maker space is an emerging concept and its viability remains unproven within the current economic context.

On the behavioural side, the absence of widely shared best practices among practitioners continues to affect print consistency and efficiency. It is not clear to what extent does the entry of individuals into the industry without a foundational understanding of textile characteristics contribute to limited awareness and comprehension of sustainable practices. A lack of opportunities for training and skill-sharing, often driven by concerns about industrial espionage within digital print bureaus, acts as a significant barrier. Additionally, designers and users of these services frequently lack the technical terminology needed to communicate effectively with print technicians. This is compounded by a limited understanding of the digital print process itself, leading to unrealistic expectations, such as assuming the final print will precisely match the on-screen image, without recognising the variables that influence print outcomes, Additionally, current models often position digital textile printing as a remote, service-oriented offering. Reframing it as a participatory tool for co-creation, where consumers and designers collaborate more directly, could unlock greater value and align the technology more closely with principles of localised, sustainable production.

3.9.3.2 How Might Design-Led Approaches Leverage This Technology to Drive Systemic Change in Fashion Production?

From a commercial perspective, SMEs and freelancers must drive growth as entrepreneurial progress (Ujiie, 2006a, b). Additionally, designers, in particular, textile designers are well equipped to problem-solve including overcoming printing variables and inconsistencies based on their understanding of textile characteristics (Carden, 2016). As print bureau customers they can drive demand towards sustainable improvements, such as bio-colouration and greater ranges of GOTS-certified fabrics prepared for digital print. However, too much inconsistency can lead to significant waste of resources. Therefore, education and the right tools are needed to enable the designer to anticipate issues and problem solve in the initial design stages (Henry, 2013; Kelly., 2014; Leak, 1998) thus reducing material wastage, colour issues and cost.

The proposed model, micro factory and makerspace aligns with the Council of Logistics Management's supply chain principles for commercial success '*Right goods, Right place, Right time, Right quality,* and *Right quantity*' (Grose & Mansfield, 2023), by enabling production on demand and consumer behaviour driving the use of sustainable and appropriate materials for real-world impact. This increases opportunities for designers, print bureaus and consumers, to collaborate directly, enabling the development of shared solutions that respond to both creative and production challenges.

To shift consumer behaviour away from the traditional *take-make-dispose* model, a micro factory approach using digital textile printing should position consumers as active co-creators. Through mass customisation, scan to fit technology, and opportunities for consumers to participate in the making process, either by engaging in cut, make, and trim services or by learning to assemble their own clothing, there is potential for wearers to develop a deeper connection to their garments. This connection, combined with better fit and personalised design, can encourage longer use and reduce the likelihood of disposal before the garment's end of life. Components including care labels, unnecessary to the garment construction, could be printed reducing cost and materials required. This model should also include digital repair services and education around garment care, furthering garment lifespans and encouraging more mindful consumption habits with potential to meet the 7R model, *reduce, reuse, recycle, restyle, rewear, redesign, reimagine* (Grose & Mansfield, 2023; Sinclair, 2014), to rethink we design, manufacture and care for our clothing (Krasny, 2020).

Printing on more sustainable substrates such as those certified by GOTS, the Global Recycled Standard (GRS), or the Recycled Claim Standard (RCS), has the potential to substantially lower the environmental footprint from production to end use. Further innovations in raw materials such as chemical-free fibres (Thomas, 2019) and bio-fibres from waste (Thomas, 2019) promise to expand the range of printable substrates, provided they can be suitably prepared for digital printing. For the UK market, these developments present valuable opportunities to utilise localised digital textile printing, especially if such materials can be sourced locally, helping to shorten supply chains and reduce environmental impact.

These design-led approaches encompass co-creation processes, efforts to influence and modify consumer behaviour, and initiatives aimed at transforming supply chains by driving demand for the availability and use of sustainable fabrics and colouration methods. Additionally, innovations in printing aesthetic, such as the integration of printed components and the development of digital repair techniques, illustrate how design strategies can harness digital textile printing technology to effect meaningful and practical changes in production practices.

Examples of two design-led initiatives that have demonstrated innovative uses of digital textile printing and have the potential to inform how it could be used for complementary manufacture are: a case study of printed denim, illustrating a pilot study working with cellulosic fibres to produce printed jeans, and a case study of a UK-based fabric manufacturer expanding operations by establishing an in-house design studio to digitally print and showcase their Irish linens. While neither case study offers an exact match to the micro factory maker space model proposed, both provide valuable insights and examples of combining design innovation with digital technology to achieve sustainable outcomes and localised production. As such, they should be regarded as important building blocks for future developments. Key to both was the utilisation of fibres and fabrics produced in the UK which for different reasons boosted the sustainable outcomes. This is of importance because the choice of substrate plays a crucial role in the success of digital textile printing as a technology (Gooby, 2023a). Thus, each study details the respective substrates and their sustainable credentials to provide additional context.

3.10 Case Study One: Printed Denim—The £27 Jean

Denim jeans are a prime example of *take-make-dispose* culture. Research by Ellams in 2017 (Ellams, 2017) provides a valuable example of how digital textile printing could be used for alternative and sustainable manufacture. Working with textile manufacturer Lenzing Tencel, based in Grimsby UK and Austria, Ellams manufactured a jeans protype, printed onto cellulosic fibre, made from eucalyptus and beech wood pulp, at a price point of £27 (Trend Hunter, 2013).

Denim jeans are always in fashion and production spans the globe. However, their design and manufacture is highly polluting and consumes substantial resources. To produce a pair of jeans it takes 2,911 L of water and emits 21 kg carbon dioxide equivalent (Levi Strauss, 2015). There are health implications for those working in the industry due to exposure to toxins and pollutants in the dyeing and finishing stages. The documentary 'River Blue' shows effluent from textile factories in Xingtang, China, the production capital of blue jeans, running into rivers, slowly infecting the rest of the world's waterways with heavy metals and highly persistent chemicals from dyeing, and chemical and abrasion finishing processes (RiverBlue, 2014). The growing of cotton from which denim is made requires vast amounts of water, fertilisers and pesticides which pollute and contributes to the loss of arable land. Overproduction sees 5 billion pairs (Thomas, 2019, p. 77) made yet 2 billion bought each year (EMF, 2023). Once purchased laundering consumers more water and energy and contributes to further aquatic pollution through dye loss as the indigo colour fades (Fletcher, 2013). Comfort stretch fit denim jeans use blended fibres for example elastane and include difficult to remove secondary components rendering them a challenge to recycle (EMF, 2023). Discarded jeans are sent to second hand clothing markets, eventually dumped slowly infecting the environment (Manieson & Ferrero-Regis, 2023). The UK Fashion and Textile market alone imports £587 M denim jeans and exports c£220 M (HMRC, 2025) a year demonstrating market demand and potential.

Ellams' method, recreate the colour of blue jeans using digital textile printing and printing to pattern, established an 80% reduction in water and energy use compared to conventional denim manufacturing (Ellams, 2017). The method importantly eliminated the need for high-impact, toxic dyeing processes involving synthetic indigo, which is a major industrial pollutant that contaminates wastewater and generates hazardous effluent, thereby offering a more efficient and improved approach to sustainable production. However, it must be acknowledged that the dyes used in digital textile printing are still classified as hazardous and contain harmful chemicals, despite being used in significantly reduced quantities compared to conventional denim colouration. Further, digital printing was used to emulate the visual texture and aesthetic appearance of denim, achieving authenticity while reducing the need for chemical and physical abrasion processes which use forever chemicals or PFAs (RiverBlue, 2014) to recreate the look of denim through an alternative colouration process.

In her research, Ellams printed onto cellulosic fibres, presenting an alternative substrate to both natural protein fibres and synthetic materials. Although cellulosic fibres had been used in garments such as underwear and casual wear, Ellams' application to denim demonstrated a novel integration with digital manufacturing techniques. Derived from wood pulp sourced exclusively from FSC® or PEFC-certified forests, TENCEL™ fibres in particular benefit from a closed-loop production system that recycles over 99.8% of solvents, drastically reducing waste and chemical use in comparison to other cellulosic fibres such as viscose rayon. Life Cycle Assessments and data from the Higg Materials Sustainability Index (MSI) show that TENCEL™ Lyocell has up to 53% lower global warming potential and up to 69% less water consumption than conventional fibres like cotton and PET, which positions it as an environmentally viable option. The fibres are also biodegradable in soil, freshwater, and marine environments, and compostable, further addressing end-of-life concerns in textile sustainability (Gooby, 2023b; Lenzing Tencel, 2025).

However, while these fibres offer a significant reduction in environmental impact at the raw material and fibre production stages, it is important to note that these assessments are based on the fibre rather than finished products. As such, the sustainability benefits rely on continued responsible practices throughout the supply chain including fabric construction, dyeing and finishing.

Several examples of printed jeans are beginning to appear on the market from brands such as Rag and Bone (Rag & Bone, 2025) and Bottega Veneta (Bottega Veneta, 2025) who have printed leather to look like denim. Although in the case of Bottega Veneta, high end and using leather an unsustainable material. Ellam's faced resistance from denim connoisseurs however due to the different feel, handle and weight of the fabric. But printed jeans present a viable and more sustainable alternative to fast fashion and ultra-fast fashion denim typically sold through mass-market retail channels such as supermarkets with potential to change consumer behaviour.

Consequently, this research proposes that building on this study, there are clear opportunities for commercial exploitation. Technology has developed considerably since Ellam's pilot, with increased speeds, improvements to print heads and colourants as well as software, bringing further potential. Coupled with supply chain support, design-led SMEs could access digital printing through micro factory and maker space setups to establish a denim printing facility, utilising developments in scan-to-fit technology and pattern generation software for better fit. Mass customisation would allow selection of distressed finishes and printed components according to preference. Consumers could engage with makerspace services for cut, make, and trim production, particularly as jean construction is complex, or participate in hands-on learning to partially construct their own garments. This offers a way to circumvent the costs associated with the labour-intensive process of sewing. Additionally, they could return for digital repair services, supporting a circular approach to denim ownership and extending product lifespans. Connections to sustainable supply chains, through facilitators such as Esce-tex, can offer designers expanded opportunities to utilise digital textile printing on a wider range of materials, including varying fabric weights, thereby enhancing the ability to more closely replicate the characteristics of traditional jeans.

This case study illustrates a design-led approach to leverage digital textile printing technology to drive systemic change in fashion production offering a viable pathway towards localised denim production. In particular, the collaboration with cellulosic fibre manufacturer TENCEL™ introduced an additional localised sustainability benefit in supplying a substrate made from fibres produced within the UK. To fully realise the potential of localised denim production, future efforts must prioritise the development and scaling of accessible infrastructure, specifically micro factories and makerspaces, that support designer-manufacturer collaboration, while also evaluating consumer demand to ensure such models are both practical and desirable. Continued innovation and commercial investment will be essential in turning these sustainable prototypes into mainstream solutions.

3.11 Case Study Two: Ancient and Modern—Digital Textile Printing in the Mill

Earthed Fabrics, the creative arm of the historic William Clark and Sons linen mill in Northern Ireland (William Clark & Sons, 2025), illustrates how digital textile printing can be strategically adopted to expand into new markets such as finished products. Established in 1736 William Clark have been producing linen for three centuries. Seeking to innovate they integrated digital textile printing into their operations by launching *Earthed* as a design-led interiors brand in 2016 (Gooby, 2023a). By leveraging their control over the entire textile process, they could print directly onto their Irish linens to produce luxury interior products such as cushions, lampshades, and decorative fabrics.

In research exploring colour variability in digital textile printing (Gooby, 2023b) the Creative Director spoke about his early exposure to digital textile printing during his university studies which sparked his interest in the technology. His subsequent industry experience expanded his expertise working with digital textile printing mechanisms from Stork (now SPG) and M&R Polaris refining his knowledge of digital workflows. At the time Earthed were using a d. Gen Arachne printer, with reactive dyes, alongside AVA design and raster image processing software. Challenges remained, particularly with maintenance issues such as clogged nozzles and, at times, ensuring consistency across the print-to-finish stages, exacerbated by the split-site operation where printing occurs in a modern facility and steaming/washing in an older building with less environmental control (Gooby, 2023a).

However, deep material knowledge enabled them to mitigate many of these issues. Because William Clark constructs the substrate from raw linen fibre through to the finished fabric, *Earthed* was able maintain a high degree of quality assurance and substrate consistency. This included developing a bespoke precoating formula tailored specifically for digital printing on linen. The ability to control the entire process, including creative strategies to mask minor inconsistencies in print, enabled the company to confidently diversify into premium textile products (Gooby, 2023b) using digital textile printing for efficient and localised production.

Linen, made from the flax plant, is considered one of the most sustainable fibres, particularly in comparison to cotton and synthetics, due to its low water requirements as it doesn't require irrigating, and has high resistance to pests (Sinclair, 2014). However, pesticides are required to prevent weeds growing in between the flax crop which produces straight stems leaving space and light for other plants which grow between (Gooby, 2025c). The fibres are strong which make it a durable fabric, historically used for a myriad of textile products including sail cloth due to its durability and hydrophobic qualities (Gooby, 2025a). Flax production has traditionally been concentrated within Europe, particularly Belgium, France and Lithuania, however production and processing has with China now a significant player (Gooby, 2025b). Due to the unique characteristics of the flax plant, industrialisation of its processing is more complex compared to other crops (Gooby, 2025c). Flax requires specialised machinery for distinct stages making mechanised cultivation and processing more labour- and equipment-intensive (Gooby, 2025a). In industrial settings, retting, a process used to separate the fibre from the plant, may involve the use of chemical agents or substantial water consumption, both of which raise environmental concerns and impact the overall sustainability of flax fibre production (Gooby, 2025b; Sharma, 1987) However, the entire flax plant can be used to make oil, food from seeds in addition to the fibre so there is minimal wastage. Flax produces a delicate blue flower and is primarily self-pollinating, though it can also be pollinated by wind (Gooby, 2025c; Sinclair, 2014). As a result, it does not rely on insect pollinators, which contributes to its ecological resilience and suitability for low-input farming systems. As a natural fibre, flax is fully biodegradable; however, like all dyed textiles, the presence of synthetic colourants can hinder composability and contribute to environmental pollution during degradation. Despite challenges, linen represents a highly sustainable fabric choice.

Although William Clark continues to offer digital printing services, their *Earthed* fabric line has, since 2024, been sold exclusively through third-party retailers (The Sustainable Marketplace, 2025), marking a change to the original business model (Canon, 2025). This suggests that while the *Earthed* venture was initially promising, it has not continued in its original form. To fully understand the factors behind this transition and assess the long-term viability of such initiatives, it would be valuable to conduct follow-up research to identifying the challenges and decisions that led to the reconfiguration of the business to draw meaningful lessons. In particular, it would be important to assess whether any market research was conducted to determine if consumers valued the option of locally printed linen as this would provide evidence that the dual approach of efficient production coupled with consumer behaviour change held measurable value in this instance.

This case presents an example of a distinctive application of digital textile printing with some notable success and value in delivering a high-quality interior product range, despite subsequent changes to the original business model. Additionally, it offers further insight into how fibre production can be integrated with digital textile printing, as described in Printed Denim, to enable localised manufacture, serving as a value-adding strategy that leverages material expertise and the use of a sustainable substrate. It offers valuable lessons to build upon, including the importance of process control for consistency, the role of design approaches in ensuring successful print outcomes, and the significance of expertise in digital printing and workflow management for overall success. Further research would be required to ascertain fully the lessons learned from the business's closure to further inform the proposal that the unique selling point of digital textile printing lies in its integration within localised production models and to determine whether any measurable consumer behavioural change was established.

3.12 Conclusion

This chapter has proposed that digital textile printing holds significant potential as a complementary and scalable solution for localised manufacturing within the fashion industry. To support this proposition, it has explored three key research questions: *What role can digital textile printing play in enabling more local and sustainable modes of production? What material, logistical, and behavioural barriers must be addressed to establish a truly sustainable digital textile workflow? And how might design-led approaches harness this technology to drive systemic transformation in fashion production?*

To address the question of *what role digital textile printing could play in fostering more local and sustainable manufacturing*, this chapter first contextualised the challenges within the existing fashion and textiles industry model. These included convoluted global supply chains, overproduction, reliance on finite resources, pollution, labour exploitation, and unsustainable consumer behaviour. While systemic change was recognised as essential, it was also understood to be a gradual process. In the meantime, accelerating immediate, process-level interventions through accessible technologies such as digital textile printing was proposed as a timely, though imperfect, step toward more sustainable production models.

Next, it further contextualised the topic through a review of research paradigms around sustainability in fashion and textiles, identifying the contributions of organisations and researchers who were addressing industry challenges, developing roadmaps and circularity strategies, and proposing design-led approaches to support systemic change. It was established that methods which improved production efficiency and encouraged consumer behaviour change were welcomed and recognised as positive developments for sustainability further supporting the idea that digital textile printing could be utilised for sustainable production. It highlighted research focused on innovating British manufacturing, particularly the role of maker spaces as existing networks that digital textile printing micro factories could leverage, especially where cut, make, and trim and garment production resources could be harnessed. Given that garment construction remains labour-intensive, it represents a key challenge in terms of affordability and scalability. Additionally, access to sustainable supply chains for improved colouration and substrate options was identified as a critical factor for enhancing overall sustainability. It examined why sustainability represents an urgent challenge, the necessity for change, and how the industry has begun to respond, demonstrating that the adoption of technology is not only desirable, but both achievable and essential for progressing towards sustainable transformation. Furthermore, it contextualised digital textile printing by surveying key areas of research in the field and identifying leading contributors, noting that current scholarship largely focuses on foundational knowledge rather than exploring business models or strategic applications for digital textile printing.

It concluded that digital textile printing presents opportunities for enabling localised manufacturing within a micro factory model, integrated into maker spaces accessible to consumers, designers, and SMEs. Such spaces could support prototyping, small-scale production, knowledge exchange, and co-creation, fostering adaptive, transparent, and lower-impact manufacturing by facilitating production closer to the point of use or market. The ability to print on demand was identified as a means to reduce the risk of overproduction. The chapter proposed a '*customise–print–make–wear*' workflow, emphasising the reduced water, dye usage, and material waste associated with digital textile printing, while also calling for greater compatibility with sustainable colouration methods and responsibly sourced substrates.

To address the question, '*What are the material, logistical, and behavioural barriers to achieving a sustainable digital textile workflow?*', the chapter first outlined the technical distinctions between inkjet and dye-sublimation printing and examined the key components of digital textile printing, including colourants, substrates, and the associated pre- and post-processing stages to understand the technical and environmental challenges inherent in the production process. It then explored examples of creative engagement, highlighting the practices of specialist print bureaus and designers who are leveraging the technology in innovative ways. In addition, it reviewed existing research on mass customisation, positioning it as a potential model for more responsive and resource-efficient production. It focused on the UK market, examining the current landscape of digital textile printing by contextualising prevailing business models through insights from print bureaus and relevant research. The analysis identified a lack of standardised best practices, frequent reprinting, and slim profit margins as key challenges. Despite these issues, it proposed that the UK possesses a strong foundation, particularly in design expertise and existing print infrastructure, on which to develop micro factories as a viable and entrepreneurial model for digital textile printing.

To assess the barriers the research identified three key areas: reduced waste and water usage in production, the impact of energy and chemical use, and the integration of the technology into existing supply chains. While digital textile printing presents clear advantages, it emphasised that a full life cycle perspective is essential. The chapter also called for further research to assess the true environmental credentials of the process and advocated for the integration of automation and appropriate AI approaches to enhance production efficiency. It also concluded that material barriers include the limited availability of sustainable fabrics and environmentally responsible colourants. Logistically, a novel business model and operational concerns such as waste management, forward planning, and identifying appropriate location conditions were identified as key considerations possibly compounded by lack of in-depth knowledge of textile characteristics. Behavioural barriers were found to involve the lack of shared best practices within the industry and the need to reposition digital textile printing as a tool for co-creation, engaging consumers directly, rather than as a remote, service-based process.

To answer the question, *How might design-led approaches leverage this technology to drive systemic change in fashion production?*, the chapter proposed that designers and SMEs have the potential to lead improvements and develop innovative, entrepreneurial business models to influence consumer behaviour. It outlined a design-led micro factory framework that positions the consumer as co-creator, strengthening their connection to clothing through active involvement in the design and production process. This approach also encourages longevity by fostering emotional durability, promoting care and repair through digital tools, and offering services such as scan-to-fit, digital alterations, and qualified cut-make-trim options, including '*make your own*' models, that support better fit, functionality, and circular practices.

It then presented two case studies in which digital textile printing served as a foundation for innovation and as potential springboards for further development, offering insights to inform future practice. Printed denim utilised cellulosic fibres sourced from the UK and Europe and circumnavigated the use of synthetic indigo dye, a known hyper-pollutant, further bypassing intensive chemical processes by digitally reproducing the aesthetic qualities of traditional jeans at a commercially viable price point. It provided examples of printed denim already on the market and proposed a denim making facility using the *customise print make wear* model.

In a second example, heritage expertise and intricate knowledge of linen production were highlighted as giving the mill significant control over its fabric, which in turn enhanced the quality and reliability of digital print outcomes. However, despite this advantage, a shift in the business model ultimately led to the venture's discontinuation. This raises important questions about what could have been done differently and underscores the need for further research to inform future businesses, particularly those adopting micro factory models, on how to avoid similar challenges. Although this case study does not fully align with the proposed micro factory model, it is included for the specific insight it offers into print setup and the critical role of textile knowledge. It highlights how designers, equipped with an understanding of fabric characteristics, could play a leading role in driving the success of such models. Since the substrate was a key factor in both case studies, an assessment of the fabrics used, focusing on their sustainability qualities and characteristics, was presented.

In both cases, the choice of substrate emerged as a critical factor influencing the success and sustainability of the application. Whilst efficient production was achieved in both instances, challenges emerged in shifting consumer behaviour, particularly due to resistance to the denim prototype from industry stakeholders and denim connoisseurs. However, this research proposes that fast and ultra-fast fashion markets may be more suitable for these digitally printed jeans. It was not possible to determine whether there was any measurable consumer behaviour change in relation to the printed linen.

In conclusion, digital textile printing holds significant potential to support more sustainable fashion and textile production, particularly through its application in localised manufacturing. Its capacity for on-demand production enables alignment with models such as concept-to-consumer and just-in-time supply chains, allowing facilities to be situated in urban areas closer to market demand. The technology is versatile, capable of printing on a wide range of fabrics and patterns. However, for this approach to be truly effective, it must be integrated with broader considerations across the supply chain, including responsible material sourcing and garment construction.

The material, logistical, and behavioural challenges associated with sustainable digital textile printing include limited access to sustainable supply chains, highlighting the need for improved substrates and colourants. Logistics would require sourcing robust waste management systems, ensuring that the location meets the environmental conditions necessary for digital printing, and testing the business model to align with both consumer and community demand. Additionally, exploring alternative funding mechanisms, including grants, in-kind support, and material sponsorship, could prove essential in supporting both the implementation and early-stage integration of the model. Bespoke training to ensure in-depth knowledge would also be crucial to mitigate lack of expertise. Profitability remains a concern, pointing to a need for better working practices and greater knowledge-sharing across the industry. Moreover, involving designers and consumers as co-creators can help reposition digital textile printing as a collaborative tool rather than a remote service. While the environmental benefits of this technology show promise, further research is needed to validate these claims, particularly through comprehensive life cycle assessments across the supply chain.

A design-led approach to digital textile printing should be realised through co-creation and community collaboration, positioning the technology as both a creative and educational tool. By embedding micro factories within underutilised urban spaces, such as vacant high street units, this model offers the potential to regenerate communities while fostering local production. The UK already has a strong foundation of textile knowledge and an existing network of micro factories and maker

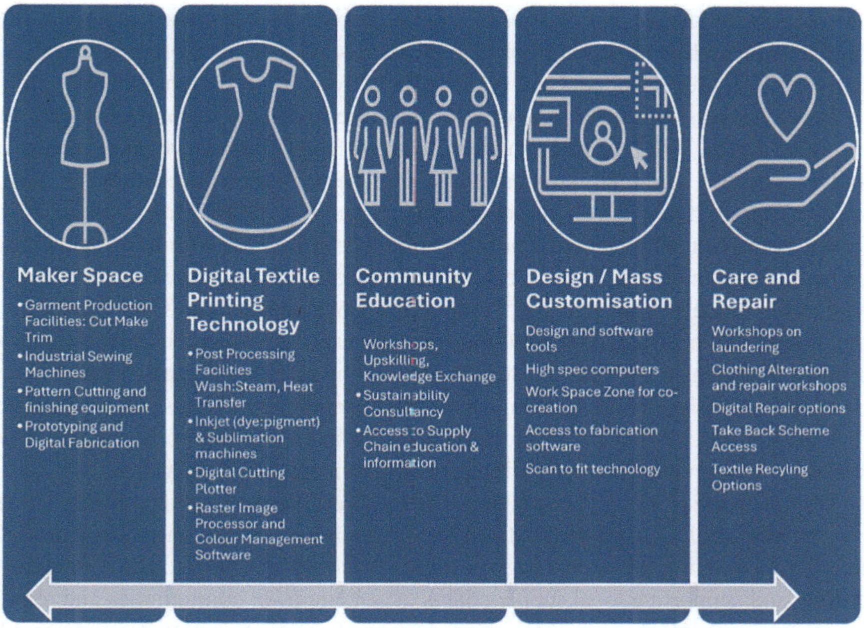

Fig. 3.2 Visualisation of the micro factory model and the five elements required for the customise-print-make-wear approach Gooby (2025a)

spaces, making it well-placed to integrate digital textile printing as a plug-in technology. Additionally, this approach supports the longevity of clothing by enabling digital repair, alterations, and reprints, while also strengthening consumer connection and awareness around garment care.

Figure 3.2 illustrates the five key elements proposed for the micro factory: the maker space, digital textile printing technology, design and mass customisation, care and repair services, and community education, all supporting the *customise–print–make–wear* model.

These elements are fundamental as they collectively enable localised, flexible production, foster consumer engagement and skill-sharing, support sustainability through garment care and repair, and drive innovation through design-led customisation.

The potential of digital textile printing lies in its ability to serve as a scalable and complementary production method for localised manufacturing, particularly through a *customise–print–make–wear* model. This technology opens up new creative possibilities for designers and supports the development of innovative, design-led business models (Gooby, 2023b). However, to realise, manufacturers must improve the sustainability of colouration processes, including exploring the viability of bio-based dyes and pigments. Advances are also needed in developing more sustainable pretreatment coatings, ideally using renewable materials such as plant-based polymers, starches, and natural gums. Additionally, printer manufacturers should consider the use of recycled metals and plastics in the construction of machinery components to further reduce environmental impact. For example, Canon have proposed that they will start using recycled steel and printing parts from 2030 (Canon, 2025).

Policy makers have a critical role to play in embedding sustainability into the future of digital textile printing through targeted legislation that regulates resource and energy use. Strategic support could include increased investment in maker spaces, funding for research and development, and innovation grants to encourage industry adoption. Digital textile printing should be embedded within national digital and circular economy strategies, encouraging community collaboration and shared ownership models. Improving access to sustainable supply chains, particularly for colourants and substrates, alongside the development of advanced digital tools such as scan-to-fit and mass customisation software, will be essential. Locating micro factories within libraries or regeneration zones, and forming partnerships with universities that already have digital textile facilities, could also expand public access and drive broader adoption, particularly during periods when academic use is low.

Digital textile printing, if reimagined through a collaborative and sustainable lens, has the potential to support localised, circular, and accessible models of production. By embedding this technology in shared maker spaces, promoting open-source design, and supporting local supply chains, it can empower communities and reduce environmental impact using efficient production. To attain this, future efforts should focus on investment in infrastructure, sustainable materials, and accessible digital tools. Importantly, embedding strategies for consumer behaviour change would be vital to make meaningful impact.

While this research is not exhaustive, it offers a foundation for further exploration and proposes that pilot schemes or prototypes of the micro factory model are now needed to test its viability in real-world settings.

Declaration The authors declare that they have no known competing financial interests or personal relationships that could have appeared to influence the work reported in this chapter. Generative AI and AI-assisted technologies have been used in the writing process to improve the readability and language of the work only. After using this tool, the authors carefully reviewed and revised the content to ensure accuracy and quality. The authors take full responsibility for the final version of this publication.

References

Association of Suppliers to the British Clothing Industry (ASBCI). (2014). *Introduction to colour in clothing and textiles*. Halifax: ASBCI.

Belsham, A. (2017). *Soft signage: The state of the market* (pp. 12–14). France: Caldera Gamut Magazine.

Bottega Veneta. (2025). *Printed denim*. Retrieved June 8, 2025, from https://www.bottegaveneta.com/en-gb/printed-leather-denim-medium-blue-denim-713844V2BC04600.html.

Britt, H. (2012). Digital textiles: An investigation into technology utilisation, teaching and significance of educator creative practice. *Research Journal of Textile and Apparel, 16*(4), 112–121. https://doi.org/10.1108/RJTA-16-04-2012-B013.

Business of Fashion Textiles and Technology. (2025). Retrieved June 8, 2025, from https://www.bftt.org.uk/.

Cahill, V. (2006) 1-The evolution and progression of digital printing of textiles. In H. Ujiie (Ed.), *Digital printing of textiles* (pp. 1–15). London: Woodhead Publishing. https://doi.org/10.1533/9781845691585.1.

Campbell, J. R. (2006) 9-Controlling digital colour printing on textiles. In J. H. Xin (Ed.), *Total colour management in textiles* (pp. 160–190). Woodhead Publishing London. https://doi.org/10.1533/9781845691080.160.

Canon. (2025). Retrieved June 8, 2025, from https://www.canon.co.uk/press-centre/press-releases/2025/01/canon-begins-using-scrap-recycled-steel-in-printing-products-and-supplying-steel-scrap/. (Published: 17th January 2025)

Carden, S. (2016). *Digital textile printing*. Bloomsbury Publishing.

Carr, W. W., et al. (2006) 4-Drop formation and impaction. In H. Ujiie (Ed.), *Digital printing of textiles* (pp. 53–68). London: Woodhead Publishing. https://doi.org/10.1533/9781845691585.1.53.

Centre for Circular Design. (2025). Retrieved June 8, 2025, from https://www.circulardesign.org.uk.

Centre for Industrial Sustainability. (2025). Retrieved June 8, 2025, from https://www.ifm.eng.cam.ac.uk/research/industrial-sustainability/.

Centre for Sustainable Fashion. (2016). Fostering sustainable practices report. Retrieved June 8, 2025, from https://www.sustainable-fashion.com/fsp.

Centre for the Understanding of Sustainable Prosperity. (2025). Retrieved June 8, 2025, from https://cusp.ac.uk/projects/arts/.

Centre for Sustainable Fashion. (2025). Retrieved June 8, 2025, from https://www.sustainable-fashion.com/.

Dawson, T. L. (2003). *The use of digital systems in textile printing*. Society of Dyers and Colourists.

Dawson, T. L. (2006) 10-Digital colour management. In H. Ujiie (Ed.), *Digital printing of textiles* (pp. 163–179). London: Woodhead Publishing. https://doi.org/10.1533/9781845691585.2.163.

De Castro, O. (2021). *Loved clothes last: How the joy of rewearing and repairing your clothes can be a revolutionary act*. Penguin.

Deraniyagala, H. (2017). Textile colour waste and sustainability. *Colour design* (pp. 653–669). Woodhead Publishing.

Ding, Y. (2016). *Color gamut comparison methodology and evaluation for textile ink jet printing*. North Carolina State University.

Ding, Y., Zhendong, W., Chuanxiong, Z., Ruobai, X., & Wenliang, X. (2021). A study on the applicability of pigment digital printing on cotton fabrics. *Textile Research Journal, 91*(19–20), 2283–2293.

Driessen Vandeijne. (2025). Retrieved June 8, 2025, from https://www.driessenenvandeijne.design/textile/.

Earley, R. (2022). Sustainable textiles for fashion. *Journal of Physics: Materials, 5*(032001), 81–84.

Ellams, D. (2017). *Environmentally conscious fashion through responsible coloration techniques applied to sustainable fabrics: Colouring outside the lines*. Doctoral dissertation, Heriot-Watt University.

Ellen MacArthur Foundation (EMF). (2019). *Fashion and the circular economy deep dive*. Retrieved June 8, 2025, from https://www.ellenmacarthurfoundation.org/fashion-and-the-circular-economy-deep-dive. (Published 16th September 2019)

Ellen MacArthur Foundation (EMF). (2023). *The jeans ReDesign*. Retrieved June 8, 2025, from https://www.ellenmacarthurfoundation.org/the-jeans-redesign/overview. (Published 18th July 2023)

Ellen McArthur Foundation (EMF). (2013). *Towards the circular economy*. Retrieved June 8, 2025, from https://www.ellenmacarthurfoundation.org/towards-the-circular-economy-vol-1-an-economic-and-business-rationale-for-an. (Published 1st January 2013)

Esce-Tex. (2025). *Update about digital textile printing*. Retrieved June 8, 2025, from https://esce-tex.com/digital-print-update. (Published 24th January 2025)

European Environment Agency. (2020). The impact of textile production and waste on the environment. Retrieved June 8, 2025, from https://www.europarl.europa.eu/topics/en/article/20201208STO93327/the-impact-of-textile-production-and-waste-on-the-environment-infographics. (Published: December 2020)

Fashion Revolution. (2022, April). *Denim to denim project report*. Retrieved June 8, 2025, from https://www.fashionrevolution.org/wp-content/uploads/2022/04/Denim-to-Denim-PROJECT-REPORT-.pdf.

Fashion Enter. (2025). Retrieved June 8, 2025, from https://fashion-enter.com/.

Feisner, E. A., & Reed, R. (2014). *Color studies*. Fairchild Books.

FESPA. (2021). Automation and the impact on the digital textile workflow. Retrieved June 8, 2025, from https://www.fespa.com/en/news-media/automation-and-customisation-%E2%94%80the-positive-impact-of-the-digital-workflow#:~:text=Debbie%20McKeegan%20discusses%20how%20digital,from%20paper%20to%20digital%20data.&text=Within%20the%20global%20textile%20market,exceed%20$5.3%20bn%20by%202026. (Published 21st February 2022)

Fletcher, K. (2013). *Sustainable fashion and textiles: Design journeys*. Routledge.

Fraser, B. M., Murphy, C., & Bunting, F. (2004). *Real world color management*. Peachpit Press.

Freire, E. M. (2006). Ink jet printing technology (CIJ/DOD). In H. Ujiie (Ed.), *Digital printing of textiles* (pp. 29–52). Cambridge, London: Woodhead Publishing. https://doi.org/10.1533/9781845691585.1.

Future Fashion Factory. (2025). *R&D calls*. Retrieved June 8, 2025, from https://litac.leeds.ac.uk/innovation/rd-funding-calls/.

Göçer Akder, D., & Bahçecik, Ş. O. (2023). The impact of brand-supplier relations on producers in the earthquake zone, Turkey 2023. Workers Rights Foundation White Paper, New York. https://open.metu.edu.tr/handle/11511/104801.

Goldsworthy, K. (2014). The case for transforming the approach to waste, and growing a circular economy; A design perspective. Discussion Paper. Environmental Audit Committee, Parliament of the United Kingdom. https://ualresearchonline.arts.ac.uk/id/eprint/6977/.

Goldsworthy, K. (2017). The Speedcycle: A design-led framework for fast and slow circular fashion lifecycles. *The Design Journal, 20*(sup1), S1960–S1970.

Goldsworthy, K., Harris, J., Hassan, M. M., & Solomon, L. (2022). *Catalyst report on circular fashion and textiles*.

Goldsworthy, K., Earley, R., & Politowicz, K. (2023). *A review of fast & slow sustainable design approaches for fashion & textile applications (2019)* (p. 142). The Textile Reader.

Gooby, B. (2020). The development of methodologies for color printing in digital inkjet textile printing and the application of color knowledge in the ways of making project. *Journal of Textile Design Research and Practice, 8*(3), 358–383.

Gooby, B. (2023a, March). *Personal interview Jim Taylor Lenzing Tencel and site visit to Tencel Grimsby textile mill.*

Gooby, B. (2023b). *An exploratory study of methodologies for digital colour printing in textile design, through an understanding of interwar historical colour palettes*. Doctoral dissertation, UWE Bristol.

Gooby, B. (2025a, April). *Personal interview with phil patterson colour connections.*

Gooby, B. (2025b, April). *Personal interview with PrinFab.*

Gooby, B. (2025c, April). *Personal workshop with FlaxLands*. https://www.flaxland.co.uk/workshops.

Greenpeace. (2011). *Dirty laundry*. Retrieved June 8, 2025, from https://www.greenpeace.org/international/publication/7168/dirty-laundry/. (Published 13 July 2011)

Grose, V., & Mansfield, N. (2023). *Fashion supply chain management: Integrating sustainability through the fashion supply chain*. Routledge.

Hatley Print. (2025). Retrieved June 8, 2025, from https://hatleyprint.co.uk/.

Hawkyard, C. (2006). 12-Substrate preparation for ink-jet printing. In H. Ujiie (Ed.), *Digital printing of textiles* (pp. 201–217). London: Woodhead Publishing. https://doi.org/10.1533/9781845691585.3.201.

Hayward, J. (2017). Cambridge investment research-global digital textiles ink-jet printing trends & market growth. In *FESPA 2017, Hamburg, 8 May*. Exhibition Presentation.

Henry, P. (2013) *The development of a new colour-picker tool*. Doctoral dissertation, University of Leeds.

Her Majesty's Revenue and Customs. (2025). *UK trade info*. Retrieved June 8, 2025, from https://www.uktradeinfo.com/.

HONGSAM. (2024). *China dyeing and printing association: 2024 China textile digital printing development report*. Retrieved June 8, 2025, from https://www.hongsamdigital.com/news/china-textile-digital-printing-development-report-1.html. (Published 26th December 2024)

Hoskins, S., & Craine, M. (2025). *Ink for printmaking: The art, craft and chemistry of ink*. Bloomsbury Publishing.

IA London. (2025). Retrieved June 8, 2025, from https://www.ialondon.com/.

Ingamells, W. (1993). *Colour for textiles*. Society of Dyers and Colourists. https://instituteofpositivefashion.com/bfcnews/4909/CIRCULAR-FASHIONINNOVATION-NETWORK-LAUNCHESREPORT-OUTLINING-CRITICAL-INSIGHTSAND-NEXT-STEPS-FOR-UK-CIRCULAR-FASHION.

Institute of Positive Fashion. (2025). *Critical insights and next steps for UK circular fashion*. Retrieved June 8, 2025, from https://instituteofpositivefashion.com/bfcnews/4909/CIRCULAR-FASHION-INNOVATION-NETWORK-LAUNCHES-REPORT-OUTLINING-CRITICAL-INSIGHTS-AND-NEXT-STEPS-FOR-UK-CIRCULAR-FASHION. (Published 15th May 2025)

Intelistyle. (2025). Retrieved June 8, 2025, from https://www.intelistyle.com/.

Katrantzou, M. (2009). *Spring summer collection*. Retrieved June 8, 2025, from https://marykatrantzou.com/pages/spring-summer-2009.

Kelly, A. (2014). *An investigation into colour accuracy and colour management issues in digitally printed textiles for higher education*. Doctoral dissertation, The University of Manchester.

Kim, Y. K. (2006) 15-Effect of pretreatment on print quality and its measurement. In H. Ujiie (Ed.), *Digital printing of textiles* (pp. 252–275). London: Woodhead Publishing. https://doi.org/10.1533/9781845691585.3.252.

King, K. M. (2006). Just-in-time printing. *Digital printing of textiles* (pp. 312–336). Elsevier.

Knowle West Media Centre. (2025). *The factory*. Retrieved June 8, 2025, from https://kwmc.org.uk/thefactory/.

Krasny, E. (2020). *Precarious commons. Commoning the city: Empirical perspectives on urban ecology, economics and ethics*. Oxford: Routledge.

Lawn, R. (2006). 10-Colour management across the supply chain. In J. H. Xin (Ed.), *Total colour management in textiles* (pp. 191–209). London: Woodhead Publishing. https://doi.org/10.1533/9781845691080.191.

Leak, A. (1998). *A practical investigation of colour and CAD in printed textile design*. Doctoral dissertation, Nottingham Trent University.

Lenzing Tencel. (2025). *Tencel LCA*. Retrieved June 8, 2025, from https://www.lenzing.com/products/brands/tenceltm/.

Levi Strauss. (2015). *The lifecycle of a jean*. Retrieved June 8, 2025, from https://levistrauss.com/wp-content/uploads/2015/03/Full-LCA-Results-Deck-FINAL.pdf.

Macchion, L., Da Giau, A., Caniato, F., Caridi, M., Danese, P., Rinaldi, R., & Vinelli, A. (2018). Strategic approaches to sustainability in fashion supply chain management. *Production Planning & Control, 29*(1), 9–28.

Manieson, L. A., & Ferrero-Regis, T. (2023). Castoff from the West, pearls in Kantamanto? A critique of second-hand clothes trade. *Journal of Industrial Ecology, 27*(3), 811–821.

Martino, G., Fera, M., Iannone, R., & Miranda, S. (2017). Supply chain risk assessment in the fashion retail industry: An analytic network process approach. *International Journal of Applied Engineering Research, 12*(2), 140–154.
McKeegan, D. (2017). *Bridging the language barrier* (pp. 21–25). France: Caldera Gamut Magazine.
McKeegan, D. (2018). *Good design is in the detail*. Retrieved June 19, 2025, from https://www.whichplm.com/good-design-detail/.
McKeegan, D. (2022). The Metaverse: Rebuilding the supply chain. *AATCC Review, 22*(1).
Miles, L. W. (2010) Production and properties of printing pastes. In *Textile printing*. Bradford: Society of Dyers and Colourists.
Miller, D., & Woodward, S. (Eds.). (2011). *Global denim*. Berg.
Moser, L. S., & Wilson, O. E. (2000). 6. Trends in preparation, dyeing, and printing. *Textile Progress, 30*(1–2), 68–83. https://doi.org/10.1080/00405160008688885.
NESTA. (2014). *Making sense of the UK collaborative economy*. Retrieved June 8, 2025, from https://www.nesta.org.uk/report/making-sense-of-the-uk-collaborative-economy/. (Published 27th August 2014)
Office of National Statistics. (2025). *Fashion and textiles*. Retrieved June 8, 2025, from https://www.ons.gov.uk/businessindustryandtrade/business/activitysizeandlocation/adhocs/12391fashionandtextileindustry.
Parraman, C. (2017). 25-Colour printing techniques and new developments in colour printing. In J. Best (Ed.), *Colour design* (2nd ed., pp. 589–618). London: Woodhead Publishing. https://doi.org/10.1016/B978-0-08-101270-3.00025-4.
Polston, K., Parrillo-Chapman, L., & Moore, M. (2015). Print-on-demand ink-jet digital textile printing technology: An initial understanding of user types and skill levels. *International Journal of Fashion Design, Technology and Education, 8*(2), 87–96. https://doi.org/10.1080/17543266.2014.992050.
Provost, J. (2017) *Textile fabric preparation in digital textile printing: More focus and education required for this key step*. Digital Textile Magazine.
Rag & Bone. (2025). *Printed denim collection*.
RiverBlue. (2014). Retrieved June 8, 2025, from https://riverbluethemovie.eco/.
Romano, R. (2017). *Textile printing 101-whattheythink*. Retrieved June 19, 2025.
Ross, D. (2018). The roles of digital print and specialist design & colour management software in sustainable textile printing. In *Textile Institute World Conference*. University of Leeds.
Scrimshaw, J. (2017) *Is digital the greener textile option?* (pp. 15–18). France: Caldera Gamut Magazine.
Sharma, H. S. S. (1987). Studies on chemical and enzyme retting of flax on a semi-industrial scale and analysis of the effluents for their physico-chemical components. *International Biodeterioration, 23*(6), 329–342.
Sinclair, R. (Ed.). (2014). *Textiles and fashion: Materials, design and technology*. Elsevier.
Stewart, H., & Tooze, J. (2015). Future makespaces and redistributed manufacturing. *Making Futures Journal*. https://makingfutures-journal.org.uk/index.php/mfj/article/view/258.
Street, P. (2024). *Fashion, revolution, and narrative*. Mediating Sustainability in the Consumer Society.
The OR Foundation. (2025). Retrieved June 8, 2025, from https://theor.org/.
The Sustainable Marketplace. (2025). *Earthed collection*. Retrieved June 8, 2025, from https://www.thesustainablemarketplace.co.uk/pages/our-brands/earthed-by-william-clark?srsltid=AfmBOorURnTgZmvz453gsG564axxNodVrVDve-Qtez10j5AbFbrJ4cr4.
Thomas, D. (2019). *Fashionopolis: The price of fast fashion–and the future of clothes*. Penguin.
Tomaney, M. (2015). Sustainable textile production. *Textiles and fashion* (pp. 547–560). Woodhead Publishing.
Treadaway, C. (2007). Digital crafting and crafting the digital. *The Design Journal, 10*(2), 35–48.
Trend Hunter. (2013). *Wood based denim*. Retrieved June 8, 2025, from https://www.trendhunter.com/trends/dawn-ellams. (Published 22nd February 2013)

Tyler, D. J. (2005). Textile digital printing technologies. *Textile Progress, 37*(4), 1–65.
Ujiie, H. (Ed.). (2006b). *Digital printing of textiles*. Woodhead Publishing.
Ujiie, H. (2006). Design and workflow in digital ink-jet printing. In H. Ujiie (Ed.), *Digital textile printing*. Cambridge, London: Woodhead Publishing.
Ujiie, H. (2021, October). Digital textile printing: Status report 2021. In *NIP & digital fabrication conference* (Vol. 37, pp. 47–52). Society for Imaging Science and Technology.
UK Government. (2019). *Libraries and makerspaces*. Retrieved June 8, 2025, from, https://www.gov.uk/government/publications/libraries-and-makerspaces/libraries-and-makerspaces. (Published 4th October 2019)
UK Government. (2025a). *All-party parliamentary manufacturing group, industrial evolution–making British manufacture sustainable*. Retrieved June 8, 2025, from https://www.policyconnect.org.uk/research/report-industrial-evolution-making-british-manufacturing-sustainable. (Published 8th January 2016)
UK Government. (2025b). *House of lords, creative industries: Growth, jobs and productivity*. Retrieved June 8, 2025, from https://lordslibrary.parliament.uk/creative-industries-growth-jobs-and-productivity/#:~:text=The%20creative%20industries%20contributed%20%C2%A3,the%20UK's%20economy%20in%202023. (Published 30 January 2025)
UKFT. (2023). *The fashion & textile industry's footprint in the UK*. Retrieved June 8, 2025, from https://unbxd.ams3.digitaloceanspaces.com/ukft.org/wp-content/uploads/2023/11/04181525/OE-Report-executive-summary.pdf. (Published October 2023)
UKRI. (2012). *UK research and innovation, montetising fashion MetaData March 2012*. Retrieved June 8, 2025, from https://gtr.ukri.org/projects?ref=100944.
UKRI. (2015). *UK research and innovation,fashion recommendation system prototype July 2015*. https://gtr.ukri.org/projects?ref=720695.
UKRI. (2020a). *UK research and innovation, pattern project March 2020*. Retrieved June 8, 2025, from https://gtr.ukri.org/projects?ref=42876.
UKRI. (2020b). *UK research and innovation,circular fashion pivot March 2022*. Retrieved June 8, 2025, from https://gtr.ukri.org/projects?ref=10022501.
UKRI. (2020c). *UK research and innovation, fashion the sustainable recovery of the fashion industry from COVID May 2020*. Retrieved June 8, 2025, from https://gtr.ukri.org/projects?ref=58923.
UKRI. (2020d). *UK research and innovation, fashion Zoa-OS: A digitally and physically integrated end-to-end Operating System to offer any fashion brand sustainable clothing rental-as-a-service November 2023*. Retrieved June 8, 2025, from https://gtr.ukri.org/projects?ref=10082300.
United Nations. (2019). *Department of economic and social affairs sustainable development*. Retrieved June 8, 2025, from https://sdgs.un.org/goals.
United Nations. (2022). *Environment program*. Retrieved June 8, 2025, from https://www.unep.org/news-and-stories/press-release/textile-producing-nations-unite-reduce-chemical-waste#:~:text=Islamabad%2C%2014%20October%202022%20%E2%80%93%20The,chemicals%20in%20their%20textile%20industries. (Published: 14 October 2022)
United Nations. (2025). *Net zero coalition, climate action*. Retrieved June 8, 2025, from https://www.un.org/en/climatechange/net-zero-coalition#:~:text=Current%20national%20climate%20plans%20%E2%80%93%20for%20195,2030%20and%20reach%20net%20zero%20by%202050.
University of Sheffield. (2025). *Maker futures*. Retrieved June 8, 2025, from https://makerfutures.sites.sheffield.ac.uk/.
UnMade. (2025). Retrieved June 8, 2025, from https://www.unmade.com.
Weffan. (2025). Retrieved June 8, 2025, from https://www.weffan.co.
Whitson-Smith, J., Rigby, E., & Harrison, L. (2019). *The laundry pile 2019*. University of Huddersfield. https://pure.hud.ac.uk/en/publications/the-laundry-pile-2019.
William Clark and Sons. (2025). Retrieved June 8, 2025, from https://wmclark.co.uk/.
World Textile Information Network. (2025). *Joseph link. Digital Textile market review 2024*. Retrieved June 8, 2025, from https://www.wtin.com/article/2025/may/12-05-25/digital-textile-market-review-2024. (Published: 14th May 2025)

WRAP. (2019). *SCAP (2012–2019) progress report*. Retrieved June 8, 2025, from https://www.wrap.ngo/resources/report/scap-2012-2019-progress-report. (Published 13th January 2021)

Chapter 4
Eco-Friendly Digital Printing of Cellulosic Fibers and Manmade Fibers

G. Krishnaraj

Abstract This chapter explores the realm of digital printing on textiles, highlighting the tangible benefits it provides to the fashion and textile industries in terms of efficiency, sustainability, and an unparalleled degree of design freedom, especially when contrasted with traditional printing methods. The growing desire for mass customisation and personalisation, especially in luxury, fast fashion trends and small-scale sectors, has propelled the development of digital textile printing, making it a moving tale of advancement. On-demand manufacturing and bespoke clothing are two areas where this technology is expanding quickly because of its increased profitability and efficiency. In the field of digital textiles, digital printing signifies a paradigm leap towards a more sustainable and all-encompassing strategy. Digital printing ensures that firms stay competitive in the ever-changing market by saving time, money, and various textile materials, in contrast to traditional printing, which frequently involves high waste and rigid design limits. In this chapter, a detailed analysis of the fundamental concepts of digital printing technologies, with particular emphasis on Transfer printing and inkjet printing along with their advantages and limitations, ink types, various natural and synthetic fibres, pre-treatment and post-treatment requirement are discussed. Further to this, the benefit of digital printing for designers and its lessened environmental impact of digital printing will be discussed. The versatility and adaptability of digital printing continue to reshape the textile industry, making it a crucial player in both innovation and sustainability.

Keywords Digital printing · Transfer printing · Digital inkjet printing · Piezo electric inkjet printing · Thermal inkjet printing

G. Krishnaraj (✉)
Department of Textile Design, National Institute of Fashion Technology, Chennai, India
e-mail: krishnaraj.g@nift.ac.in

A. Athalye and H. Patil (eds.), *Digital Printing*,
https://doi.org/10.1007/978-981-95-2593-5_4

4.1 Introduction

In light of the heightened competition from emerging textile markets worldwide, it is imperative to enhance quality and reduce response time. Additionally, the market requires high productivity, high-resolution printing of delicate patterns, a wider colour gamut for colour reproduction, increased strike-through to the back surface of products like scarves, high fastness, and low operating costs. The textile industry has seen considerable changes as a result of the digital revolution. This entails delivering cutting-edge design concepts, sample fabrics, and production-printed textiles in a wide range of colour variations in a timescale that has never been seen before. With shorter lead times, lower inventory levels, and quicker reactions to fashion trends, textile printers have benefited greatly from the introduction of textile inkjet technology for sample and proofing (Alomari et al., 2015; Azizi et al., 2019; Fang et al., 2020;Gao et al., 2014; Shah et al., 2021; Evans et al., 2021). Over the past decade, inkjet printing technologies have undergone substantial advancements and have been increasingly adopted in the textile and fashion industries (Uddin et al., 2022). It was initially limited to small sample creation and proofing. Nonetheless, a lot of extensive research has been done to advance digital textile printing to the level of mass manufacturing. Moreover, digital textile printing has continued to evolve, as new materials have emerged and new models of production rooted in sustainability, like the "neo cottage industry approach" have flourished. It has become an essential tool for digital industries seeking more sustainable, transdisciplinary, and widely applicable solutions in the fabric printing domain.

The textile printing industry is extensive and intricate, necessitating the production of prints on many textile substrates, ranging from cotton to an array of synthetic fibres. To give each of these fibres the excellent colour fastness qualities that the consumer wants, a particular class of dyestuff must be applied through a chemical or physical procedure. Because colour correction printing requires excellent resolution and colour reproduction, A technique called inkjet makes it easier to apply liquid ink to a surface while guaranteeing that only the ink droplets comes into contact with the substrate. Three essential parts make up an inkjet printer, and each one needs to work well to provide output that is satisfactory. The substrate, ink, and print head are the constituents (Shah et al., 2021). The latest developments in digital inkjet printing for textiles based on the above constituents will be reviewed in this chapter, along with a focus on the technology's potential applications in textile printing.

4.2 Evolution of Digital Printing of Textiles

The production of textiles has been revolutionised by digital printing, which offers an affordable, adaptable, and sustainable alternative to traditional printing methods (Pekarovicova & Husovska, 2016a). This is due to the increased competitiveness in textile print markets in terms of quick response, short-run production, faster delivery, more innovation and new fashion concepts, and development of designs for the customers by forming alliances with major retail groups, etc. (Le, 1998). Digital printing uses digital files to print directly onto fabrics, as opposed to screen printing, which requires the creation of physical screens for every design. This makes it possible to create incredibly intricate patterns, vivid colours, and elaborate designs with reduced waste and faster turnaround times. Additionally, this digital printing method is economical for short runs, as there is no requirement for preparing printing screens. It is no longer necessary to store the cylinders and screens. Additionally, production time can be reduced. It takes at least three to five weeks to prepare the screens and colour finalisation for conventional printing. Hence, It is considerably simpler to create samples, short to medium runs in a short amount of time as compared to the existing conventional system.

4.3 A Tool for Designer's Creativity: Design Freedom and Innovation

Initially, there was a significant amount of opposition from the renowned designers. While it is possible to create many beautiful and creative things by hand, a computer allows creativity to reach new heights and even generate ideas and images that were previously unimaginable. Human engagement was limited before the invention of this technology. Contemporary technology has eliminated the difficulty for creating the designs on the fabric from the imagination. Although the replicated design is acceptable, the outcome can be more effectively generated by the system's applications than by hand-drawn images (Tkalec et al., 2024). Hence, it is essential to comprehend the potential inherent in the digital printing systems. This innovative technique enables designers to maintain control over the final outcome, preventing alterations during the production process. Designers, including those in fashion and interior design, can collaborate with clients in a more effective and cohesive manner (Tyler, 2005). Digital technologies offer advantages such as rapidity, effective idea communication, last-minute innovations, and competitive superiority on the catwalk with distinctive products. The process may commence with verbal concepts, followed by a compilation of ideas, after which the designer can refine them digitally, experimenting with several iterations, confident that the ultimate outcome would be promptly available in the market. Designers possess complete authority over the final product regardless of yardage specifications. Limited edition clothes can be manufactured without the necessity of engraved screens, maintaining inventory, or manufacturing thousands of items. Consequently, the novel technologies provide authentic creation on demand. Innovative advancements have emerged that connect the creation of textile designs with the patterns, cuts, and overall design of garments.

Fashion designers and producers encounter numerous problems, like the creation of high-quality, sustainable, and exclusive clothes while maintaining a competitive edge in the market (Masod, 2024). As a result, the digital printing solutions are tailored to fulfil the distinct requirements of the high-end fashion industry, providing unmatched precision, variety, colour vibrancy, and rapidity in the creation of innovative design items. Additionally, it helps designers in producing bespoke designs or facilitating small-batch production runs. In luxury fashion, where design exclusivity, fabric quality, colour precision and customization are paramount. The digital printing has addressed the aforementioned parameters as well as created new opportunities for innovation. Further, producing a similar print with high resolution image in screen printing requires 100–300 mesh size. Hence, the designers can produce elaborate, custom prints on materials such as silk, wool, cotton, and leather, without sacrificing texture or tactile quality. Further, the need to rely on a distant factory has been eliminated by the elimination of several layers of industrial processes. This will enhance speed, competitiveness, and creativity. The designer now has direct access to the whole range of alternatives. It is no longer required to depend on intricate production procedures. Additionally, It's time for mass customisation. Moreover, newer generations exhibit not as many concerns towards these technology and methodologies. Furthermore, there is no necessity for photocopying, cutting, pasting, or struggling with repeats. Instead, additional time can be allocated to the genuinely creative aspects of a designer's work.

4.3.1 Difference Between Conventional Printing and Digital Printing

The sample length may be produced from two to five metres, contingent upon the design type. The conventional screen engraving technique is employed to create screens for generating samples that may also be utilised directly in mass production. Sample preparation is a lengthy, intricate, and costly process (Ujiie, 2006). As seen in Fig. 4.1, a collection commences with a consensus among design stylists, driven by factors such as sales potential, trend forecasts, and their expertise in the contemporary fashion market. Art-work is the initial requirement for commencing screen printing procedure. Hence, the initial phase involves the production of textile artworks by external studios or freelance textile designers. The subsequent process is deconstructing this artwork into its constituent colours for engraving a screen for each hue. This means that if a textile design incorporates 24 colours, then 24 screens must be created, in contrast to the spot colours utilised in the graphic arts sector. It is essential to guarantee a proper registration among each screen. Traditionally, it is executed by proficient colour separation artists. Traditionally, separation of colour would be a highly specialised and challenging task.

Moreover, it is increasingly difficult to locate individuals possessing the requisite expertise, as time advances. Currently, CAD systems are employed to scan drawings, minimise the number of colours, design manipulation, create repeats, apply colour, and produce separations. The digital information generated through design processes can subsequently be employed for direct screen development using

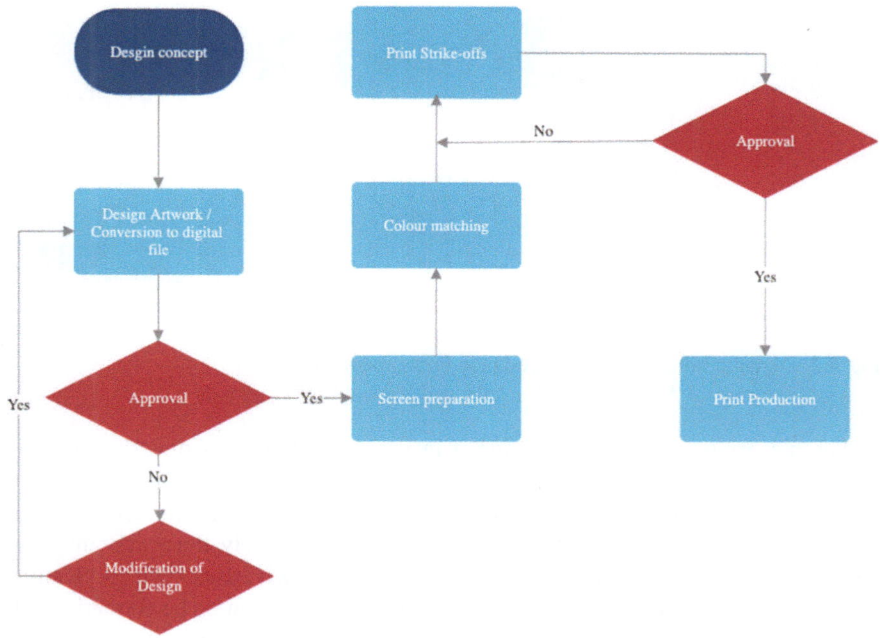

Fig. 4.1 Conventional printing process for Design to Print production

advanced laser engraving technologies or through conventional methods involving computer-generated films. In the traditional approach, the artwork is separated into its constituent colours to produce film positives. Each film is aligned with a screen that has been coated with a light-sensitive emulsion. Later, the prepared screen is placed in the exposure unit where UV light is used for exposing. Following the designated exposure period, the screen undergoes a washout process using a washer with high pressure. This step removes the emulsion which has not been exposed to UV light, thereby revealing the positive or design area. The screen is then placed in a drying cabinet for drying. Ink is subsequently applied and pushed through the open areas of the screen to the textile substrate.

Unlike CMYK four-colour process printing, this technique allows for the independent manipulation of each screen colour, a feature that is essential in textile collections to facilitate varied colourways or options during sample production (El-Kashouti et al., 2019). A significant challenge in this workflow lies in the simultaneous preparation of multiple design samples under stringent deadlines, such as those imposed by biannual trade fairs. This is further complicated by the inherent complexity of fashion industry operations, frequent last-minute design modifications, and the ongoing development of colourways during the sampling phase. Digital printing technologies can substantially mitigate these bottlenecks, as illustrated in Fig. 4.2. Such advancements reduce the time required for colourway and sample selection, while also greatly improving the efficiency with which printers can translate design concepts into tangible outputs (Castrejon Pita et al., 2013).

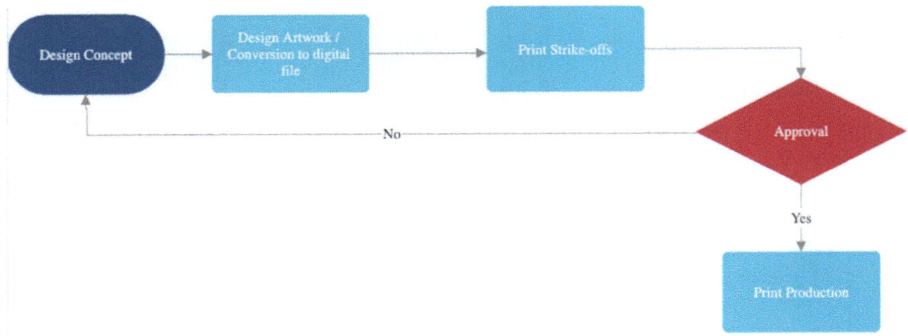

Fig. 4.2 Digital printing process for Design to Print production

Sample prints can now be produced in various colourways directly on textile substrates for client approval. Since the adoption of this methodology, clients have been able to submit new designs or digital files generated through their own computer-aided design (CAD) systems to the textile printer. Alternatively, the printer may scan the physical pattern into a CAD environment, where processes such as cleaning, setting repeat patterns, colour separation, and recolouring can be performed—tasks that modern digital technologies are increasingly capable of automating. A crucial advantage of this approach is the ability of the textile printer to use CAD-generated data to drive inkjet printers for sample production. The resulting printed sample then undergoes standard post-printing procedures, including drying, curing or steaming, and washing. This enables the rapid production of realistic, high-quality samples without the need for traditional screen engraving or sample print preparation.

Furthermore, the creation of colour atlases specifically calibrated for inkjet fabric printing introduces new avenues for precise colour specification and communication. These atlases can facilitate the translation of sample print colours into bulk production recipes. Such recipes may be electronically transmitted to automated dispensing systems for accurate and efficient production of colours at scale. This capability supports the seamless integration of sampling and production processes within a unified digital workflow. By leveraging digital and inkjet printing technologies, designers are empowered to develop and implement innovative textile designs more efficiently and with greater creative freedom.

4.4 Transfer Printing

Transfer printing refers to the printing technique of fabrics with patterns or motifs that are released from a paper medium through the combined application of heat, pressure, and controlled dwell time (Xue et al., 2012). This method generally involves the initial application of a disperse dye onto the transfer paper, followed by the sublimation and diffusion of the dye into a hydrophobic fabric under elevated temperatures

(Yin et al., 2013). Compared to conventional printing techniques, transfer printing provides enhanced production efficiency, greater manufacturing flexibility, and finer pattern detail. Moreover, it is advantageous in terms of operational simplicity, reduced cost, minimal maintenance requirements, and lower energy consumption (Dawson & Hawkyard, 2000). As a result, no additional fixing or wash-off is necessary. Because the machines were initially made for the graphics sector, jet printers that print directly onto fabric may also produce transfer paper. A transfer calendar equipment is used to transfer the design continuously onto fabric rolls with the support of blanket. Although transfer printing is a desirable substitute for ink-jet printing on polyester substrates due to thermoplastic nature. Further to this, it is not practical for natural textiles like cotton, linen, rayon, silk and wool as the disperse dyes are not suitable for the aforementioned fibres (Miles, 1977). In the transfer printing process, disperse dyes are employed at elevated temperatures, with the fabric placed in close contact with heated calendar rolls for a dwell time of approximately two minutes. The specific dwell time is generally determined by the volatility of the dye and the thermal tolerance of the substrate, in order to prevent fabric damage or melting during the transfer stage (Rattee, 2010). To prepare the design for transfer, a mirror image of the intended pattern is first printed onto a paper. Disperse dyes used in this process are capable of sublimation at temperatures ranging from above 100 °C to below 220 °C, making them suitable for application on synthetic fibers (DeVries, 1977; Lewis et al., 1984). The referenced patent describes a dry-release sublimation transfer method for textile decoration, which includes a temporary backing sheet, a sublimation transfer design layer composed of sublimation transfer inks, and a polymeric layer in direct contact with the design layer.

The most widely employed transfer printing techniques in commercial applications are sublimation transfer and film release. Additional methods include the use of meltable inks (melt transfer) and transfer through an aqueous film (Cie, 2015a). Sublimation transfer operates on the principle that certain disperse dyes are capable of transitioning directly from a solid to a gaseous state upon the application of heat, without undergoing a liquid phase (Cie, 2015b). This technique is suitable for use on synthetic fibers, natural fibers, and blended fabrics composed of both natural and synthetic materials, provided that disperse dyes are utilized (Pinto & Pinto, 2013). The film-release technique involves applying a thermoplastic film over printed paper, which, under heat and pressure, adheres fully to fabric while releasing from the backing paper (Cie, 2015a). This system uses strong films and coated paper to ensure efficient transfer, unlike sublimation which leaves some dye behind (Rattee, 1977). The adhesive film activates at low temperature and pressure, minimizing fabric distortion and energy consumption (El-Halwagy et al., 2001). Unlike melt-transfer methods, film-release relies on a tacky film layer for adhesion rather than melting the print, making it suitable for high-speed, high-efficiency applications, including printing disperse dyes on natural fibers (Miles & Leslie, 2010).

Wet-transfer printing refers to the process of transferring images from one surface to another in an aqueous state. It is synonymous with techniques such as hydrographics, immersion printing, water-transfer printing, dip coating, and cubic printing. This method is primarily used to apply printed designs onto three-dimensional surfaces (Harris & Anderson, 2013), hence the term "cubic printing." Wet-transfer printing has been widely applied to a variety of substrates, including metal, plastic, glass, hardwoods, and other materials. In contrast, melt-transfer printing is a technique commonly employed in the textile industry, particularly for wool (Ikeda & Tomiki, 2013). This process involves mixing traditional wool dyes with chemicals and a meltable transfer medium, which is then printed onto a transfer paper. The printed transfer paper is subsequently pressed against pre-treated wool fabric and subjected to heat in a calendering process. Under controlled pressure and a temperature range of 100–120 °C, the dye and transfer medium are transferred to the wool substrate (Bisutti, 2011). A critical preparatory step in wool-transfer printing is the chlorination of the wool (Bowles et al., 2005). This treatment enhances the wool's receptivity to the dye, ensuring better color penetration and durability of the printed design. During the transfer, the fabric and paper are passed between a heated, polished metal roller and a nylon roller, applying extreme nip pressure to achieve effective dye migration.

4.5 Digital Inkjet Printing

In inkjet printing, ink is precisely ejected from fine nozzles onto a designated location on a treated substrate, under the control of a computer. The deposited ink is then subjected to heating to evaporate the solvent, thereby forming a predetermined pattern. Pre-treating the cloth before printing is a problem since it involves applying chemicals and thickening agents onto the fabric for fixing the dyes and avoid excessive wicking and absorption. These must be removed after the fixation stage, along with any unfixed dye. This contrasts with sublimation or transfer printing, in which the chemicals and thickener stay on the paper after the dispersion dyes are permeated into the fabric during the printing process (Sarkodie et al., 2018).

4.5.1 Printhead

Printhead carriages are typically positioned up to 10 mm from the drive belt to accommodate the use of thicker fabrics. During initial installation, or whenever a printhead is replaced, a precise alignment procedure must be conducted (Castrejon Pita et al., 2013; Lee et al., 2002). Alignment is carried out in two stages: first, printhead-to-printhead alignment must be performed for each individual carriage; subsequently, a carriage-to-carriage alignment procedure is executed to ensure overall system accuracy. The replacement and alignment of a printhead are designed to be completed within one hour. The printhead is engineered for industrial applications and offers an extended operational lifespan. Failures are most commonly attributed by nozzle contamination or external physical damage. Under normal operating conditions, the printhead is capable of withstanding over 4 billion ink drops per nozzle. The printhead utilizes shared-wall piezoelectric technology, wherein ink drop is ejected through a nozzle by squeezing the piezoelectric walls. Each chamber shares its piezo walls with adjacent chambers. While this design ensures compactness and efficiency, it also means that when one nozzle is firing, its neighbouring nozzle cannot be activated simultaneously. Nonetheless, individual nozzles can be fired without interfering with adjacent ones. Electrodes responsible for activating the piezo elements are embedded within the ink chambers themselves, with alternating chambers carrying opposite electrical polarity (Kwon et al., 2020).

4.5.2 Ink Carriage

The ink delivery system consists of a rigid plastic cartridge that encloses and safeguards an internal ink bag, which functions as the primary ink reservoir. Before being filled into the bag, the dissolved air in the ink is degassed. This procedure is critical for optimizing the performance of printhead jetting function. However, the dissolved air present in the ink may be discharged at the time of jetting process which is known as rectified diffusion (Ujiie, 2006). It has resulted the misfiring or misdirected ink drops. This challenge is inherent to all piezoelectric inkjet technologies. In the absence of proper degassing, the printhead must operate at reduced firing frequencies, adversely impacting the printer's overall productivity (Hoath, 2016). To prevent air ingress, the ink bag is meticulously engineered using a multi-layer laminated structure composed of polyethylene, nylon, aluminium, and polyester. Among these, the aluminium layer serves as the primary barrier to atmospheric gases, while the remaining layers contribute to mechanical strength and resistance to physical stress.

4.5.3 Principle of Inkjet Technology

In the inkjet technology, ink is delivered to a substrate where only ink drops are made to contact with it (Hoath, 2016). It is a reprographic technique that enables the precise and controlled deposition of minute ink droplets on a substrate (Boland et al, 2006; Saunders & Derby, 2014). The print head, ink, and substrate are the three key components of an ink jet printer that must function properly to produce a satisfactory result as mentioned in Fig. 4.3. The print head comprises of ink reservoir, nozzle and piezoelectric element. The two general categories into which ink jet technologies are classified, namely Continuous Ink Jet (CIJ) and Drop-on-Demand Ink Jet (DOD) as shown in Fig. 4.4. In CIJ, a constant pressure is applied to nozzles to spray ink at a steady speed. The ink jet quickly fragments into droplets owing to its inherent instability while exiting the nozzle. Depending on the picture being printed, the droplets are either deposited on the substrate or directed to a gutter for recirculation (Khan et al., 2019). Typically, an electric field is applied to regulate the trajectory and electrically charge the drops to create the deflection. The term "continuous" refers to the constant ejection of droplets. In DOD method, drops are released only when the image is to be printed. The two primary drop ejector techniques are Piezoelectric inkjet (PIJ) and thermal inkjet (TIJ). In the Piezoelectric Inkjet (PIJ) method, a piezoelectric actuator is employed to mechanically deform the ink chamber, thereby forcing an ink droplet through the nozzle (Wijshoff, 2010). In contrast, the Thermal Inkjet (TIJ) method utilizes an electrical heater located within each nozzle to rapidly increase the temperature of the ink to its bubble nucleation point. The resulting vapor bubble generates sufficient pressure to eject an ink droplet from the nozzle. This material-conserving patterning technique, commonly implemented through piezoelectric inkjet printing, is used for liquid-phase material deposition which contains solutes in a solvent (Ellen, 2017). Ejection of ink droplets is regulated by a piezo-ceramic actuator, which, upon receiving an electric impulse, deforms to generate a pressure wave, forcing the droplet out through the nozzle. A diaphragm is employed to prevent undesired interactions between the ink and the actuator. The droplets travel toward and settle on the substrate due to their acquired momentum, after which the solvent evaporates (Singh et al., 2010).

Before drying commences, the droplet reaches thermal equilibrium through heat loss and evaporation into the warmer surroundings (Boel et al., 2020; Farid, 2003). Droplet drying occurs in two distinct stages. In the constant-rate stage, evaporation is limited by the heat required to vaporize the solvent, with solvent molecules moving to the surface via diffusion, convection, or capillary flow. This stage persists as long as solvent is readily available at the surface and ambient temperature remains stable (Vehring, 2008; Singh & Van den Mooter, 2016; Mezhericher et al., 2011). In the falling-rate stage, the surface concentration of solute increases due to continued evaporation, eventually leading to crust formation at the droplet surface—referred to as the locking or critical point. This crust restricts mass transfer, causing the drying rate to decline. Further evaporation proceeds through diffusion or capillary

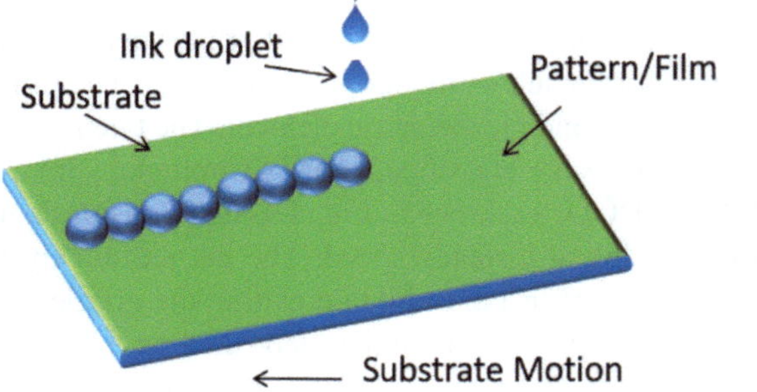

Fig. 4.3 Digital inkjet printing technology (Khan et al., 2019)

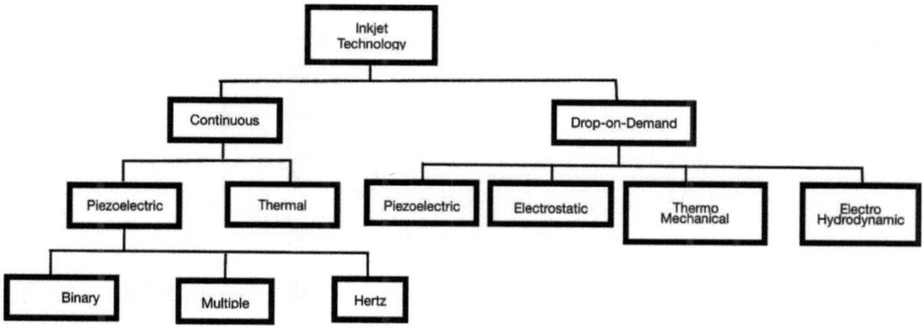

Fig. 4.4 Classification of Digital inkjet printing (Uddin et al., 2022)

transport from the wet core through the crust, which progressively thickens and increases resistance. The drying kinetics during this stage vary depending on material properties and environmental conditions (Boel et al., 2020; de Souza Lima et al., 2020). In summary, the inkjet printing mechanism consists of three primary stages: (1) ink ejection and droplet formation, (2) liquid–solid interaction following droplet deposition on the surface of the fabric (3) Stable printed structure formulation by drying of droplets and solidification (Machekposhti et al., 2020; Sohrabi & Liu, 2018).

4.5.4 Continuous Inkjet Method

Continuous inkjet (CIJ) printing is widely employed in applications such as textile printing, labelling, and other high-speed graphic production tasks (Alaman et al., 2016). Inkjet printing (IJP) technologies can be further categorized based on their actuation mechanisms into two primary types: (a) piezoelectric and (b) thermal (Kumar et al., 2021; Alaman et al., 2016; Murphy & Atala, 2014; Calvert, 2007). In both methods, the printing material is stored within a chamber and ejected through a nozzle in the printhead as shown in Fig. 4.5. However, the two techniques differ fundamentally in the mechanism by which droplets are generated (Gaisford, 2017; Ozkol et al., 2009; Mannerbro et al., 2008). In these methods, the ink jet is fragmented into droplets after it leaves the nozzle. The fragmentation would happen at random and produce droplets of varying sizes in the absence of any additional intervention. This is often fixed by giving the nozzle a time-domain periodic stimulation that causes a distortion in the ink fluidity. The size of the droplets are adjusted to a very high degree of precision by combining the frequency of excitation and jet velocity.

In the piezoelectric approach, a piezoelectric transducer is integrated with the printhead to generate periodic actuation. These actuations are mechanical oscillations, which deform the ink chamber and facilitate droplet ejection. The quantity of electrical charge applied on the drops are determined by the image that has to

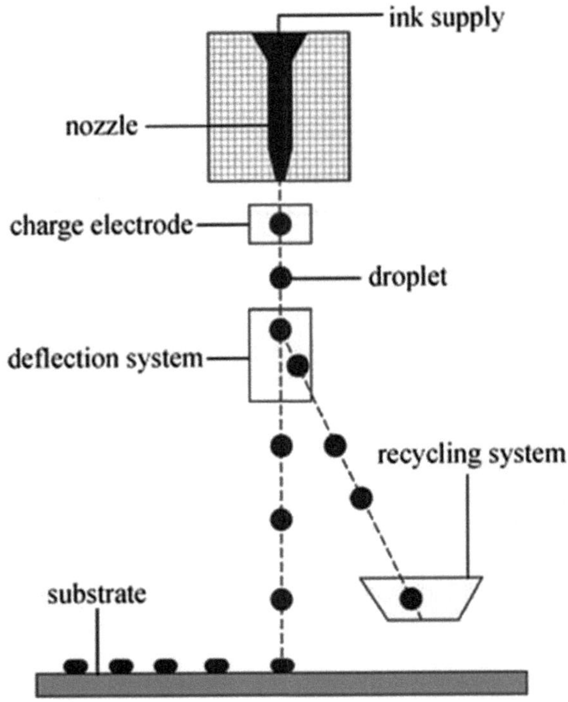

Fig. 4.5 Inkjet printing—continuous mode (Lan et al., 2017)

be produced. After that, the drops deflect as they move through an electric field. In piezoelectric-driven CIJ method, the drops can be deflected in three different ways, namely binary deflection, multiple deflection and hertz method. In the binary deflection method, the ink droplets are either deposited as a single pixel spot on the fabric or directed to the recirculating gutter whereas the drop deposition is varied in several pixels in the multiple-deflection method. In hertz method, the quantity of ink deposited per pixel varies by producing extremely tiny drops (approximately 3 pL) at 40 m/s and stimulation frequencies greater than 1 MHz. The drops are charged and diverted to a gutter or recycling system if they are not meant to reach the substrate. To avoid merging in flight, a lower charge is applied to the printing drops (Lan et al., 2017).

In thermal based CIJ approach, thermal pulses are applied to the inkjet for breaking up uniformly. An annular electrical heater embedded in each nozzle is pulsed at a specific frequency to elevate the temperature of the ink in the vicinity of the nozzle. It helps to reduce the viscosity of the ink. The resultant jet splits into uniformly sized droplets in a repeatable manner because heating pulse recurs at regular time intervals. The print heads of traditional CIJ is expensive due to various complexity issues like charge and deflection, ink recirculation, and pressurisation, etc. Furthermore, the operational frequency of these devices generally more than that of DOD systems by an order of magnitude. Since the positive pressure action actively refills the nozzles.

These factors contribute to the widespread adoption of CIJ systems in industrial settings.

4.5.5 Drop-on-Demand: Piezoelectric Ink Jet Method

Generally, lead zirconate titanate (PZT) is used for making the piezoelectric element for generating droplets as shown in Fig. 4.6 (Lan et al., 2017). The piezoelectric transducer could be integrated either as a component of the chamber wall or as the entire ink chamber. In both cases, ink is expelled from the nozzle by applying a voltage to the electrodes of the piezoelectric element, resulting in a reduction of the chamber volume (Hoath, 2016). The PIJ print heads in the DOD method ink ejection process are regulated by a variety of drop ejector geometry, such as shear mode, bend mode, push mode, squeeze mode, nozzle excitation, and porous layer feed. In shear mode operation, an electric field is established perpendicular to the poling direction of electrode to produce shear displacement. The poling and electric field directions are parallel in the bend mode. The membrane is orientated in the expansion direction of electrode when in push mode. On the other hand, the poling and electric field directions are parallel in push mode. In the squeeze mode configuration, applying an electric field to the hollow piezoelectric tube forming the drop ejector causes a decrease in its internal volume. In the case nozzle excitation class, the nozzle plate is fitted with piezoelectric elements, which offers a cost-effective alternative to air bubbles in the ink bath. In porous layer class configuration, ink flows into the chamber equipped with a porous metal layer. Additionally, the amount of ink in the front channel can be changed by bulging the nozzle meniscus in or out according to the pre-pulses which resulted the smaller or larger droplets. The drop volume is adjusted across an order of magnitude using all of the aforementioned strategies (Ujiie, 2006).

4.5.6 Drop-on-Demand: Thermal Ink Jet Method

The nozzle structure of thermal inkjet (TIJ) technology usually incorporates an electric heater, which is frequently made utilising microelectronic fabrication methods. In order to raise the ink temperature in the vicinity above 300 °C, a pulse of electric current is delivered to the heater (Lan et al., 2017). An ink droplet is violently squirted out the nozzle aperture owing to the formation of vapour bubble and quickly expands due to the abrupt temperature increase. Water-based inks are especially well-suited for TIJ applications since water is more prone to producing stronger bubble formation than other solvents.

The controlled micro-explosion method is quite similar to the TIJ process. Since the vapour bubble is a poor heat conductor, there is little point in keeping the heater powered on once bubble nucleation and expansion have started. As a result, the

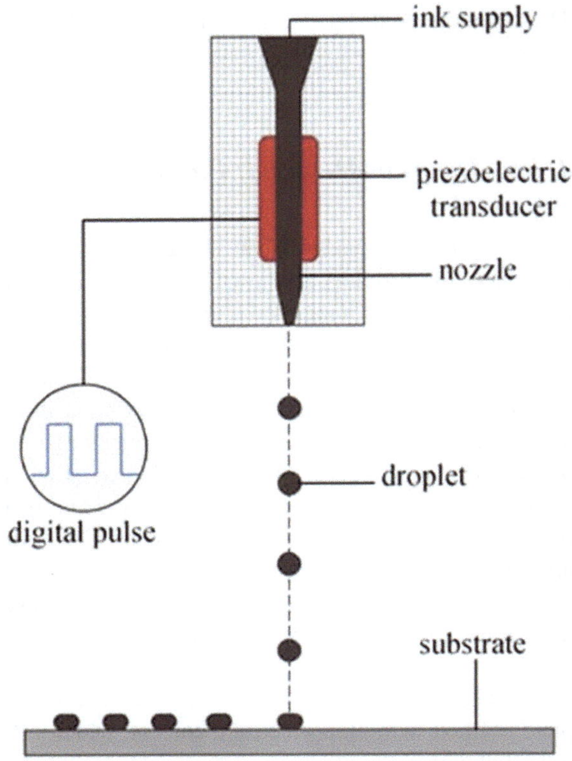

Fig. 4.6 Inkjet printing—DOD mode (Lan et al., 2017)

current pulse usually ends soon after nucleation. The bubble attains its largest size before collapsing because it quickly cools and its internal pressure drops. The ink meniscus is drawn back into the channel by this collapse. Capillary activity then replenishes the channel, readying it for the subsequent firing cycle. The duration and power of the heating pulse are the only factors that can be precisely controlled because of the explosive dynamics of the process. In certain situations, the ink may be preheated using a brief pulse, which gives the user some control over the amount of ink that is ejected.

Though various types of TIJ drop ejectors are available, the most common models are roof-shooter, side-shooter, and back-shooter models as shown in Fig. 4.7 (Christie, 2015). In roof-shooter arrangement, the heater's plane is placed parallel to the nozzle plane whereas in the side-shooter arrangement, a heater plane is placed perpendicular to the nozzle plane. In the back-shooter configuration, the heater is located on the back side of the nozzle plate. Furthermore, energy-efficient designs have been created in which the heater is suspended inside the ink. Compared to setups where the heater is embedded in a substrate, this design increases thermal transfer efficiency by permitting a larger percentage of generated heat to be directly absorbed by the ink. Despite having hundreds of nozzles, commercial TIJ printheads are usually powered by just a few dozen electrical lines. This leads to low-cost,

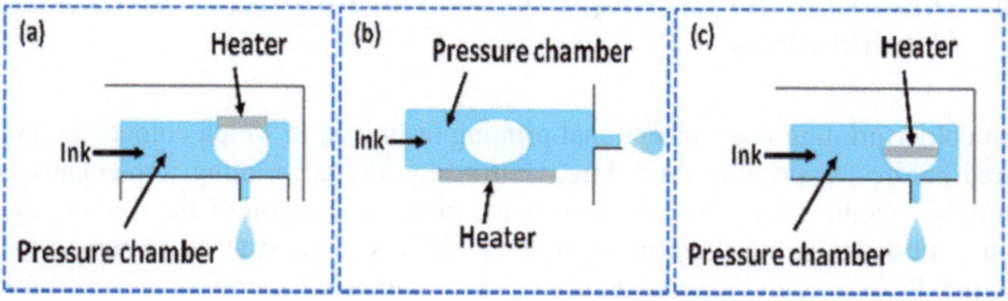

Fig. 4.7 **a** Roof shooter, **b** side shooter, and **c** back shooter system (Christie, 2015)

multi-nozzle printhead arrays with the cost advantages of batch processing that are inherent in integrated circuit (IC) fabrication techniques.

4.5.7 Drop-on-Demand: Electrostatic, Thermo Mechanical and Electro-Hydrodynamic Extraction Method

There are a number of distinct approaches to ink droplet ejection besides thermal inkjet technology that make use of various actuation mechanisms and physical concepts (Cai, 2021). Similar to piezoelectric transducer, electrostatic technique uses an electric field directly to actuate the ink chamber's membrane. In this arrangement, the electric field causes the membrane to move mechanically, which propels the ejection of ink droplets. This technique is enabled by Micro-Electro-Mechanical Systems (MEMS) technology, which allows for precise micro-scale actuation and integration with fluidic components.

In thermo-mechanical inkjet systems, droplet ejection occurs through the rapid movement of a composite structure caused by differential thermal expansion. The materials' different thermal expansion coefficients result in abrupt mechanical movement when an integrated electric resistor is heated. In one use, the paddle behind the nozzle that is submerged in the ink chamber receives this motion. Droplet generation begins when the paddle is activated, dislodging ink. In another method, ink ejection is directly facilitated by the nozzle structure itself moving inward. The Electro-Hydrodynamic (EHD) Extraction process is another unique approach where extraction electrode is placed in front of the nozzle (Cai, 2021). In this case, The static electric field from the extraction electrode and the negative pressure applied at the ink inlet remain balanced during standby or non-printing mode. A increased electrical potential is given to the electrode to cause the equilibrium to be upset and for an ink droplet to be ejected when droplet ejection is needed.

4.6 Digital Ink Jet Printing Process: Ink-Fibre Combinations

Spot colour printing is a conventional printing technique in which colours are pre-mixed before the printing procedure. Conversely, digital printing implements an alternative methodology, which involves the direct generation of the desired hues on the substrate during the printing process. This is referred to as process colour printing. Process colour is accomplished by superimposing microscopic dots of four primary inks—cyan, magenta, yellow, and black (CMYK). In order to enhance the quality of print and broaden the colour spectrum, it is possible to integrate another four to eight colours in addition to the aforementioned four colours. A virtually unlimited range of tints and tonal variations can be achieved by combining these inks in various proportions. Ink droplets with diameters typically spanning from 50 to 60 microns are ejected onto the substrate during the printing process. The size and position of these particles are precisely regulated to guarantee the reproduction of high-resolution images. The timing and trajectory of ejection are influenced by high-frequency electrical signals, which regulate the placement of particles. The ink stream is subjected to regulated pressure, which enables the stream to be consistently divided into individual droplets, thereby further controlling drop formation.

4.6.1 Fabric

The fabric must be fed into the printer with dimensionally stable and should be free from any tension. The fabric is advanced beneath the printheads by the printer after each carriage pass. It is imperative to maintain precision in this motion in order to reduce dot placement errors and guarantee high print quality. The fabric movement and belt movements are regulated by the motion control system for attaining sub-pixel accuracy. This necessitates positional accuracy that exceeds 12 microns at a resolution of 540 pixels per inch.

A belt with an adhesive surface, which is technically referred to as an adhesive print blanket or commonly known as a sticky belt, is used to maintain the fabric's position during printing. The fabric's dimensional stability is guaranteed by this adhesive belt during the printing procedure. It is essential that the fabric is not stretched or wrinkled when it is introduced onto the belt. The fabric roll is unwound by a specialised feed mechanism that ensures minimal tension, a critical factor for elastic materials like Lycra, which are frequently employed in sports apparel and swimwear. The printed image may become distorted after the fabric is removed from the printer if tension control is not meticulously maintained.

The fabric is carefully placed onto the adhesive belt after passing over a de-wrinkling roller prior to printing. A pressure roller is then used to secure the fabric in position as it enters the print zone by pressing it onto the belt's surface. After the printing, the fabric is dried by passing it over a heated plate, which removes moisture from the ink and allows it to be rewound without ink transfer. It is crucial to recognise that the ink is not intended to be cured by this curing process. Lastly, the printed fabric is rolled back onto a tension-controlled roll to prevent elongation. A cleansing system consists of a water-based brush to remove residual ink and fibres from adhesive print blanket regularly, followed by a squeegee to dry the surface in preparation for the next section of fabric.

4.6.2 Fabric Pre-treatments

A preparatory process is required for all textile substrates that are intended for colouration, whether through dyeing or printing, to guarantee that they are clean and appropriately responsive to aqueous treatments (Campbell, 2006). This requirement is applicable to both natural and synthetic fibres; however, natural fibre based fabrics are requiring an additional rigorous and expensive preparation process. The flow process sequence for cotton woven fabric preparation is mentioned below in Fig. 4.8.

The sequence of preparatory processes for textile substrates may vary little depending on the specific requirements, and certain stages may be combined to increase efficiency. Hot alkaline scouring is a critical step in the preparation of cotton fabrics, as it effectively removes the hydrophobic materials of the cotton fibres. Mercerisation is a process that is frequently implemented to improve the absorption of dyes and to add lustre to cotton fabrics. This process involves the application of a concentrated caustic soda solution. Conversely, regenerated cellulose fibres, including viscose rayon and lyocell, undergo a gentler causticization process.

The batched fabric should be assessed for wettability subsequent to its preparation. The fabric is deemed prepared for dyeing or printing if it exhibits adequate wetting characteristics. In the context of printing, this fabric is frequently referred to as "P for P" (Prepared for Print). Wettability and moisture absorbency are essential parameters in screen printing. Inadequate absorbency, as seen in water-repellent fabrics, can lead to the initial printed pigment remaining on the surface, which may subsequently be smudged upon contact with subsequent screens. In contrast, fabrics that are excessively absorbent may permit the print paste to penetrate too deeply, resulting in a diminished surface colour yield and dulled hues. Thus, it is imperative to obtain a balance in the absorbency of the fabric. This principle is also applicable to textiles that are designed for inkjet (jet) printing.

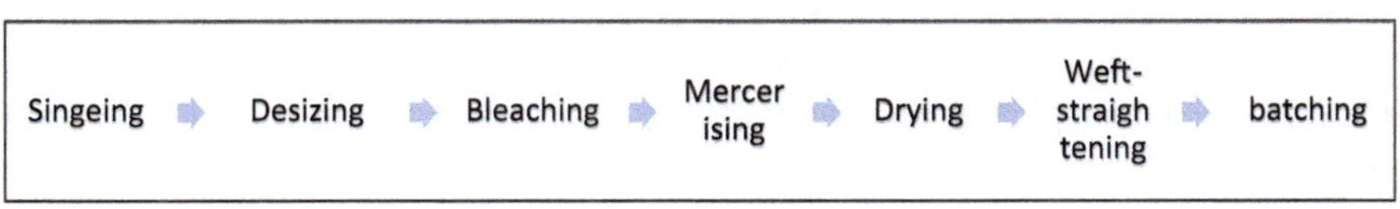

Fig. 4.8 Flow process sequence for cotton fabric preparation

Fabrics of varying types contain a variety of finishing agents that must be eliminated during the preparation process. Sizing agents are typically applied to woven fabrics, knitting oils to knitted fabrics, lubricants to nonwovens, and similar additives to carpets. Uniform printing performance necessitates the complete elimination of these substances. In traditional textile printing, reactive dyes are employed in conjunction with alkali and other essential auxiliaries in a print paste formulation. The printed fabric is subsequently steamed to facilitate dye fixation to the hydroxyl groups in cellulose. Subsequently, it is subjected to thorough washing to remove any remaining dye, chemical residues, and thickening agents. This multi-stage process is considerably more intricate than inkjet printing on paper.

Adaptations in printing technology have been required due to the inherently low volume of ink applied in inkjet systems, the reactivity of reactive dyes under alkaline conditions, and the necessity for additional chemicals to repair them. Introduction of chemical pre-treatments is one example of such an adaptation (Tawiah et al., 2016). The direct incorporation of functional compounds and thickeners into the low-viscosity inks used in digital printing has presented numerous challenges, necessitating the pre-treatment of textiles before inkjet printing. Consequently, the inkjet printing process for textiles is closer to the "two-phase" method of conventional printing than the traditional "all-in" approach. The print paste contains all components, including pigments, chemicals, and thickeners, in the all-in method. The two-phase method involves the application of specific ingredients, particularly reactive substances, separately either before or after the printing stage as presented in Fig. 4.9.

There are numerous compelling reasons to separate dyes from compounds and thickeners and apply them directly to the substrate: All-in-one ink formulations are generally less stable and have a shorter expiration life. For instance, reactive dyes are susceptible to hydrolysis in the presence of alkali. Sodium chloride is a well-known example of a corrosive agent on steel, and the incorporation of reactive chemicals in ink formulations can contribute to nozzle corrosion. Additionally, continuous inkjet systems that operate on the "charged drop" principal necessitate inks with low electrical conductivity. Ink formulations frequently fail to exhibit the desirable rheological characteristics due to the inclusion of thickeners. Sodium carbonate, which is employed as an alkali for reactive dye fixation, can be efficiently applied to the fabric surface during pre-treatment. However, if it is incorporated into the formulation, it may destabilise the ink. The solubility of dyes is diminished by the presence of higher salt concentrations in aqueous inks. Concentrated inks are necessary to obtain the necessary level of colour intensity, as inkjet printing is characterised by its small droplet size.

Although low-viscosity inks are essential during the jetting process, higher viscosity inks are preferred after deposition to regulate lateral ink dispersal (Tawiah et al., 2016). This typically necessitates the use of thickeners that are highly pseudoplastic. Nevertheless, the use of such formulations can result in unfavourable effects during the printing process, such as the formation of satellite particles and tailing.

Fig. 4.9 Flow process sequence for digital printing

Conversely, pigments that exhibit Newtonian flow characteristics are more consistent during jetting. The capacity to customise the fabric's wettability and penetration behaviour is an additional benefit to thickener and other auxiliaries application. Wettability is the term used to describe the ability of the liquid to spread when it comes into contact with a fibrous surface. This phenomenon is primarily driven by capillary forces existing in the limited spaces between the fibrous and yarn structures.

4.6.3 Ink Systems

The ink utilized for material deposition, along with its physical properties, represents a critical component of inkjet printing technology (De Gans & Duineveld, 2004; McManus et al., 2017). The overall printing process which includes generation of droplet and its movement, interaction with the fabric, and drying, etc. is highly dependent on the quality and various ink characteristics such as viscosity, density, surface tension, conductivity, and pH (Thakker et al., 2022). Among these characteristics, viscosity and surface tension are particularly influential, as they significantly affect the resolution, uniformity, and overall printed pattern quality (Sohrabi et al., 2018). These two parameters play a decisive role in governing the three fundamental stages of the inkjet printing process (Machekposhi et al., 2020). An increase in ink viscosity can lead to reduced speed and accuracy in droplet ejection. At high viscosity and drop rates, the ink may fail to adequately replenish the cartridge between jetting events, resulting in printing failure. Viscosity is commonly adjusted using excipients such as glycols, including glycerol, propylene glycol, and polyethylene glycol (Genina et al., 2012, 2013; Sandler et al., 2011). In contrast, surface tension governs the ink's ability to detach from the nozzle and form droplets, and is typically modulated through the addition of surfactants (Ihalaninen et al., 2015).

To ensure reliable inkjet printing, an optimal balance among several physical properties, namely surface tension (γ), viscosity (η), density (ρ), and nozzle diameter or mesh aperture (l)—is essential (Saunders, 2014; McManus et al., 2017). These parameters contribute to the derivation of a dimensionless quantity (Z), based on fluid dynamics principles such as the Reynolds, Weber, and Ohnesorge numbers. For effective inkjet printing without the undesirable formation of satellite or secondary droplets, Z must fall within the range of 1–10 (i.e., $1 < Z < 10$) (Kumar et al., 2021; Alaman et al., 2016; Murphy & Atala, 2014). Satellite droplets can interfere with the trajectory and placement of primary droplets, thus compromising the precision of droplet deposition. Table 4.1 present the various ink properties and their functions (Thakker and Christie, 2015;Christies, 2015). Therefore, achieving accurate, uniform, and well-positioned droplet placement is vital for the overall success of inkjet printing (Hossain & Lu, 2023). Water-based inks are advantageous in inkjet printing due to their ease of cleaning from printheads and reduced likelihood of nozzle clogging, particularly when the ink exhibits Newtonian rheological behaviour (Tawiah et al., 2016; Żołek-Tryznowska, 2016). The development of water-based inkjet inks for textile applications is meticulously tailored to ensure

Table 4.1 Ink properties and the effect on inkjet printing

Physical property	Purpose and function
Viscosity	Influence the drop formation and ink fluidity
Density	Decides the viscosity
Surface tension	Influencing the satellite formation, substrate interaction
Conductivity	Assists in accurate droplet positioning on the substrate and regulates the waveform, which impacts droplet characteristics such as volume, velocity, and trajectory
pH	Affects ink stability

compatibility with the chemical and physical properties of the textile substrates (Le, 1998). Generally, the water-based inkjet ink used for textile printing comprises of water soluble solvents, deionized water, dye or pigment, surfactant, biocide, pH controller, chelating agents, binder, defoaming agents, etc.

Pigment, reactive dye, acid dye, and disperse dye are the four forms of ink that can be used on a wide variety of textiles (Yang et al., 2022; Madhukar Thakker & Sun, 2024; Pekarovicova & Husovska, 2016a). Dye-based inks necessitate pretreatment and post-treatment processes as presented in Table 4.2 (Tawiah et al., 2016; Le, 1998; Fespa, 2020). The benefits of pre-treatment with specific auxiliaries based on the colourant include controlled ink penetration, higher colour build-up, greater levelness, and less bleeding. In the post-treatment, the unfixed pigments will be removed through wash-off, which is followed by steam fixation with saturation condition or high temperature fixation.

4.6.4 Pigments

Digital printing of a wide variety of textile substrates, such as 100% cotton, cotton/polyester mixes, and the majority of other fabric types, frequently uses pigment inks. In contrast to dyes, pigments need a binding process in order to stick to the fabric surface because they have no inherent affinity for textile fibres (Pekarovicova & Husovska, 2016b). In order to accomplish this, pigment particles are dispersed in an aqueous media combined with a polymeric, self-crosslinking binder that encapsulates the pigments and promotes their mechanical adherence to the fabric. The four main dispersion techniques in pigment printing are surface modification through hydrophilic functional groups, microencapsulation, surfactant dispersion, and polymer dispersion. When it comes to ink compatibility, jetting performance, and textile print quality, surfactant-based dispersions have demonstrated especially positive outcomes. In order to ensure proper fixing and durability, binders are usually added directly to pigment ink formulations. An ideal pigment-to-binder ratio of 1:2 has been determined for all formulations.

Table 4.2 Digital printing Inks, fibre—interaction and finishing treatment

Colourant	Ink type and interaction	Fibres	Pre-treatment	Post-treatment	Printing applications
Pigment	Dispersion and no interaction, complex surface formulation with binder	All type of fibres	Not required	Dry heat fixation	Home furnishing/ low quality apparel/ certain children's wear
Reactive	Water Soluble and covalent bonding	Cotton, linen, viscose rayon, silk, wool	Alkali	Steam fixation, rinse, wash-off, hot rinse, rinse, dry	Apparel/ fashion
Acid	Water Soluble and electrostatic and hydrogen bonding	silk, wool, nylon, cashmere, angora	Acid donor	Steam fixation, rinse, wash-off, rinse, dry	Apparel/ fashion/ swimwear
Disperse	Dispersion and hydrophobic, solid state	Polyester, nylon, acrylic, and acetate rayon	Thickener	High temperature steam fixation, rinse, reduction clearing, rinse, dry	Apparel/ fashion/ automotive (high wet fastness)/ automotive
Disperse Sublimation	Dispersion and hydrophobic, solid state	Polyester, poly lycra blends	Thickener	Heat transfer fixation only	Apparel/ flag/banner/ signage/ Sportswear

The three main steps in the pigment printing process are printing, drying, and curing (Li et al., 2023). One of pigment printing's biggest benefits is that it does not require rigorous fabric preparation or post-printing process, which makes it a very cost-effective and ecologically friendly process. Due to their ease of use and wide range of substrate compatibility, pigment inks now make up around 50% of all textile printing worldwide. For blended textiles, like polyester/cotton, which pose technical difficulties when printed using traditional dye-based techniques, pigment printing is particularly beneficial. Because of their greater processability and decreased susceptibility to changes in fibre content, pigment inks are widely chosen in these situations.

Initially, the extensive usage of thermal drop-on-demand (DOD) printheads in early inkjet systems has posed technological challenges to the application of pigment inks in digital textile printing (Marie et al., 2013). These systems use heated ink to create a vapour bubble that pushes droplets from the nozzle. This method did not work well with pigment inks, especially those that contained binders, because heating made problems like nozzle clogging and ink instability worse. These difficulties have been greatly lessened by recent developments in piezoelectric printhead technology. Piezo printheads work by not heating the ink, which improves compatibility with formulations that use pigment. The economic viability of pigment inkjet printing on fabrics has been made possible in large part by this invention.

These days, a wide range of performance parameters, including as jetting reliability, shelf life, image quality, image permanence, and regulatory compliance, are used to evaluate the quality of pigment inks created for inkjet printing. To obtain the best fastness qualities, thermal fixation—usually done in a fabric oven or heated calendar—remains crucial. Contemporary pigment inks have outstanding lightfastness and strong resistance to rubbing and cleaning. However, there is still a significant disadvantage such as the binder forms a film, the printed cloth may have a harder hand, or less softness of feel. Previous drawbacks including tendency of nozzle block and dull colour reproduction have been resolved by manufacturers' continuous advancements, which have been made possible by intensive research and development. As a result, modern pigment inkjet systems offer a useful and progressively better textile printing solution for a variety of uses.

4.6.5 Reactive Dyes

Reactive dye inks are primarily employed to print on cellulosic based materials like cotton and viscose because they create vibrant colours with outstanding light and washing fastness (Zhao et al., 2022). Under alkaline conditions, these dyes create covalent bonds with the fibre, guaranteeing robust fixation and longevity. Although reactive dyes are mainly employed on cellulosic fibres, they can also be used on wool, silk, and nylon, however printing on these materials usually calls for extra process modifications. Reactive dyes come in a variety of classes, but not all of them are appropriate for traditional or inkjet textile printing. The functional groups of the reactive dyes used in textile printing are particularly selected because they can react with the hydroxyl groups in cellulose to create stable covalent connections. Reactive dyes for textile printing have been divided into two main classes:

1. Dye monohalogenotriazine, which is nucleophilically substituted with fiber's hydroxylgroups.
2. Vinyl sulfone-based dyes are made from sulphuric acid esters of β-hydroxyethylsulfones.

Which change into vinyl sulfone groups in alkaline environments and then nucleophilically attach to the fibre. Monohalogenotriazine dyes are the most commonly utilised in textile printing because of their varying reactivity and the need for long-term ink stability in printing applications. On cellulosic textiles, they make up about 85% of traditional reactive printing. Reactivity and substantiality are the two main factors that affect the choice of reactive dyes for inkjet applications. Despite being extremely reactive, dichlorotriazine dyes have poor storage stability in aqueous environments, making them inappropriate for inkjet printing. They are susceptible to hydrolysis due to their high reactivity, which shortens their shelf life and degrades the consistency of their print quality. On the other hand, monochlorotriazine (MCT) dyes are the best option for inkjet ink compositions because they provide the best balance between stability and reactivity. Textile printing requires dyes with minimal affinity to the substrate to prevent unwanted absorption or spreading, which means that before fixation, it has little affinity for the fibre. This feature reduces bleeding or accidental dye migration and guarantees crisp print definition. Usually, bulky side groups that sterically prevent premature contacts between the dye and fibre are included into the dye structure to produce low substantivity. Unlike exhaustion dyeing, where great substantivity is preferred, this method fits the requirements of continuous dyeing processes.

Inkjet systems cannot use liquid reactive dyes, which are used in traditional screen printing. Significant levels of inorganic salts, such as sodium sulphate or sodium chloride, are frequently present in these commercial formulations. These salts help with dye solubility and fibre penetration during padding, but they cause issues when printing digitally. Particularly in the low-viscosity conditions needed for inkjet delivery, high salt concentrations cause corrosion of jet nozzles and decreased dye solubility. Reverse osmosis (RO), a membrane filtration method that extracts ionic components from the dye solution, is used to significantly reduce salts from reactive dye inks in order to address these problems. Dye solubility is still an issue in spite of this. Lithium salts are occasionally used in place of sodium salts in dye formulations to increase solubility and decrease electrical conductivity. A range of auxiliary agents are also included in ink formulations:

- Diethylene glycol, propylene glycol, and diethylene glycol monobutyl ether are examples of hygroscopic (or hydrotropic) additives that stop ink from drying inside the nozzles.
- Surfactants improve dispersion and wetting.
- Buffers based on phosphoric acid preserve pH stability.

These speciality ingredients add to the high price of reactive dye inks, which can cost more per weight than precious metals like gold. The substrate must be pretreated for inkjet printing; this procedure is similar to that for acid dyes. A solution including moderate alkali, urea (to encourage moisture retention), and sodium alginate (as a thickening) is used to treat the fabric beforehand. After the printing application, the fabric is dried. Inkjet printing usually uses a "two-phase" procedure as opposed to the "all-in" approach used in traditional screen printing, when all chemicals and colours are included in the print paste:

- The ink, which solely contains pure colours and wetting ingredients, is applied after the substrate has been pre-treated.

Despite producing very high-quality prints, this method has extra expenses because of the pad/dry preparation stage. Furthermore, reactive inkjet printing is only economically viable for short-run or sample applications due to the need to stock pre-treated fabric. Reactive dye inks are highly prized for creating rich, vivid colours and are

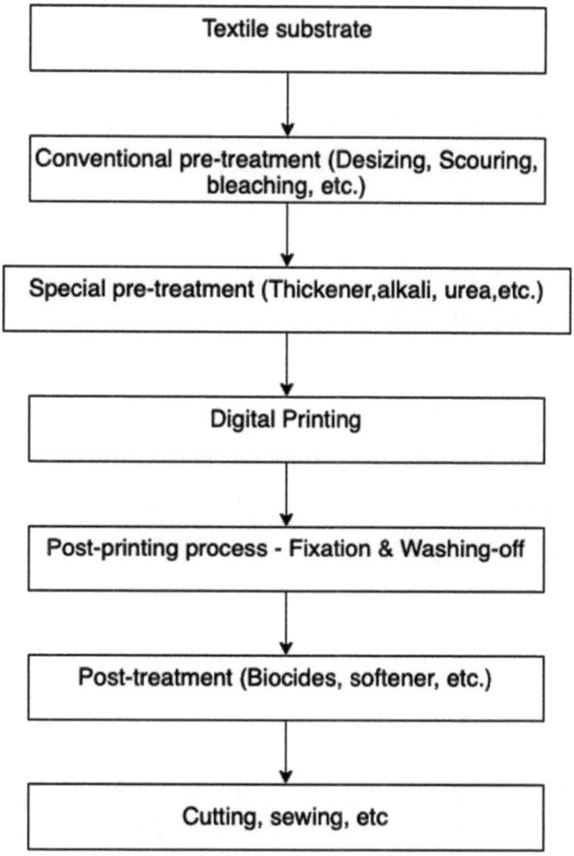

Fig. 4.10 Flow process sequence for digital printing with reactive ink

mostly used on cotton, linen, viscose rayon, and wool textiles. However, the reactive printing requires two more steps in contrast to pigment-based methods such as Fixation i.e., steaming and washing as mentioned in the flow process in Fig. 4.10.

For effective dye fixation via covalent bonding, the fabric must be steamed at 100 ± 20 °C for 10 min. A steaming process at 100 ± 20 °C for 10 min is required to achieve covalent bonding between the dye and the fibre. Thorough washing, including boiling and rinsing in both cold and hot water, is then necessary to get rid of any remaining pretreatment chemicals and unfixed dye. High-quality sampling and fashion applications, where colour gamut and vibrancy are crucial, are the main uses for reactive inks. Reactive inkjet printing is still the go-to technique when outstanding print quality, softness, and wash endurance are required, despite the process's complexity. The recipe and process parameter for pre-treatment, printing and post-treatment is given below.

Specific pre-treatment recipe for Cellulosic fabric by using Reactive Ink

Sodium alginate thickener: 60–100 gpl

Urea: 100–200 gpl

NaOH: 30–40 gpl

Resist salt: 20–30 g/kg

Water: 1000 g

pH $\geq$ 11

Padding pick-up: 70%

Drying: 100 °C for 12 min.

Digital printing

Post treatments

Steam fixation: 102 °C for 8 min in loop ager

Cold rinse, Washing 98 °C 2 g/l anionic washing agent for 7–10 min

Hot and cold rinsing

Drying.

4.6.6 Acid Dyes

Nylon is frequently printed with water-soluble anionic dyes due to their outstanding affinity for these fibres (Faisal et al., 2020; Hassabo et al., 2024). They are drawn to positively charged dye spots on the fibre surface rather than forming covalent connections with the fibres like reactive dyes do. When compared to reactive dye systems, these dyes frequently yield deeper and more vibrant hues. For maintaining the right moisture content and ink stability, hygroscopic compounds like glycerine, diethylene glycol, and triethylene glycol, monobutyl ether are most commonly used in ink formulations along with a surfactant. The fabric must be pretreated before printing for ensuring efficient dye fixation. In order to help the dye adhere to the fibre, a paste containing an acid donor must be applied uniformly. Once the printing and drying processes are complete, the fabric is steam-treated for 30–60 min at about 100 ± 2 °C. By causing fibre swelling and redissolving the dye, this procedure improves dye fixation and penetration. In order to ensure fastness and get rid of any unfixed dye, a second washing step is required. The recipe and process parameter for pre-treatment and post-treatment is given below.

Specific pre-treatment recipe for Nylon fabric by using Acid ink

Thickener: 60–80 gpl

Urea: 100–150 gpl

Ammonium sulphate: 20–40 gpl

Padding pick-up: 70%

Drying: 120 °C.

Digital printing

Post treatments

Steam fixation 105 °C for 20 min in loop ager

Cold rinse, Washing with soap, acetic acid 2–3 gpl at 80 °C for 15–20 min

Drying.

4.6.7 Disperse Dyes

Disperse dye inks is primarily used for polyester material (Kosolia et al., 2011). Due to their extremely low water solubility, these colours are finely distributed particles. Two sets of dispersion dye inks are often produced by several ink manufacturers: one for transfer printing and the another for printing the polyester materials directly. Former one is referred as Indirect digital printing and later one is referred as Direct digital printing. Compared to the former, the latter have more volatile dyes with a smaller molecular weight. Higher molecular weight dyes may be required to meet fastness standards, as in the case of some clothing end applications and materials used for car seat covers, where light-fastness is essential. As the small molecular weight dyes require shorter durations or lower temperatures for fixation compared to those with higher molecular weight, it is possible to print directly onto polyester fabric instead of paper using transfer printing inks. Jet printing directly into fabric produces results comparable to the high definition and tone work made available by transfer printing, without the need for additional paper or a second printing step. Pre-treatment of the fabric with a specific solution is essential to enhance the effectiveness of the subsequent steaming process. While dry heat at approximately 190 °C for one minute or pressurized steam at around 160 °C for 20–30 min can be employed to fix disperse dyes, steaming generally yields deeper colour shades. To get rid of any unfixed dye, including a reduction clear, washing is necessary as shown in flow process in Fig. 4.11. Disperse dye on polyester has a wide range of uses, including banners, flags, and soft signage. The recipe and process parameter for pre-treatment and post-treatment is given below.

Generally, direct print is better than sublimation as it provides good colour build-up, brilliant colours, and fastness properties. Initially, the polyester fabric is given heat setting treatment and specific pre-treatment like thickener as shown in the flow process Fig. 4.10 and mentioned below (Evans et al., 2021). Specific pretreatment is carried out in order to control the ink penetration, higher colour build up, better levelness, better quality image resolution and colour fastness. Post-treatment can be limited to a simple thermal fixation which is greater than 200 °C. However, time, water, and energy consumption is required for post-treatments like steaming,

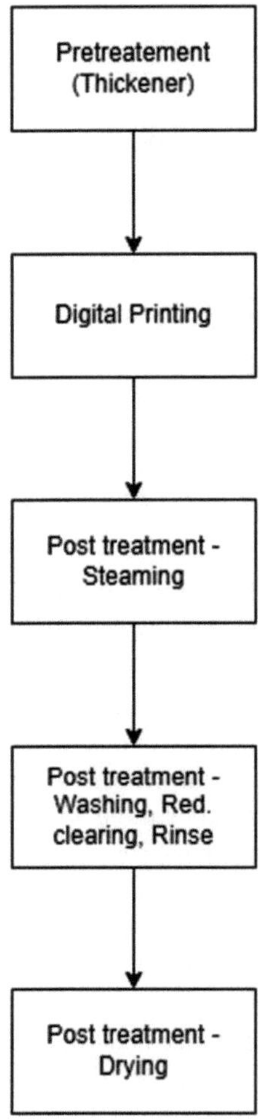

Fig. 4.11 Flow process of direct digital printing—disperse ink

washing, and final drying. Further, high fixation temperature limits the use of DP inks for polyester with thermal sensitive fibres like elastane. The other problems are nozzle clogging, limited stability, and lower print resolution compared to sublimation process.

Specific pre-treatment recipe for Polyester fabric by using Disperse Ink

Thickening agent: 50–80 gpl

Acetic Acid: 2–3 gpl

Padding pick-up: 70%

Drying: 120 °C.

Digital printing

Post-treatments

Steam fixation: 180 °C for 12 min in loop ager or 130 °C for 30 min in 30 psi

Cold rinse,

Reduction clear at 65–70 °C

Cold wash with acetic acid

Drying.

4.6.8 Disperse Ink- Indirect Digital Printing Through Sublimation

Sublimation printing does not require pre-treatment of fabric. Initially, it is printed on transfer paper which is the additional cost for the printing process. Afterwards, it is transferred to the fabric under heat 190–210 °C for the duration of 30–45 s with 2 bar pressure as shown in Fig. 4.12. This technique is mainly limited to polyester fibres which is the main limitation. As like the Direct printing method, high fixation temperature limits the use of polyester with thermal sensitive fibres like Lycra, Dorlastan and other problems such as nozzle clogging, limited stability. Further it is essential to ensure the uniformity of maintaining temperature and pressure of the calendar over the whole width and whole process.

4.6.9 Post-treatments in Textile Printing

After the pre-treated fabric has been dried and jet printed, the use of a separate drying station is generally not required. This is attributed to the relatively slow pace of the printing process, which allows the fabric to dry naturally by the time it is batched onto a roll, owing to prolonged exposure to the ambient warmth of the surrounding environment (Hossain & Lu, 2023). In some instances like catalogue, the fabric is printed only for photographic purposes. In such cases, fixation of the print may not be necessary. However, for the majority of dyes applications, fixation and washing are essential. These steps ensure that the dyes achieve their full fastness properties, as well as enhancing the colours of the print effectively.

4.6.10 Fixation Process

The fixation of printed textiles is typically achieved through steaming [80]. The fixation of reactive and acid dyes is typically carried out through steaming under

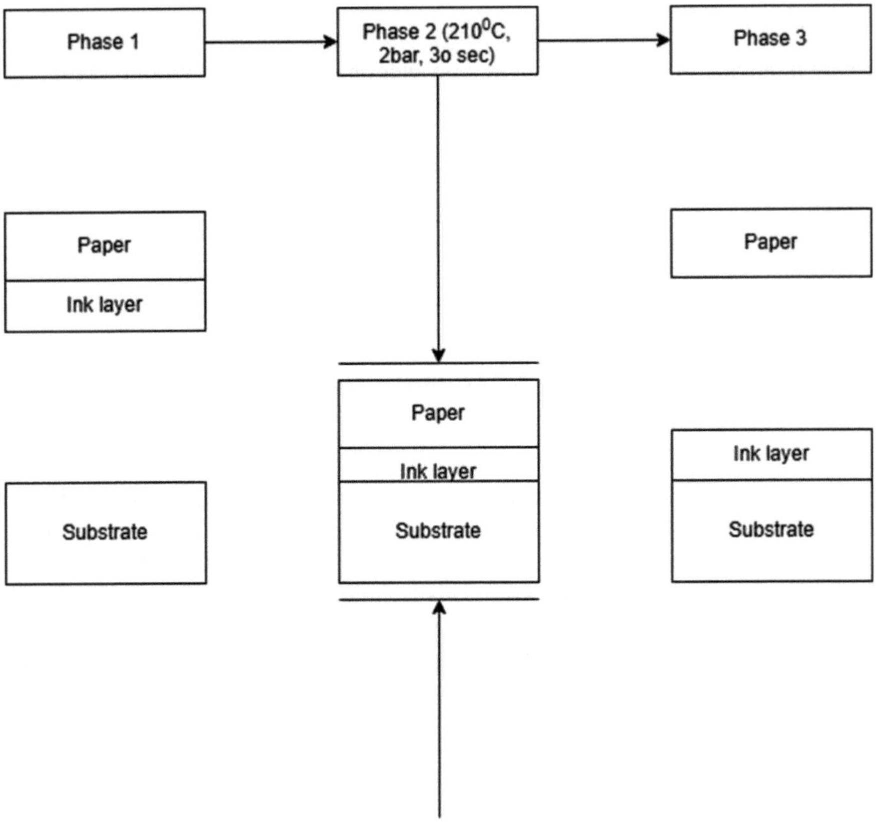

Fig. 4.12 Flow process of indirect digital printing—disperse ink

atmospheric pressure at temperatures marginally exceeding 100 °C. In this stage, steam condenses on the fabric surface and is subsequently absorbed by the thickener and hygroscopic agents present in the printed regions. This leads to the dissolution of dyes and chemicals, resulting in the formation of concentrated dyebaths confined within the thickener matrix. Due to the exceptionally low liquor ratio (approximately 1:1), dye fixation proceeds considerably faster than in traditional exhaustion dyeing methods.

Fixation of disperse dyes on polyester is typically carried out using high-temperature steam, owing to the reduced glass transition temperature (Tg) of polyester in steam compared to dry air, which enhances the fixation process. The steaming process is commonly conducted at atmospheric pressure, with temperatures ranging from 170 to 180 °C. In some cases, the temperatures range between 130 and 150 °C is also utilized for the steaming process. Pigment prints, on the other hand, are cured using hot air with the temperature around 150 °C in a stenter.

4.6.11 Washing-Off Process

Washing-off of fixed jet-printed fabrics is typically undertaken in batch operations, as the lengths of the printed substrates are frequently limited (Hossain & Lu, 2023). This process can be done either batchwise or continuously and generally involves several stages. The initial stage consists of a cold rinse, which is necessary to remove thickeners, auxiliaries, and any loose dye under conditions where there is little risk of staining the white or unprinted areas of the fabric. The addition of reagents that retain unfixed dye within the bath can further minimize the risk of staining. For example, The inclusion of 1 g/L sodium carbonate in the initial rinse bath during the post-dyeing process of acid dyes serves to maintain an alkaline environment, thereby preventing the protonation of amino groups on fibers such as nylon, wool, and silk, and consequently minimizing the risk of dye re-adsorption. After the first cold rinse, the primary requirement for washing off reactive dye prints is that the subsequent hot wash attains a temperature of no less than 95 °C to ensure thorough removal of unfixed dye. This is essential to ensure the removal of hydrolyzed dye. A reduction clearing treatment with alkali and sodium dithionite ($Na_2S_2O_4$) is often necessary to remove surface dye from fabrics printed with disperse dyes. It is crucial to maintain the bath temperature below 70 °C during this process, as sodium dithionite is unstable at higher temperatures.

4.7 Cotton and Cellulosics

The reactive dyes are typically printed on cotton, viscose rayon, and lyocell fabrics using the two-phase jet printing method. In this procedure, the ink composition comprises the dye, and the cloth is pre-treated with an alkali and thickening (Zhao et al., 2022). Although screen printing can also be used, a pad mangle is typically used to provide pre-treatment. Whatever the technique, the fabric needs to be dried until there is only around 5–7% moisture left on it before printing. A typical aqueous pad liquor formulation for reactive dyes of the monochlorotriazine (MCT) type contains 20–30 gpl sodium carbonate, 100 gpl sodium alginate, and 100 gpl urea. These substances serve as alkali, hygroscopic agents, and thickeners, respectively. In textile printing, thickeners are essential. By reducing ink wicking into the substrate, they aid in preserving the sharpness of printed edges and outlines. Furthermore, thickeners hold onto moisture, which helps chemicals and colours dissolve and diffuse into the fibres during the post-print steaming stage. The rheological characteristics of inks and print pastes are also influenced by thickeners, which has an impact on print penetration and definition.

The perfect thickener should be chemically inert to the dye and other ingredients in the composition. When thickeners and dyes react, insoluble residues are usually left behind. These residues cannot be washed out and might give the fabric an unwanted rigidity. Natural thickeners are very hydrophilic; they are mostly polysaccharides with many hydroxyl groups. Their use is restricted, nevertheless, by their reactivity with reactive dyes (much like cellulose). For reactive dye printing, sodium alginate, a polysaccharide made from brown seaweed, is the recommended thickener. Since it is a polyanionic molecule, mutual electrostatic repulsion prevents it from reacting with anionic reactive dyes. There are several grades of sodium alginate, with varied molecular weights. High molecular weight grades produce low-solids formulations with extremely pseudoplastic (shear-thinning) behaviour, whilst high-solids thickeners are produced by less molecular weight grade. Higher thickener loadings help maintain design detail during drying and steaming, therefore solids content is given precedence over flow characteristics when the thickener is applied during fabric pre-treatment.

The thickener creates a microporous, sponge-like structure while adding to the aqueous phase of an oil-in-water emulsion for retaining moisture. Afterwards, the material is dried. Excellent wash fastness is ensured using reactive dyes, which create covalent bonds with cellulose fibres in alkaline environments. For the best fixation, different reactive dye classes—such as vinyl sulphone and monochlorotriazine—need different alkali strengths. Because sodium bicarbonate has a lesser tendency to cause dye hydrolysis during storage, it is frequently suggested for "all-in" pastes and inks. However, sodium carbonate is typically appropriate for pre-treatment applications due to its strength and cost-effectiveness. In certain formulations, 25 gpl of sodium bicarbonate is used in place of sodium carbonate. A common ingredient in print pastes, urea functions as a hygroscopic agent (humectant) and a dye solvent. Urea is still the most used option, despite environmental concerns about the nitrogen load in wastewater. Urea concentrations in pad liquors are frequently raised to 200 gpl for viscose rayon in order to improve moisture retention and dye solubility. Pad liquors also frequently contain sodium meta-nitrobenzene sulphonate. During the steaming process, it acts as a moderate oxidising agent, preventing reactive colours from being reduced and then decolorised.

4.7.1 Nylon

Nylon is typically printed by using acid dyes, following a pre-treatment process as mentioned in Table 4.1. Optimization of print quality may require experimental adjustment of the concentrations of urea and acid in the formulation (Kosolia et al., 2011). It has been reported that polyamide substrates are treated with bisphenol derivative solution for thirty minutes at 90 °C, followed by washing, drying, and subsequent inkjet printing with acid dye inks. This technique is found to yield high colour and excellent colour fastness.

4.7.2 *Polyester*

Polyester fabrics are typically printed using disperse dye inks. Effective dye diffusion into the fibre requires temperatures exceeding the polymer's glass transition point. While disperse dyes exhibit minimal reactivity with thickening agents, it is essential that the selected thickener remains stable under the elevated fixation temperatures involved in the process. Sodium alginate, commonly used at a concentration of 100 gpl, is a preferred thickening agent due to its role as a movement inhibitor during both drying and steaming stages. In addition to sodium alginate, synthetic thickeners have also been considered as viable alternatives. Furthermore, the same anti-bleeding agent employed in the pre-treatment of nylon has been incorporated into pre-treatment formulations for polyester textiles, enhancing print definition and colour sharpness. However, the inclusion of urea in such formulations poses environmental and health concerns, as it may generate noxious fumes during the high-temperature dye fixation process. To mitigate these issues, alternative thickeners have been investigated. One such example, cited in a patent, involves a pre-treatment composition containing 1% sodium salt of polyacrylic acid and 2% polyvinyl alcohol, offering an effective and potentially less hazardous approach.

4.8 Benefits and Scope of Digital Printing

Digital textile printing offers the significant advantage of enabling direct output from a digital file, thereby eliminating the need for setup procedures. The printing cost for short runs are lower (Khatri et al. 2015). Moreover, the cost per square meter is almost flat with volume of production. In the case of screen printing, the setup requirements and screen expenses are typically spread over the full length of the production run. It has been revealed that screen printing will be economical for the printing job quantity of above 1000 m^2. But, the digital printing tends to offer greater cost efficiency at lower print volumes. Generally, print design has 40–150% coverage in the pattern. At 100% coverage, ink usage averages 17.5 ml/m^2. However, based on print density and design characteristics, this can range from approximately 7–26.3 ml/m^2. It is also required to consider the waste ink for the calculation. It is mainly produced when the printer has been kept idle for long time and periodic maintenance of printhead cleaning activity. It has been revealed that 2.6 ml/m^2 ink will be incurred as a waste. Further to this, printing cost is mainly influenced by ink cost. Hence, it can be addressed by producing more volume in the market which also facilitate the faster market growth (Tyler, 2005).

Digital printing has profoundly influenced the fashion industry, especially within niche markets, due to its capacity for customisation or small runs, rapid production, and flexibility, thereby enabling designers to develop innovative and exclusive bespoke collections in apparel and non-apparel markets such as niche fashion market, swim suits, home furnishing, wall hangings, flags, banners, etc. In addition, there is a continual expectation from customers for enhanced speed, superior quality, and greater cost efficiency which are also assist the aforementioned markets. Improvements in image quality can be achieved through the use of smaller ink drop sizes. Similarly, tonal or gradations and colour gamut are enhanced by the improvement of gray scale capability and ink technology, respectively. Despite persistent obstacles, especially regarding fabric compatibility and durability, continuous technological developments are mitigating these concerns.

Today's markets are evolving at an unprecedented pace, and customer expectations are higher than ever. In response, digital textile printing has emerged as a powerful solution, enabling the print production which are tailored to individual customer preferences. Customization allows consumers to express their uniqueness, and applications are broad and diverse—ranging from customized automotive upholstery and interior furnishings to exclusive wall coverings and window treatments. Digital printing enables distinctive and memorable designs. For instance, casino gaming tables can feature artwork specific to a particular convention or event, while pool table covers can display targeted advertising messages. The possibilities for customization and innovation are expanding rapidly.

Digital inkjet printing addresses these growing opportunities and unlocks access to entirely new markets (Tkalec et al., 2024). It supports the creation of intricate and creative designs that are not feasible with traditional screen printing methods. While even the most advanced screen printers are generally limited to 12 spot colours, digital printing using process color technology allows for an almost unlimited colour palette within a single design. This advancement also brings significant operational advantages. Design cycle times are significantly reduced, and samples can be produced instantly. The ability to economically manage short production runs enables companies to maintain smaller inventories and respond swiftly to market demand. For example, popular apparel items can be replenished quickly without requiring retailers to reduce prices, thanks to the flexibility offered by digital printing. Hence, inkjet printing offers several advantages over other existing printing techniques such as lower cost, higher efficiency, improved resolution, compatibility with polymeric materials, process simplicity, flexible operation, minimal environmental constraints, and reduced material consumption (Kholghi et al., 2017; Yin et al., 2010). In summary, digital textile printing is not only meeting the needs of today's dynamic market but is also reshaping the future of design, manufacturing, and customer engagement.

4.9 Sustainability Aspects of Cellulosic and Manmade Fibres in Digital Printing

In textile printing, cellulosic (natural or semi-synthetic) and synthetic (man-made) fibres present different environmental difficulties over all phases of their lifecycles, and sustainability varies greatly between them. The dye affinity and printing methods are two important differences. Disperse dyes are usually used for synthetic materials like polyester, whereas acid dyes are used for nylon. Both dyes usually require high temperatures and chemical fixatives. Even while these methods typically use less water than natural fibres, the effluent that is produced could still contain harmful materials. Often used on polyester, sublimation printing uses a lot of energy but produces very little water waste. Reactive dyes, on the other hand, are used for cellulosic fibres like cotton, rayon, and lyocell because they form an efficient bond with cellulose. Despite being widely used for cotton and lyocell, reactive dyes need a lot of water and salt, which significantly pollutes wastewater (Khatri et al., 2015). However, compared to screen printing, digital inkjet printing method is more water-efficient when applied to cellulosic materials. Man-made fibres usually require less water for coloration process, but they need more energy to heat-set dyes, and they can cause long-term pollution because they release microplastics when they are washed (Ellen, 2017; Boucher & Friot, 2017). Although cellulosic fibers—particularly cotton—consume more water overall, their long-term environmental impact is lessened by their biodegradable nature. Because of its closed-loop production method and exceptional dye absorption, Lyocell (Tencel™) is a more environmentally friendly cellulosic choice that reduces dye waste (Lenzing, 2021).

4.10 Market Scenario of Digital Printing Based on Material and Process

Owing to the direct application of designs on the fabric surface and other advantages such as rapid production, design flexibility, and reduced environmental impact compared to conventional methods like screen printing, the market for digital inkjet printing technology has been growing rapidly now a days. In addition, its cost-effectiveness and adaptability make it especially valuable for the various textile sectors, viz. fashion, home décor, and personalised merchandise. The growing shift toward digital solutions is being influenced by rising demand for precision, sustainability, and product customisation. The boom in e-commerce and direct-to-consumer (DTC) channels is also a major catalyst. In the recent IOSR report in 2024 of sales in retail sector, global e-commerce sales represented 19% in 2023. Moreover, the projections showing a 39% increase by 2027, surpassing USD 8 trillion. In line with these trends, the market share of inkjet technology was assessed at USD 5.64 billion in 2024 and is predicted to grow at a compound annual growth rate (CAGR) of 12.4% to reach USD 18.22 billion by 2034 (Digital Textile Prinitng Sarex, 2025; [90]). Changes in consumer behaviour and advancements in technology are driving this increase.

From a regional perspective, the market is divided into North America, Europe, Asia Pacific, Latin America, and the Middle East & Africa. In 2024, North America topped the market among them owing to its sophisticated manufacturing infrastructure, strong R&D investments, and early technology adoption. The region's high demand for customised apparel and premium textiles, combined with a mature e-commerce ecosystem, continues to push demand for efficient, short-run digital printing solutions. In contrast, the fastest growth is expected in Asia Pacific region through 2034. The expansion is being propelled by its robust textile manufacturing base, increasing disposable incomes, and surging consumer interest in fashion-forward and personalized products. Rapid urbanisation and rising internet penetration in countries like China, India, and Southeast Asia are further accelerating adoption of affordable, high-performance digital printing technologies.

Design software, ink formulas, and printhead speed improvements are all greatly increasing manufacturing efficiency while lowering manufacturing expenses and waste. These innovations support the market's move toward mass customisation, meeting consumer expectations for personalized products. Digital textile printing enables businesses to produce custom items on demand, without large inventory or long lead times—providing agility to adapt to fast-changing trends and seasonal shifts. The inkjet technology is used with several ink types such as sublimation for polyester fibre materials, reactive for cotton, rayon and linen fibre materials, acid for silk and nylon fibre materials, direct disperse for polyester fibre materials, pigment inks for any fibre materials. Among these, sublimation inks dominated in 2024 which is mainly used for polyester-based fabrics applications in fashion, home décor products, and sportswear. Sublimation delivers bright colors, durability, and wash resistance. The waterless nature of the process and low waste output also make it a sustainable choice. The widespread use of polyester in many different industries has resulted sublimation inks' leading position.

For natural fibers like cotton, rayon, and linen, reactive and pigment inks are most frequently used due to their compatibility and ability to deliver high-quality prints on cellulosic materials. Its popularity stems from rising demand for lightweight, vivid, and wrinkle-resistant polyester fabrics, commonly seen in fashion, activewear, and signage. This method offers superior color vibrancy and sharpness, supports customization, and enables fast, environmentally conscious production for both small and large-scale operations.

4.11 Conclusion

This chapter has shown that the textile industry is currently undergoing a profound transformation, propelled by the escalating demand for high-quality, cost-efficient, and rapid production solutions. Digital inkjet printing has emerged as a revolutionary technology that effectively addresses these requirements by facilitating the creation of intricate designs, vibrant colour reproduction, and on-demand manufacturing capabilities. This advanced methodology eliminates the necessity for physical screens inherent in traditional screen printing, thereby enabling swift design modifications, substantial inventory reduction, and more sustainable operational practices. Such technological progression is particularly crucial in the contemporary marketplace, which is increasingly characterized by fast fashion trends and growing consumer preference for personalized products. Digital printing technologies present numerous strategic advantages that align with current market expectations, including dramatically compressed production timelines that allow brands to expedite product development from initial concept to retail availability within days. The inherent flexibility of these systems supports both limited-edition productions and fully customized collections, while their exceptional resolution capabilities ensure photorealistic print quality. Furthermore, digital processes demonstrate superior environmental performance through significantly reduced water consumption and decreased chemical usage compared to conventional methods.

The integration of advanced piezoelectric and thermal inkjet systems, combined with specialized ink formulations—including pigment, reactive, acid, and disperse variants—ensures broad compatibility with diverse textile substrates and enables versatile applications across multiple sectors, including fashion apparel, home furnishings, and industrial textiles. As environmental sustainability considerations increasingly intersect with evolving market demands, digital textile printing is positioned for substantial market expansion. Industry valuations indicate the sector was worth USD 5.64 billion in 2024, with projections suggesting robust growth at a compound annual rate of 12.4%, potentially reaching USD 18.22 billion by 2034. Regional analysis reveals North America maintains leadership in customized production adoption, while the Asia–Pacific region demonstrates the most rapid growth, driven by its established manufacturing infrastructure and expanding e-commerce ecosystems. Continuous technological advancements in printhead engineering, ink chemistry, and automated systems are not only revolutionizing production methodologies but are also establishing digital inkjet printing as the definitive sustainable and innovative solution for future textile manufacturing.

References

Alamán, J., Alicante, R., Peña, J. I., & Sánchez-Somolinos, C. (2016). Inkjet printing of functional materials for optical and photonic applications. *Materials, 9*, 910.

Alomari, M., Mohamed, F. H., Basit, A. W., & Gaisford, S. (2015). Personalised dosing: Printing a dose of one's own medicine. *International Journal of Pharmaceutics, 494*, 568–577.

Azizi Machekposhti, S., Mohaved, S., & Narayan, R. J. (2019). Inkjet dispensing technologies: Recent advances for novel drug discovery. *Expert Opinion on Drug Discovery, 14*, 101–113.

Bisutti, G. (2011). WO Patent 2011027375.

Boel, E., Koekoekx, R., Dedroog, S., Babkin, I., Vetrano, M. R., Clasen, C., & Van den Mooter, G. (2020). Unraveling particle formation: From single droplet drying to spray drying and electrospraying. *Pharmaceutics, 12*, 625.

Boland, T., Xu, T., Damon, B., & Cui, X. (2006). Application of inkjet printing to tissue engineering. *Biotechnology Journal, 1*, 910–917.

Boucher, J., & Friot, D. (2017). *Primary microplastics in the oceans: a global evaluation of sources* (Vol. 10). Gland, Switzerland: Iucn.

Bowles, R. J., Jr., Epling, J. P., Hand, J. P., Ruffner, S. C., & Walker, J. W. (2005). US Patent 6,953,511.

Cai, S., Sun, Y., Wang, Z., Yang, W., Li, X., & Yu, H. (2021). Mechanisms, influencing factors, and applications of electrohydrodynamic jet printing. *Nanotechnology Reviews, 10*(1), 1046–1078.

Calvert, P. (2007). Materials science. Printing cells. *Science, 318*, 208–209.
Campbell, J. R. (2006). *Controlling digital colour printing on textiles* (pp. 160–190). Woodhead Publishing Ltd.
Castrejon Pita, J. R., Baxter, W. R. S., Morgan, J., Temple, S., Martin, G. D., & Hutchings, I. M. (2013). Future, opportunities and challenges of inkjet technologies. *Atomization and Sprays, 23*(6).
Christie, R. M. (2015). *Colour chemistry* (2nd ed.). Cambridge: The Royal Society of Chemistry.
Cie, C. (2015a). Heat transfer and sublimation printing. *Ink jet textile printing* (pp. 125–137). Woodhead Publishing.
Cie, C. (2015b). Inks for digital printing. In *Ink jet textile printing* (pp. 85–97). Cambridge, UK: Woodhead Publishing.
De Gans, B. J., & Duineveld, P. C. (2004). Schubert, U.S. inkjet printing of polymers: State of the art and future developments. *Advanced Materials, 16*, 203–213.
De Souza Lima, R., Ré, M. I., & Arlabosse, P. (2020). Drying droplet as a template for solid formation: A review. *Powder Technology, 359*, 161–171.
DeVries, R. F., & William, H. S. I. (1977). US Patent 4,021,591.
Dawson, T., & Hawkyard, C. (2000). *Review of Progress in Coloration and Related Topics, 30*(1), 7–20.
Digital Textile Printing. (2025). *Pretreatment for digital printing.* Sarasol-1101-Sarex chemicals. Retrieved 19, April, 2025 from https://www.sarex.com/textile/product.
El-Halwagy, A. A., El-Sayad, H. S., & El-Molla, M. M. (2001). *Macromolecular Materials and Engineering, 286*(10), 618–623.
El-Kashouti, M., Elhadad, S., & Abdel-Zaher, K. (2019). Printing technology on textile fibers. *Journal of Textiles, Coloration and Polymer Science, 16*(2), 129–138.
Ellen MacArthur Foundation. (2017). *A new textiles economy.*
Evans, S. E., Harrington, T., Rodriguez Rivero, M. C., Rognin, E., Tuladhar, T., & Daly, R. (2021). 2D and 3D inkjet printing of biopharmaceuticals—a review of trends and future perspectives in research and manufacturing. *International Journal of Pharmaceutics, 599*, Article 120443.
Faisal, S., Tronci, A., Ali, M., Lin, L., & Mao, N. (2020). Pretreatment of silk for digital printing: Identifying influential factors using fractional factorial experiments. *Pigment & Resin Technology, 49*(2), 145–153.
Fang, M., Li, T., Zhang, S., Rao, K. V., & Belova, L. (2020). Design and tailoring of inks for inkjet patterning of metal oxides. *Royal Society Open Science, 7*, Article 200242.
Farid, M. (2003). A new approach to modelling of single droplet drying. *Chemical Engineering Science, 58*, 2985–2993.
Fespa. (2020, 10, 27). Inks for textile printing. Retrieved 07, November, 2022, from https://www.fespa.com/en/news-media/features/inks-for-textile-printing.
Gaisford, S. (2017). 3D Printed pharmaceutical products. In *3D Printing in medicine* (pp. 155–166).
Gao, Y., Shi, W., Wang, W., Leng, Y., & Zhao, Y. (2014). Inkjet printing patterns of highly conductive pristine graphene on flexible substrates. *Industrial and Engineering Chemistry Research, 53*, 16777–16784.
Genina, N., Fors, D., Vakili, H., Ihalainen, P., Pohjala, L., Ehlers, H., et al. (2012). Tailoring controlled-release oral dosage forms by combining inkjet and flexographic printing techniques. *European Journal of Pharmaceutical Sciences, 47*, 615–623.
Genina, N., Janßen, E. M., Breitenbach, A., Breitkreutz, J., & Sandler, N. (2013). Evaluation of different substrates for inkjet printing of Rasagiline Mesylate. *European Journal of Pharmaceutics and Biopharmaceutics, 85*, 1075–1083.
Harris, J. D., & Anderson, B. J. (2013). US Patent 8,372,478.
Hassabo, A. G., Gouda, N. Z., Khaleed, N., Shaker, S., Abd El-Salam, N. A., Mohamed, N. A., & Othman, H. (2024). Enzymes in digital printing of polyamide fabric. *Journal of Textiles, Coloration and Polymer Science, 21*(1), 149–160.
Hoath, S. D. (Ed.). (2016). *Fundamentals of inkjet printing: The science of inkjet and droplets.* John Wiley & Sons.

Hossain, S. S., & Lu, K. (2023). Recent progress of alumina ceramics by direct ink writing: Ink design, printing and post-processing. *Ceramics International, 49*(7), 10199–10212.

Ihalainen, P., Määttänen, A., & Sandler, N. (2015). Printing technologies for biomolecule and cell-based applications. *International Journal of Pharmaceutics, 494*, 585–592.

Ikeda, W., & Tomiki, A. (2013). EP Patent 2666646 A1.

Khan, S., Ali, S., & Bermak, A. (2019). Smart manufacturing technologies for printed electronics. In *Hybrid nanomaterials: Flexible electronics materials.*

Khatri, A., Peerzada, M. H., Mohsin, M., & White, M. (2015). A review on developments in dyeing cotton fabrics with reactive dyes for reducing effluent pollution. *Journal of Cleaner Production, 87*, 50–57.

Kholghi Eshkalak, S., Chinnappan, A., Jayathilaka, W. A. D. M., Khatibzadeh, M., Kowsari, E., & Ramakrishna, S. (2017). A review on inkjet printing of CNT composites for smart applications. *Applied Materials Today, 9*, 372–386.

Kosolia, C. T., Tsatsaroni, E. G., & Nikolaidis, N. F. (2011). Disperse ink-jet inks: Properties and application to polyester fibre. *Coloration Technology, 127*(6), 357–364.

Kumar, P., Ebbens, S., & Zhao, X. (2021). Inkjet printing of mammalian cells—theory and applications. *Bioprinting, 23*, Article e00157.

Kwon, K. S., Rahman, M. K., Phung, T. H., Hoath, S. D., Jeong, S., & Kim, J. S. (2020). Review of digital printing technologies for electronic materials. *Flexible and Printed Electronics, 5*(4), Article 043003.

Lan, L., Zou, J., Jiang, C., Liu, B., Wang, L., & Peng, J. (2017). Inkjet printing for electroluminescent devices: Emissive materials, film formation, and display prototypes. *Frontiers of Optoelectronics, 10*, 329–352.

Le, H. P. (1998). Progress and trends in ink-jet printing technology. *Journal of Imaging Science and Technology, 42*, 49–62.

Lee, S. W., Kim, H. C., Kuk, K., & Oh, Y. S. (2002). A monolithic inkjet print head: DomeJet. *Sensors and Actuators, a: Physical, 95*, 114–119.

Lenzing, A. G. (2021). *Life cycle assessment of TENCEL™*.

Lewis, D. M., Brady, P. R., Cookson, P. G., & Fincher, K. W. (1984). US Patent 4,455,147.

Li, Y., Zhang, X., Yang, L., Zhang, R., & Li, R. (2023). Study of color reproduction in pigment digital printing. *Textile Research Journal, 93*(11–12), 2718–2737.

Machekposhti, S. A., Movahed, S., & Narayan, R. J. (2020). Physicochemical parameters that underlie inkjet printing for medical applications. *Biophysical Reviews, 1*, Article 011301.

Madhukar Thakker, A., & Sun, D. (2024). Inks for digital printing of textiles. *The Journal of the Textile Institute, 115*(12), 2377–2390.

Mannerbro, R., Ranlöf, M., Robinson, N., & Forchheimer, R. (2008). Inkjet printed electrochemical organic electronics. *Synthetic Metals, 158*, 556–560.

Marie, M. M., El-Hamaky, Y. H., Maamoun, D., Ibrahim, D. F., & Abbas, S. M. (2013). Pigment ink formulation for inkjet printing of different textile materials. *International Journal of Innovation and Applied Studies, 4*(1), 239–247.

Masod, M. Y. M. (2024). Digital printing in art: Enhancing narrative and expression on contemporary works/Muhammad Yusuf Masod. *International Journal of Art and Design (IJAD), 8*(2/SI), 1–8.

McManus, D., Vranic, S., Withers, F., Sanchez-Romaguera, V., Macucci, M., Yang, H., et al. (2017). Water-based and biocompatible 2D crystal inks for all-inkjet-printed heterostructures. *Nature Nanotechnology, 12*, 343–350.

Mezhericher, M., Levy, A., & Borde, I. (2011). Modelling the morphological evolution of nanosuspension droplet in constant-rate drying stage. *Chemical Engineering Science, 66*, 884–896.

Miles, L. (1977). *Journal of the Society of Dyers and Colourists, 93*(5), 161–164.

Miles, C., & Leslie, W. (2010). Textile printing (2nd ed.). NC, USA: AATCC, Research Triangle Park.

Murphy, S. V., & Atala, A. (2014). 3D Bioprinting of tissues and organs. *Nature Biotechnology, 32*, 773–785.

Özkol, E., Ebert, J., Uibel, K., Wätjen, A. M., & Telle, R. (2009). Development of high solid content aqueous 3Y-TZP suspensions for direct inkjet printing using a thermal inkjet printer. *Journal of the European Ceramic Society, 29*, 403–409.

Pekarovicova, A., & Husovska, V. (2016a). Printing ink formulations. In S. T. Joanna Izdebska (Ed.), *Printing on polymers* (pp. 41–45). William Andrew Publishing.

Pekarovicova, A., & Husovska, V. (2016b). Printing ink formulations. In S. T. Joanna Izdebska (Ed.), *Printing on polymers. Fundamentals and applications* (pp. 41–55). William Andrew Publishing.

Pinto, G. R., & Pinto, C. (2013). US Patent 8,485,657.

Rattee, I. D. (1977). *Journal of the Society of Dyers and Colourists, 93*(5), 190–194.

Rattee, I. D. (2010). Transfer printing. In L. W. C. Miles (Ed.), *Textile printing, revised* (2nd ed.). Bradford, UK: Society of Dyers and Colourists.

Sandler, N., Määttänen, A., Ihalainen, P., Kronberg, L., Meierjohann, A., Viitala, T., & Peltonen, J. (2011). Inkjet printing of drug substances and use of porous substrates-towards individualized dosing. *Journal of Pharmaceutical Sciences, 100*, 3386–3395.

Sarkodie, B., Tawiah, B., Agbo, C., & Wizi, J. (2018). Status and development of transfer printing in textiles—a review. *AATCC Journal of Research, 5*(2), 1–18.

Saunders, R. E., & Derby, B. (2014). Inkjet printing biomaterials for tissue engineering: Bioprinting. *International Materials Reviews, 59*, 430–448.

Shah, M. A., Lee, D. G., Lee, B. Y., & Hur, S. (2021). Classifications and applications of inkjet printing technology: A review. *IEEE Access, 9*, 140079–140102.

Singh, A., & Van den Mooter, G. (2016). Spray drying formulation of amorphous solid dispersions. *Advanced Drug Delivery Reviews, 100*, 27–50.

Singh, M., Haverinen, H. M., Dhagat, P., & Jabbour, G. E. (2010). Inkjet printing—process and its applications. *Advanced Materials, 22*, 673–685.

Sohrabi, S., & Liu, Y. (2018). Modeling thermal inkjet and cell printing process using modified pseudopotential and thermal lattice Boltzmann methods. *Physical Review E, 97*, Article 033105.

Tawiah, B. E. N. J. A. M. I. N., Howard, E. K., & Asinyo, B. K. (2016). The chemistry of inkjet inks for digital textile printing-review. *BEST, 4*, 61–78.

Thakker, A. M., Sun, D., & Bucknall, D. (2022). Inkjet printing of plasma surface–modified wool and cotton fabrics with plant-based inks. *Environmental Science and Pollution Research*. https://doi.org/10.1007/s11356-022-20659-3.

Tkalec, M., Glogar, M., Penava, Ž., Forte Tavčer, P., Kuščer, D., & Stojanoska, I. (2024, February). The complexity of colour/textile interaction in digital printing as an integral part of environmental design. In *Arts* (Vol. 13, No. 1, p. 29). MDPI.

Tyler, D. J. (2005). Textile digital printing technologies. *Textile Progress, 37*(4), 1–65.

Uddin, M. J., Hassan, J., & Douroumis, D. (2022). Thermal inkjet printing: Prospects and applications in the development of medicine. *Technologies, 10*(5), 108.

Ujiie, H. (2006). *Design and workflow in digital ink-jet printing* (pp. 337–356). Woodhead Publishing.

Vehring, R. (2008). Pharmaceutical particle engineering via spray drying. *Pharmaceutical Research, 25*, 999–1022.

Wijshoff, H. (2010). The dynamics of the piezo inkjet printhead operation. *Physics Reports, 491*, 77–177.

Xue, M., Li, F., & Cao, T. (2012). *Nanoscale, 4*(6), 1939–1947.

Yang, X., Lin, Y., Wu, T., Yan, Z., Chen, Z., Kuo, H. C., & Zhang, R. (2022). An overview on the principle of inkjet printing technique and its application in micro-display for augmented/virtual realities. *Opto-Electronic Advances, 5*(6), 210123–210131.

Yin, Z. P., Huang, Y. A., Bu, N. B., Wang, X. M., & Xiong, Y. L. (2010). Inkjet printing for flexible electronics: Materials, processes and equipments. *Chinese Science Bulletin, 55*, 3383–3407.

Yin, Y., Wang, C., Shen, Q., Zhang, G., & Galib, C. M. A. (2013). *Industrial and Engineering Chemistry Research, 52*(31), 10656–10663.

Zhao, H., Zhang, K., Fang, K., Shi, F., Pan, Y., Sun, F., et al. (2022). Insights into coloration enhancement of mercerized cotton fabric on reactive dye digital inkjet printing. *Rsc Advances, 12*(17), 10386–10394.

Żołek-Tryznowska, Z. (2016) Rheology of printing inks. In S. T. Joanna Izdebska (Ed.), *Printing on polymers* (pp. 87–99). William Andrew Publishing.

Chapter 5
Sustainable Inkjet Printing Inks to Enhance Aesthetic Appeal of Wool and Wool Blend Fabrics

Mrinal Choudhari, Swati Mahadik, and Dinesh Marewad

Abstract The textile industry is currently experiencing a paradigm shift, driven by an increasing demand for sustainability and environmentally friendly practices. As consumers become more conscious about the ecological impact of their purchasing decisions, there is a growing emphasis on materials that fulfill both aesthetic and functional needs. Among these materials, wool stands out due to its unique qualities, including breathability, moisture management, and thermal insulation, as well as its long-standing reputation as an eco-friendly fiber. However, traditional dyeing and printing processes employed in the industry often rely on hazardous chemicals and involve excessive water usage, which poses significant environmental challenges. This research aims to address these pressing issues by developing sustainable inkjet printing inks that enhance the aesthetic appeal of wool and wool blend fabrics. We formulated these inks using biodegradable biopolymer carboxymethyl cellulose (CMC), combined with various other constituents, specifically tailored for effective printing on wool substrates. The study involved the preparation of both acid and pigment inks, followed by a comprehensive evaluation of their physical and performance properties compared to commercially available inks. The findings highlighted that the acid inks achieved a measured viscosity of 7.47 cSt, while the pigment inks recorded a viscosity of 8.24 cSt, indicating their suitability for various printing applications. Furthermore, the formulated inks demonstrated impressive UV resistance and biodegradability, affirming their eco-friendly nature and potential for reducing the environmental footprint commonly associated with textile printing. This innovative approach to digital fabric printing not only enhances the visual appeal and absorbency of woolen textiles but also represents a significant stride towards sustainable practices in the textile industry.

Keywords Wool · Natural ink · Wool blends · Acrylic · Reactive ink · Acid ink

---

M. Choudhari (✉) · S. Mahadik · D. Marewad
Wool Research Association (Approved By Ministry of Textiles, GOI), Thane, India
e-mail: drmrinal@wraindia.com

D. Marewad
e-mail: dineshsmarewad@wraindia.com

A. Athalye and H. Patil (eds.), *Digital Printing*,
https://doi.org/10.1007/978-981-95-2593-5_5

5.1 Introduction

Textile printing is a distinct technique that entails applying colorants to fabrics in a controlled and consistent manner. In contrast to dyeing, which uniformly colors the entire textile surface, printing is a targeted coloring method that selectively applies pigments or dyes to designated sections of the fabric, resulting in elaborate designs and patterns. These motifs typically adhere to a predefined repeat, serving as the basis for both aesthetic and functional textile design in various sectors, including fashion and industrial use (Miles, 2003).

5.1.1 Early Origins and Historical Development

The exact beginnings of textile printing are hard to determine accurately, but archaeological evidence suggests that it has been in practice for many centuries. Particularly, discoveries of printed clothing and woody printing blocks from the Coptic era have been found in Upper Egypt, indicating a notable level of skill and technical expertise even among early textile civilizations (Dawson & Hawkyard, 2000). Moreover, India and ancient China had advanced block printing methods, with surviving examples that date back to as early as the 13th century.

By the 14th century, textile block printing had reached Europe, where it gained popularity in countries such as France, Italy, and Germany. The technique involved intricately crafted wooden blocks with raised patterns, which were immersed in dye or pigment and then manually pressed onto fabric. Although the concept is straightforward, block printing was a labor-intensive process demanding highly skilled artisans to ensure uniformity across repeated designs (Miles, 2003).

5.1.1.1 The Rise of Industrial Textile Printing

During the Age of Exploration, the growth of global trade prompted Portuguese merchants to recognize the economic potential of brightly colored cotton fabrics in European markets. This insight spurred increased trade between Europe and India, resulting in heightened demand for printed textiles and a subsequent need for more efficient manufacturing techniques. A significant advancement in technology occurred in 1752 when Francis Nixon unveiled copperplate printing in Dublin. This technique enabled the creation of much finer and more intricate designs compared to traditional block printing. However, due to the expense and complexity involved in preparing the copper plates, it never completely replaced the conventional block printing method (Miles, 2003).

A significant advancement took place in 1783 when James Bell received a patent for the engraved copper roller printing method. This innovation culminated in the development of the first mechanized continuous textile printing machine in 1785.

By employing a series of engraved copper rollers, the machine was capable of printing multiple colors—up to six—simultaneously in a single ongoing operation. This dramatically enhanced production speed and uniformity, positioning the United Kingdom as a leader in textile printing technology and large-scale manufacturing (Miles, 2003).

5.1.1.2 Advancements in Screen Printing Techniques

Screen printing, a significant sector of textile printing, originated from stencil printing methods first used for writing and ornamenting. Its commercial growth started in the mid-19th century, especially around 1850, when it was enhanced by Lyons, who utilized silk gauze as the basis for stencils. During the hand screen printing period (1930–1954), textiles were printed manually using screens that required precise positioning and considerable labor. Technological progress resulted in the development of semi and fully automatic screen-printing machines, significantly improving production speed and reducing costs. A pivotal moment came in the mid-1950s with the introduction of rotary screen printing, which employed cylindrical screens to apply twenty or more colors at once. Rotary screen printing presented multiple benefits, such as increased production rates, improved alignment, and the ability to create intricate multicolor designs with high resolution. These advantages contributed to rotary screen printing becoming the preferred method in the textile printing industry due to its efficiency and consistent quality of output.

5.1.1.3 The Emergence of Digital and Inkjet Printing

The 20th century marked the emergence of digital technologies, with inkjet printing becoming a groundbreaking technique for coloring textiles. Initially used in limited applications, inkjet processes gained recognition after the introduction of various advanced inkjet printers at the International Textile Machinery Association (ITMA-2003) exhibition (Moser, 2003; Glover, 2003; Lewis, 1982a, b). These systems showcased the competencies of digital printing for detailed, high-resolution textile designs. Inkjet printing differs fundamentally from traditional screen printing because it is a non-contact, data-driven process. It doesn't use physical stencils or screens; instead, it digitally deposits tiny droplets of dye or pigment directly onto the fabric surface. This method enables mass customization—the ability to create highly personalized or variable designs on a large scale, which is essential in today's fashion and home textiles industries (Dawson & Hawkyard, 2000; Tyler, 2005).

Although inkjet printing currently accounts for a small portion of the total textile printing market (Fryberg, 2005), its influence is rapidly growing. Advantages such as reduced waste, faster prototyping, design flexibility, and lower inventory costs make it an increasingly viable alternative to traditional methods. Industry experts expect that once current technical and financial hurdles—particularly in fabric pretreatment, color durability, and printing speed—are overcome, inkjet printing will likely surpass screen printing as the preferred method for textile coloring (Ujiie, 2010).

5.1.2 Wool and Its Blends

5.1.2.1 Wool

Wool is the most versatile textile fiber. It can produce items like billiard table cloths and high-quality woven and knitted fabrics. Wool products are highly comfortable due to their insulating and moisture-wicking properties wear. Moreover, wool's chemical composition enables it to be dyed easily in a range of colors, from gentle pastels to rich, vivid shades.

5.1.2.2 Fibre Morphology

Wool is a key fiber from sheep fleece breeds. It is composed of multilayered, protein-based hairs that gradually narrow from base to tip. Fiber lengths vary from 4 to 40 cm, with diameters between 17 and 40 μm. Wool fibers have a complex structure with the scaly outer layer, cortex, and cell membrane complex (CMC) that surrounds cells in both the cuticle and cortex, forming a continuous phase. Coarser fibers may also contain a medulla or inner core (Fralix, 2001). The schematic diagram of fine wool and the complex physical structure of cuticle cells are illustrated in Figs. 5.1 and 5.2.

Microscopic analysis reveals that the scales present on the surface of the fiber are typical of most animal fibers. The scales overlap and point in the direction of the fiber's tip. In fine wool, these scales account for approximately 10% of the fiber. The friction caused by these scales slows down the movement of the wool fiber toward

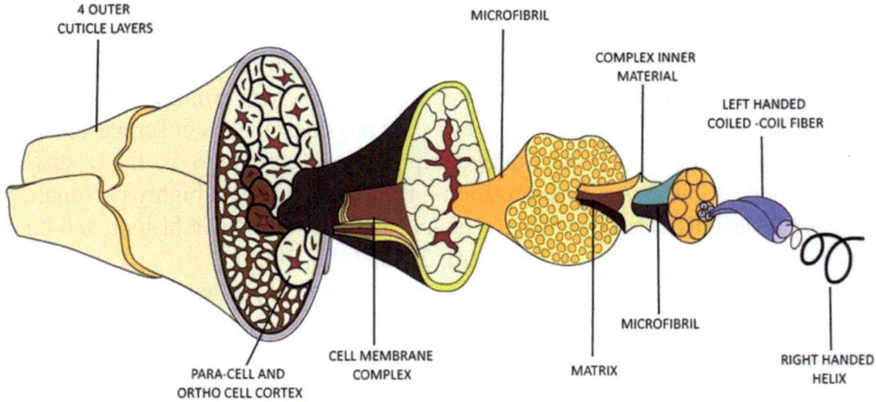

Fig. 5.1 Structure of fine merino wool fiber (Broadbent, 2001)

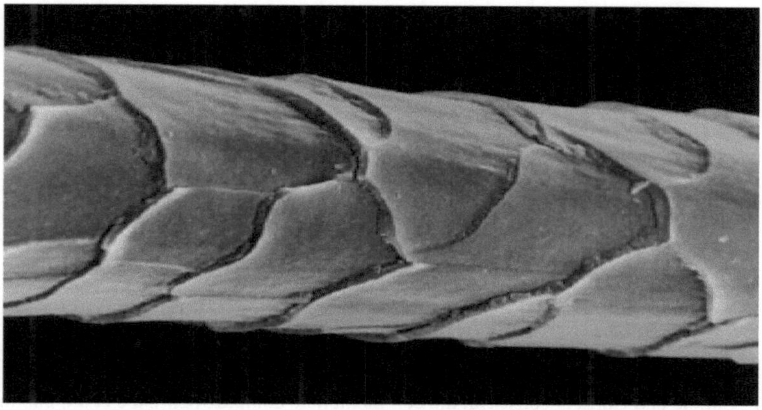

Fig. 5.2 SEM of wool fiber with overlapping cuticle cells (Broadbent, 2001)

the tip, a phenomenon known as the directional friction effect, which contributes to the felting and shrinking of wool items during washing. These scales have multiple layers protected by a hydrophobic outer layer called the epicuticle, part of the cell membrane complex, surrounding the fiber except at the tip, which is worn away by environmental factors. The cortex consists of spindle-shaped cells from keratin fibrils in a high-sulfur protein matrix, enclosed by the cell membrane. It makes up about 90% of wool fiber. In fine wools, the cortex has two regions resembling half-cylinders joined along their axes.

The two areas illustrated in Fig. 5.3 are referred to as the orthocortex and the paracortex. These regions follow the crimp pattern along the fiber, with the orthocortex facing outward from the crimp wave. This arrangement resembles that found in bicomponent synthetic fibers. The crimp results from uneven keratin hardening in hair follicles, occurring at different levels on each wool hair side. The two cortical regions have distinct structures and reactive properties.

The cell membrane complex, as depicted in Fig. 5.4, predominantly consists of proteins and lipids and encases all cell types within a continuous network throughout

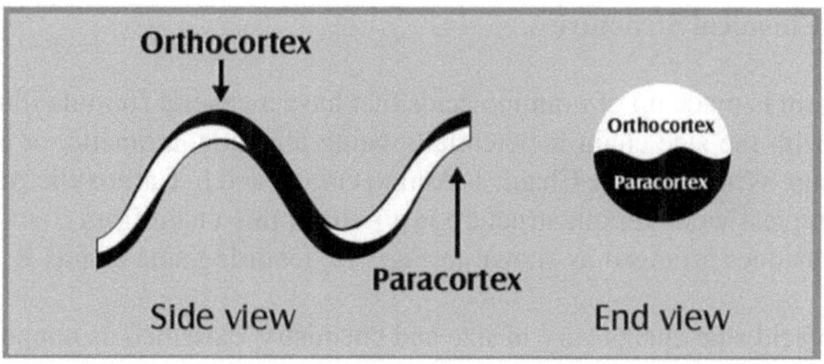

Fig. 5.3 Ortho-cortex and para-cortex region in wool (Christoe et al., 2000a)

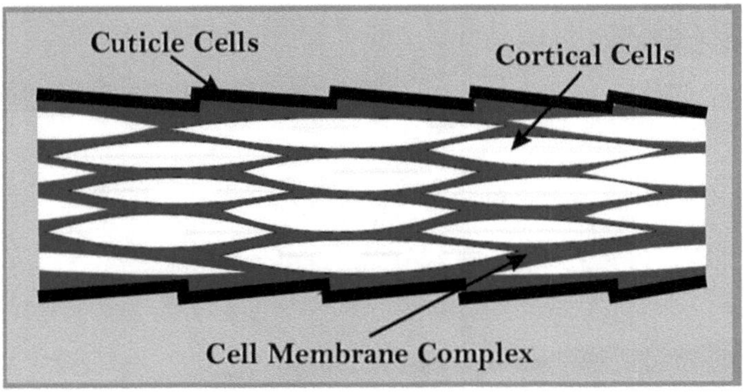

Fig. 5.4 Cuticle and Cortical cells with the cell membrane complex (Christoe et al., 2000b)

the fiber. It defines a region mechanically weaker than the rest of the fiber, which is the first to fail during abrasion, causing disintegration into cortical cells or protein fibrils. The membrane complex becomes compromised following extended dyeing in an acidic solution at elevated temperatures.

The medulla, when present, is a core of cells that may contain pigments. It doesn't run the full length of the fiber and might have empty spaces. Wool's protein structure varies: some microfibril proteins are coiled like springs, making wool flexible, elastic, resilient, and wrinkle-resistant. Others in the matrix around microfibrils are disordered, enabling wool to absorb water (up to 30% of its dry weight) and dyes. The matrix proteins also help wool hold dyes effectively.

5.1.2.3 Wool Composition

Wool is classified as a protein keratin. Pure wool contains approximately 82% keratin-based proteins, 17% non-keratin-based proteins, and approximately 1% non-protein substances such as waxy lipids (CSIRO, 2006).

5.1.2.4 Chemical Structure

Wool keratin is made up of α-amino acids that have a general formula illustrated in Fig. 5.5, with the side chain R potentially being aliphatic, aromatic, or a different cyclic group. Where R- Side Chain; a- Amino Group, and b- Carboxylic Acid group.

The simplest wool keratin structure is a polypeptide chain from α-amino acids, with the residues arranged as shown in Fig. 5.6, featuring side chains R1, R2, and R3.

Amino acid side chains vary in size and chemistry, classified as nonpolar, polar, acidic, basic, or sulfur-containing. Wool's peptide chains are linked by covalent bonds and noncovalent interactions, shown in Fig. 5.7.

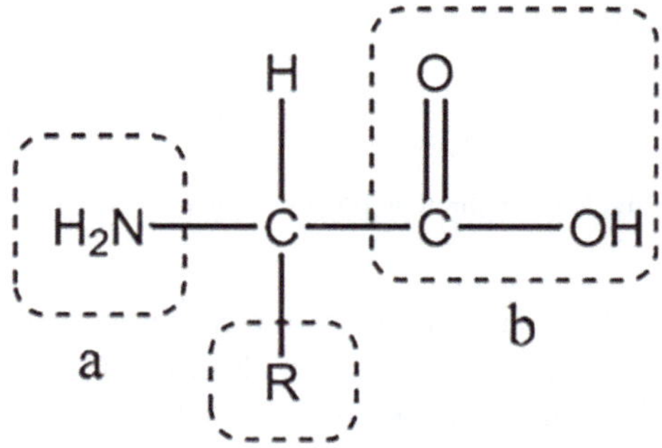

Fig. 5.5 Typical structure of an amino acid

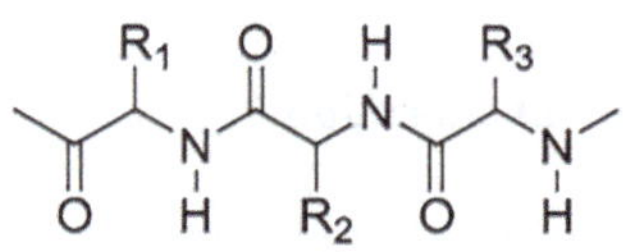

Fig. 5.6 Overall structure of wool keratin polypeptide

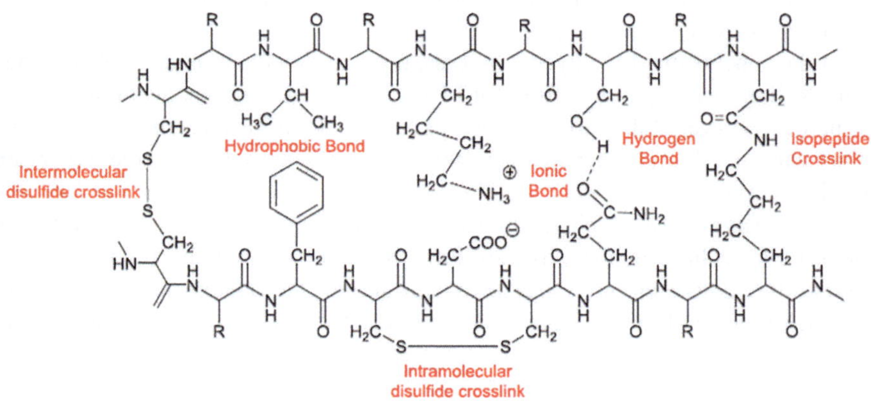

Fig. 5.7 Covalent and noncovalent interactions in wool (Broadbent, 2001)

Disulfide bonds, involving sulfur, are key covalent bonds formed during keratinization, making keratin fibers water-insoluble and more resistant to challenges. These bonds are also involved in fabric finishing, where their rearrangement helps wool fabrics dry smoothly without ironing. Another covalent bond, the isopeptide bond, forms between lysine and aspartic or glutamic acid.

Besides covalent bonds, various other interactions also contribute to the stability of the fiber in both washing and drying conditions. These interactions come from amino acid side chains in wool keratins. Hydrophobic interactions occur between hydrocarbon groups, and ionic interactions occur among groups capable of proton exchange. Among the noncovalent interactions, the ionic or salt linkages between acidic (carboxyl) and basic (amino) side chains are the most crucial.

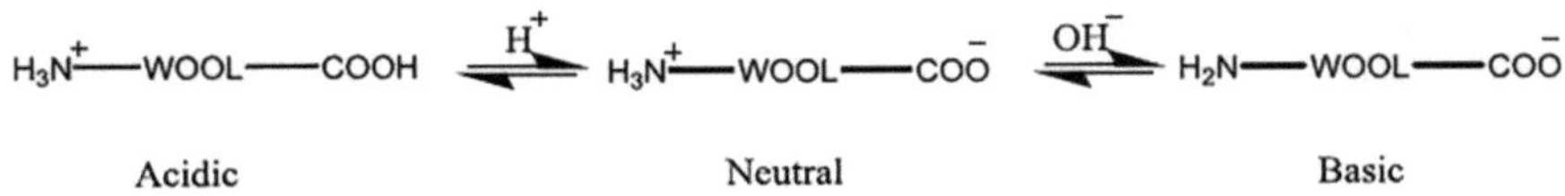

Fig. 5.8 Amphoteric properties of wool (CSIRO, 2006)

The carboxyl and amino groups in wool give the fiber amphoteric or pH buffering properties, enabling it to absorb and release acids and bases, as shown in Fig. 5.8. The ionic groups also influence dye interactions due to their charge interactions molecules.

5.1.2.5 Mechanism of Wool Coloration

Wool coloration involves dye transfer, adsorption, and diffusion into the fiber as shown in Fig. 5.9. Dyes penetrate at cuticle joints, diffusing through endocuticle, CMC, and intermacrofibrils. Over time, dye moves from nonkeratinous areas into sulfur-rich proteins surrounding microfibrils and from endocuticle into the exocuticle, especially the A-layer.

At the conclusion of the dyeing process, if the dyes remain in the nonkeratinous regions, they will rapidly diffuse out of the fiber, leading to poor wash fastness properties. A-layer of the exocuticle acts as a barrier to diffusion across cuticle cells. In untreated wool, dyes enter the fiber's interior mainly through an intercellular route (CSIRO, 2006).

Reactive dyes may exhibit variations in their distribution between non-keratinous and keratinous fiber regions, as they form covalent bonds with proteins in non-keratinous regions at an early stage on. At equilibrium, they tend to be more abundant in the CMC and endocuticle compared to non-reactive dyes. Dye absorption impacts

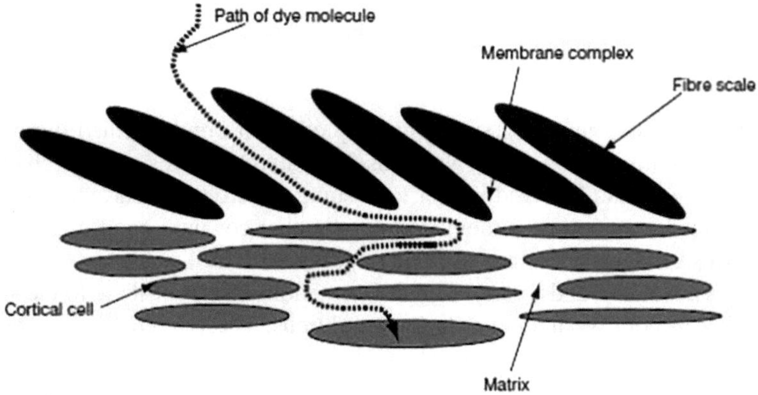

Fig. 5.9 Dye molecules diffusing into the wool cortex (Fralix, 2001)

printing more than dyeing, as dye diffusion in wool is slow. This can cause inconsistent printing results if steaming takes too long. Wool fabric is usually pre-treated to ensure proper dye penetration during printing within a reasonable time with gentle steaming.

5.1.3 Wool Blends

Blends of wool with other fibers are common in printed products for various reasons.

- By incorporating synthetic fibers, a more affordable product can be manufactured.
- A synthetic fiber can achieve a wool-like texture and look by being mixed with small quantities of wool.
- Technical benefits may necessitate the addition of synthetic fibers (for example, to enhance durability).
- Highly luxurious products are made by combining wool with premium fibers like silk or cashmere.

5.1.3.1 Wool/Cotton

Wool and cotton blends have remained important despite synthetic fibers' advances. Union fabrics with a cotton warp and wool weft are used in shirtings, pajamas, and workwear, though lighter, better-laundered synthetic blends are replacing them. Wool/cotton fabrics are used in blazers, coats, shirts, blouses, knitwear, women's outerwear (often with viscose), leisure and sportswear. Wool/linen blends are also noted. Demand for washable, durable garments has increased the need for improved wet-fastness, and for over 30% wool blends, shrink-resist treatments are often necessary, applied with resins or chlorine-treated wool using pad-batch techniques.

5.1.3.2 Wool/Polyester

The proportions of wool to polyester in blends can vary widely, ranging from 5 to 45% wool. Lower wool content is generally utilized to improve the texture and visual appeal, and these blends are typically dyed using dispersed dyes through heat-transfer methods and labeled as polyester. However, for blends that contain more than 20% wool, it is essential to consider both types of fibers; failing to do so can lead to inadequate coverage of the wool, resulting in an unattractive, skittery look (Broadbent & Rigout, 2013).

5.1.3.3 Wool/Acrylic

Acrylic fibers closely resemble wool among synthetic fibers. Wool-acrylic blends, with 30–60% wool, mimic wool's look, feel, and durability, with higher acrylic content (>60%) improving washability. These blends boost strength, bulk, water absorption, and aesthetics.

Wool blends have long been important in fashion and non-fashion sectors, and their relevance is expected to persist. Rapid manufacturing needs will increase dyeing during yarn and fabric production, with a shift towards dyeing fiber components together rather than separately. Environmental concerns will push dyers to lessen dye release into wastewater, especially for cotton and blends where dye bath efficiency is often poor. Carriers are vital for dyeing wool/polyester mixes at lower temperatures, protecting the wool. Regulations on carriers like biphenyl and halocarbons will drive the development of eco-friendly alternatives (Lewis, 1989, 2013; Walter, 1987). For wool/nylon carpets and pure wool that require insect-repellent treatment, new methods must minimize the release of dye agents or produce low-effluent volumes for safe disposal (Barton, 2000).

5.2 Inkjet Printing on Wool Using Reactive Inks

Historically, wool fabric printing has been a niche, high-end activity, making it difficult for wool printing to adapt to high-speed rotary screen-printing machines in print mills, which can reach speeds of 50 m/min. In contrast, cotton fabric printing often involves production runs of 500 m, a standard that is likely to decrease further. As demand for unique designs grows, the era of mass-produced printing with specific colorways and patterns may be coming to an end. A potential solution is the development of textile inkjet printing technologies. Successful implementation of these systems would benefit wool and wool-blend fabric manufacturers focusing on small production runs. Traditional printing typically uses reactive printing with an all-in approach, but the purity requirements for inkjet inks pose significant challenges, as standard chemicals are unsuitable for ink formulations (Yuen et al., 2010). Usually, wool inkjet printing employs a two-phase process, where inks contain only purified dyes, while thickeners and other supporting chemicals are applied during pre-treatment of the fabric.

5.2.1 Reactive Inks for Wool

Reactive dyes form covalent bonds in an acidic environment using –SH, =NH, and –NH2 groups, not salt-like bonds with protein fibers polypeptide. These products offer excellent fastness, vibrant shades, and enhance market choices. The high demand for Woolmark-labeled products and the usability of these fibers justify their higher costs dyes. Vinyl sulfone and monochlorotriazine dyes with low to medium reactivity dominate the inkjet market (Provost, 1994). Reactive inks usually fix about 70% on cotton. To ensure high wash fastness, it's important to effectively remove unfixed or hydrolyzed dye essential. Lengthy and costly washing-off procedures, like traditional fiber reactive dyeing, have a significant environmental impact from removing large dye and chemical quantities and using vast water. Unfixed reactive dyes in wastewater may also pose risks. Reactive dyes, especially high and medium reactivity ones, often suffer from reactive group hydrolysis. When this happens, the dye can't bond properly with the substrate, resulting in poor color development. Additionally, unbonded dye adds to effluent waste and may need additional treatment. Furthermore, reactive inks, due to their high hydrolysis tendency, have a limited shelf life (Kanik & Hauser, 2002; Magdassi, 2012).

Inkjet printing with reactive dyes is widely accepted for natural fibers such as cotton, wool, and silk; however, it often experiences inadequate fixation rates and requires a significant amount of auxiliary chemicals during the pretreatment and post-treatment stages. These issues lead to low dye utilization, substantial wastewater production, and higher operational costs, which ultimately affect the sustainability and efficiency of digital textile printing. To address these challenges, researchers have focused on two primary approaches:

- Chemical modification of reactive dyes to improve their reactivity, solubility, and fixation.
- Surface modification of textile substrates to enhance dye affinity and minimize the necessity for excessive chemical pretreatment.

5.2.1.1 Reactive Dye Modifications for Wool Inkjet Printing

In 1989, Gisler created trisazo dyes that incorporated fluoropyrimidine reactive groups, which were intended to enhance the reactivity with the amino groups of wool in milder, non-alkaline environments. These heterocyclic groups improve electrophilicity, allowing for effective covalent attachment to protein-based fibers like wool without causing damage to the fibers (Gisler, 1989).

Eltz and Russ (1993) incorporated cyanamide groups into established commercial reactive dyes, enhancing their hydrolysis resistance. These dyes preserve their reactivity in buffered conditions, which is vital for wool, as it cannot withstand the alkaline fixation usually applied to cellulosic fibers (Eltz & Russ, 1993).

Clark and colleagues (1994) enhanced the solubility and longevity of reactive inks by altering Drimarene K dyes with phenolsulfonic acid. This modification guaranteed prolonged stability of the ink and minimized clogging, while maintaining high reactivity on wool at a neutral pH (Clark et al., 1994).

Maheshwari (2020) developed a modified form of Reactive Yellow 160 by adding sulfonated secondary amines, producing a dye that exhibits great water solubility, compatibility with protein fibers, and high fixation yield in the inkjet printing process on wool (Maheshwari, 2020).

A medium reactivity dye, 2-sulphophenoxy-4-chloro-s-triazine, was developed from a dichloro-s-triazine precursor and showed excellent hydrolytic stability over six months, suitable for inkjet printing. The inks had proper surface tension and viscosity. Applied to wool, it showed greater color strength and fixation than the dichloro-s-triazine dye. Steaming at 102 °C improved fixation and print quality. Lightfastness increased by 0.5 grade; wash fastness was excellent with no shade change or staining. The new dye ink offers better durability and eco-efficiency for wool digital printing (Faisal & Lin, 2020).

Four different cationic reactive dye inks were developed for inkjet printing on wool/acrylic fabric blends. Stability testing showed that inks D1, D2, and D4 maintained better particle size stability than D3. Except for D3, the inks achieved good dye fixation, vibrant colors, and balanced tones on the fabrics. The best formulation contained 10% dye, 25% non-ionic dispersant IW, 15% ethylene glycol, 1.5% surfactant XP-70, and 0.02% triethanolamine. Dispersants, polyols, surfactants, and pH regulators played key roles in defining ink properties. The printed fabrics exhibited excellent to moderate color fastness. This study confirms that multifunctional cationic reactive dyes with dual fiber affinity can effectively be used in inkjet printing on wool/acrylic blends (Xiao et al., 2021).

Reactive dyes derived from p-sulphophenoxy-s-triazines were produced and utilized for inkjet printing on wool fabrics. The effectiveness of fixation, which reflects the covalent bonding of the dye to the fiber, was evaluated using various post-print development techniques. The most successful technique involved layering the printed wool with damp cotton and allowing it to batch for 3 h, resulting in the highest fixation rates. Steam fixation was found to be less effective due to an increase in dye hydrolysis at elevated temperatures. A model DCT dye (dye 2) underwent mono-substitution with HBSA at 20 °C for 60 min at a pH of 7.0. Increasing the temperature to 40 °C for a further 90 min enabled di-substitution, leading to the formation of dye 4. Structural verification was performed using FT-IR and ^{1}H NMR, while capillary electrophoresis (CE) confirmed that dye 4 was the predominant species. The stability of the ink was essential, with phosphate-buffered inks at pH 6.0 displaying minimal degradation (an 8.9% reduction in CE peak after 10 days) in comparison to those at pH 6.5 and 7.0. Fixation was enhanced to 76% with batching at 65 °C for 2 h, whereas alkaline conditions resulted in a decrease in fixation. Steam fixation only achieved a fixation rate of 59%. The high solubility of dye 4 eliminated the necessity for co-solvents. Buffering at pH 6.5 improved both ink stability and fixation, achieving a fixation rate of up to 91%, which exceeded previous research, including findings by Lewis and Seltzer that indicated 80% fixation for comparable dyes using cold-pad-batch methods (Yang et al., 2019).

Chloropyrimidine reactive dyes are well-known for their effectiveness in textile printing; however, their synthesis through nucleophilic aromatic substitution with chromophores that contain amino groups generally occurs at a slow rate and yields lower amounts. In this research, a new and effective method for creating a chloropyrimidine-based reactive dye utilizing microwave irradiation was developed, and its performance was compared to traditional heating methods. Key parameters, including reaction duration and yield, were assessed for comparison. The reactions were monitored using Capillary Electrophoresis (CE), and dye purity was evaluated through thin-layer chromatography (TLC). The synthesized trichloropyrimidine dye's identity was verified by Fourier-transform infrared spectroscopy (FTIR) and elemental analysis. It was particularly noted that the reaction conducted under microwave irradiation was roughly four times faster than the one using conventional heating, although the enhancement in product yield was modest. These findings confirm that microwave-assisted synthesis significantly improves reaction efficiency for chloropyrimidine-based reactive dyes. The synthesized dye was incorporated into an inkjet-printable ink and utilized on wool fabric, followed by steaming at 102 °C for times ranging from 5 to 25 min in 5-min increments. The printed fabrics showed an increase in color intensity and dye fixation with longer steaming durations, along with satisfactory light and wash fastness properties. Overall, this research illustrates that chloropyrimidines are adequately reactive under microwave conditions, providing a swift and practical synthetic pathway for developing high-performance reactive dyes appropriate for digital inkjet printing on protein-based fibers like wool. This development has the potential to broaden the use of chloropyrimidine dyes in efficient and sustainable textile printing methods (Faisal et al., 2020).

5.2.1.2 Wool Fiber Surface Modification

Modifying the surface of wool fibers is an essential process for improving dye absorption, wettability, and printability, especially in digital textile printing. The inherent hydrophobic characteristics of the wool cuticle and the lipid-rich epicuticle layer limit the affinity and retention of conventional reactive dye systems on untreated wool. To address this issue, a variety of both physical and chemical methods for surface modification have been utilized. Inkjet printing provides high-quality, environmentally friendly printing options for wool textiles, but the scales on the fiber's surface impede dye penetration, resulting in subpar print quality. This research introduced a sustainable air plasma pretreatment aimed at improving the printability of wool when using reactive dye inks. Plasma treatment enhanced the surface area of the fibers and added oxygen-containing functional groups, which increased wettability by lowering the contact angle from 135° to 27.5°. The oxidation of –SH and –S–S– groups to –SO–, along with the cleavage of C–C and C–N bonds, generated additional sites for dye attachment. These modifications led to improved ink absorption and stronger dye bonding to the fiber. The plasma-treated wool exhibited a significantly higher color yield (K/S 22.34) compared to untreated samples (K/S 14.64), validating the effectiveness of the method and its environmental advantages (Wang et al., 2022).

The ecological modification of wool fabric using plasma treatment notably enhanced its performance in inkjet printing while avoiding emissions or wastewater. The application of air plasma treatment decreased the wool's surface contact angle from 145.84° to 41.35°, indicating a significant increase in hydrophilicity. FTIR analysis showed changes in the 1000–2500 cm^{-1} range, which aligned with a rise in polar functional groups such as hydroxyl and carbonyls. Observations of morphological changes on the wool surface revealed increased scaliness, contributing to better absorbency. Wool fabrics that were printed with herbal red ink reached a maximum color difference value of 14.26 and demonstrated outstanding wash and rub fastness. The wash fastness regarding color change improved to a very good level, although the light fastness showed inconsistency, possibly due to uneven herbal pretreatments, suggesting that shade drying might be an effective solution. Statistically, wool that underwent plasma treatment exhibited a minimal standard error of 0.08 and a high significance level ($p = 4.630 \times 10^{-4}$, power = 0.99), confirming the effectiveness of the treatment. This environmentally friendly, waterless approach promotes climate action and sustainable manufacturing, in line with the Sustainable Development Goals (Thakker et al., 2022).

In the pretreatment process for inkjet printing on wool fabrics, two environmentally friendly agents, hydrogen peroxide (H_2O_2) and papain, were utilized. This research introduces an eco-conscious and effective method for preparing wool fabrics for digital inkjet printing by employing these two green reagents. The oxidative properties of H_2O_2 break down disulfide bonds in the keratin structure of wool, leading to softening and partial removal of the surface scales, while papain works enzymatically to hydrolyze the exposed protein chains, further smoothing the fiber surface. This combined treatment enhances the wettability of the fabric and lowers its negative zeta potential, which promotes deeper dye absorption and reduces electrostatic repulsion when using reactive dyes. Structural changes and chemical alterations on the fiber surface were confirmed through analytical techniques such as FT-IR and XPS. When treated wool fabrics were printed with Reactive Red 218 dye, they demonstrated a substantial increase in color strength (K/S value rose from 20.54 to 25.30) and resulted in more vibrant, uniform patterns with improved dye application. Significantly, the entire pretreatment process avoids generating hazardous by-products, aligning with green chemistry principles and providing a cleaner, cost-effective approach for high-quality digital printing on wool textiles (Shi et al., 2021).

A new eco-friendly pretreatment enhances wool fabric print quality with reactive dye inks via a two-step process: first, using the protease enzyme Savinase 16L to remove surface scales, then applying sodium alginate (SA) to create a hydrophilic layer. This combination improves print richness, brightness, and detail. Analytical methods confirmed scale removal and increased surface hydrophilicity, with a water contact angle of 70.1° and wetting in 3.5 s. The treated fabrics showed good washing, rubbing fastness, and acceptable tensile strength. This approach enhances dye penetration and fixation on wool, with no AOX emissions, minimal strength loss, and sustainable processing (An et al., 2019).

This investigation examined how the rheological characteristics of pretreatment solutions, specifically Carboxymethyl hydroxypropyl cellulose, sodium Carboxymethyl cellulose, and sodium alginate, influence the inkjet printing efficiency on wool fabrics using reactive dyes. The rheological assessment revealed that the CMHPC solution exhibited the highest zero-shear viscosity, improved fluidity, and predominantly elastic behavior. These attributes facilitated the creation of uniform and continuous pretreatment films, resulting in a reduction in fabric thickness. Microscopic (optical and FESEM) and XPS analyses verified that

CMHPC-treated fabrics significantly decreased ink droplet dispersion and penetration, resulting in the highest color strength (K/S value) and the sharpest printed edges. On the other hand, SA exhibited poor fluidity, marked viscous behavior, and the lowest zero-shear viscosity, resulting in discontinuous films, increased fabric thickness, and reduced print quality. Wettability assessments reinforced the idea that the structure of the pretreatment film, closely linked to its rheological properties, is crucial in determining the results of inkjet printing. Among the three thickeners assessed, CMHPC yielded the most effective and sustainable outcomes, necessitating lower concentrations, less urea, and shorter steaming durations, while still achieving improved color strength and satisfactory fastness and mechanical properties. Therefore, CMHPC emerges as a promising environmentally friendly pretreatment agent for high-quality digital printing on wool (An et al., 2020).

5.3 Inkjet Printing of Wool with Acid Inks

Acid inks are a unique subset of dye-based inks crafted specifically for application on protein-based fibers like silk, wool, and polyamide (nylon). These fibers possess amino groups ($-NH_2$) in their structure, which enable them to create ionic bonds with acid dyes. In the textile printing sector, acid inks are valued for their vibrant color accuracy, exceptional durability, and suitability for intricate designs, making them perfect for high-end fashion, swimwear, scarves, and upholstery uses. The term "acid" in acid inks comes from the necessity of an acidic pH range (generally pH 4–6) during the dye fixation process. Following the inkjet printing phase, the fabric is subjected to a steaming process that enhances dye infiltration and adhesion. During this phase, heat and moisture promote the diffusion of acid dyes into the fiber matrix, where they establish strong electrostatic (ionic) bonds with the protonated amino groups found in protein fibers (Cristea & Vilarem, 2006). The outcome is a remarkably resilient color with outstanding visual clarity on both the front and back of the fabric.

5.3.1 Key Advantages of Acid Inks

- **Exceptional Color Quality and Reproduction**: Acid inks provide an extensive range of colors, yielding vivid, saturated, and intricate images that are highly valued in fashion and home décor fabrics. The precise definition and deep absorption of dye molecules produce prints that are not just lively but also demonstrate smooth tonal shifts and photographic sharpness (Ghoranneviss et al., 2014).
- **Superior Wash Fastness**: The ionic bonding process provides remarkable durability, particularly with household washing. In contrast to pigment-based methods that reside on the surface, acid dyes delve deeply into the fiber's structure, reducing color fading or bleeding after numerous washes. This characteristic makes them

ideal for clothing items that require regular cleaning, like scarves, underwear, or activewear (Bechtold & Mussak, 2007).
- **High Resistance to Chlorine and Salt Water**: A key characteristic of acid inks is their ability to withstand harsh conditions, including chlorinated pool water and seawater. This durability makes acid inks especially ideal for swimwear, beachwear, and performance fabrics. Their chemical stability in these environments is due to the robust attachment of dyes within the fiber structure and the minimal solubility of the bonded dye molecules (Schindler & Hauser, 2004).

5.3.2 Acid Inks for Digital Printing

The successful creation, structural alteration, and analysis of five new perylene-based acid dyes resulted in the formulation of high-performance aqueous-based inks tailored for high-speed digital textile printing. These dyes were designed to fill existing gaps in the color range of digital inks, particularly to achieve scarlet and high-lightfastness red (H.Red) shades, which are typically challenging to produce with traditional azo dye systems. By incorporating large and hydrophilic substituents, the researchers notably enhanced the aqueous solubility of the dyes, ensuring the inks remained stable over time without issues of sedimentation or precipitation. Spectroscopic evaluations indicated high molar extinction coefficients, with pronounced absorption peaks in the 500–550 nm spectrum. Importantly, dyes 4b and 5b exhibited bathochromic shifts attributable to phenoxy substituents at bay positions, which improved absorption at longer wavelengths, providing an optical advantage for broader color coverage in textile applications (China Patent CN101899786A).

The synthesized dyes exhibited outstanding thermal and photostability in comparison to the reference azo dye. Thermogravimetric analysis revealed degradation temperatures ranging from 400 to 550 °C, while the reference dye had a degradation temperature of 291 °C, demonstrating superior resistance to heat-intensive processes like steaming and curing (China Patent CN101899786A). Testing for photoaging demonstrated that perylene dyes outperformed their competitors, maintaining 90% of their initial absorbance even after 1200 min of continuous exposure, whereas the reference dye's absorbance dropped to around 50%. Furthermore, the higher extinction coefficients of the perylene dyes allowed for vivid colors at lower concentrations, supporting economical dye usage and minimizing environmental impact. Improved water solubility also facilitated stable ink formulations, which is crucial for ensuring consistent jetting performance and avoiding nozzle blockages. In practical applications, the perylene-based inks displayed excellent rheological and physicochemical characteristics, with viscosity, surface tension, and pH levels falling within ideal ranges for digital inkjet printing. Trials with inkjet printing indicated smooth ejection, low occurrences of nozzle obstruction, and high precision in pattern printing, verifying their compatibility with commercial high-speed printers. Notably, the inks achieved accurate color matching in accordance with industry standards for scarlet and H.Red, with printed samples exhibiting sharp, stable dot patterns and excellent pattern control. Collectively, these features establish the synthesized perylene acid dyes as advanced alternatives to traditional azo dyes, providing improved color vibrancy, durability, print consistency, and a wider chromatic range suitable for contemporary digital textile printing (China Patent CN101899786A).

The invention presents a specialized formulation of acid dye ink designed for applications involving textile inkjet printing. The ink formulation consists of 0.5–10% acid dye, 0.5–10% viscosity modifier, 5.0–20% humectant, 0.2–5% osmotic agent, 0.1–2% pH control agent, and 0.1–1% disinfectant, with the balance made up of deionized water. This meticulously crafted mixture aims to provide optimal performance, including vibrant color, print stability, and compatibility with various textile materials like silk, wool, polyamide (chinlon), and leather. The formulation process entails combining all ingredients at a rotational speed of 100–300 revolutions per second for 2 h while regulating the temperature between 40 and 60 °C. This procedure guarantees complete dissolution and uniformity of all components. After mixing, the solution is allowed to cool to room temperature and is then sequentially filtered through microporous filters with pore sizes of 0.45 microns and 0.20 microns. This two-step filtration process eliminates impurities and ensures that the ink attains the necessary purity and consistency required for high-resolution inkjet printing. The final ink demonstrates superb jetting performance, print clarity, and long-term stability, making it ideal for commercial textile printing endeavors (Fu, 2006).

5.4 Sustainable Development for Digital Printing of Wool and Wool Blends Fabrics

An increasing amount of research and industry data underscores the urgent necessity for sustainability in textile wet processing, especially in the textile printing area. Traditional printing techniques often make extensive use of volatile organic compounds such as ethanol, methanol, toluene, and various humectants including urea, propylene glycol, polyethylene glycols, and diethylene glycol. These substances, widely utilized to aid in print formulation and application, present significant risks to environmental health and human well-being. The prevalent use of these compounds has triggered concern among environmental agencies and public health organizations. According to Fu, these chemicals are found in considerable amounts in conventional printing inks, and their release into the environment, mainly via air and water discharge, has been linked to various ecological and health-related problems (Fu, 2006).

A study conducted by the United Nations Environment Programme (UNEP) found that almost 99% of harmful emissions from the textile printing industry are released into the air, with only 1% entering water and land systems equally. The impact of this is extensive, contributing to the enhancement of the greenhouse effect, depletion of the ozone layer, acid rain, and various health problems in humans, including nausea, respiratory irritation, neurological issues, organ damage, and potential genetic mutations (Christie et al., 2000a). Additionally, the Ellen MacArthur Foundation approximates that the textile sector produces around 21 billion tons of waste each year, with 20% consisting of hazardous wastewater generated from dyeing and printing facilities. These effluents are not only challenging and costly to manage but also result in prolonged pollution of groundwater, making it unusable for many years and causing a ripple effect on public health and the sustainability of agriculture (Ellen MacArthur, 2021).

Compounding the seriousness of the problem is the type of materials typically present in conventional inks. A number of these inks have phthalate-based plasticizers, such as N-ethyl and N-methyl toluene sulphonamides and tris (2-ethylhexyl) taramellite, which are recognized for their tendency to bioaccumulate and remain in ecosystems (Ethical, 2021). The worldwide textile sector is currently acknowledged as the second-greatest source of industrial pollution, following the oil industry. Water contamination is responsible for 1.8 million deaths each year and affects more than a billion people, surpassing the consequences of global armed conflict and violence (Christie et al., 2000b). Considering the intricate and resource-demanding process of cleaning polluted water sources, the urgency for preventive measures has intensified. In light of these difficulties, digital inkjet printing has surfaced as a viable and eco-friendly option. Unlike conventional wet processing and rotary screen printing, which depend on large volumes of water and toxic chemicals, inkjet printing functions as a clean, electro-mechanical method that significantly lowers emissions and discharges. This technology allows for the precise application of colorants with minimal waste. Research indicates that digital printing generates a considerably smaller carbon footprint—approximately 74.26 kg CO_2 equivalent per unit—compared to 123.76 kg CO_2 equivalent associated with rotary printing. Furthermore, the transition to waterless technology can greatly reduce water consumption and enables the use of concentrated botanical extracts, thereby alleviating the pressure on land and water resources typically required for dye plant cultivation. However, despite the ecological advantages, the predominant inks utilized in digital printing today are still synthetic and comprised of petrochemical-derived ingredients such as 2-pyrrolidone, propanediol, and surfactants like polyethylene glycols, along with non-ionic agents such as Surfynol® and rheology modifiers. These substances are known to be environmentally persistent and often toxic. In particular, pigment-based inks have been found to be more detrimental than their water-based alternatives, raising concerns regarding their extensive use and impacts at the end of their lifecycle (Fu, 2006).

A sustainable method for coloring and enhancing polyamide, silk, and wool fabrics using curcumin-based inks has been developed through inkjet printing technology. Three unique ink formulations were crafted and examined: curcumin combined with Triton X-100 (CT), curcumin mixed with Triton X-100 and tannin acid (CTT), and tannin acid paired with sodium dodecyl sulfate (TS). These inks were thoroughly characterized to evaluate their stability, viscosity, and surface tension, which are crucial for effective inkjet application. Alongside the ink formulations, three treatment strategies were created to improve fabric performance. These strategies included Method 1 (M1), which incorporated printing with CT ink followed by a post-treatment bath in tannin acid; Method 2 (M2), which involved direct printing using the CTT ink; and Method 3 (M3), entailing sequential printing with CT ink followed by TS ink.

The treated fabrics were assessed based on several factors, including color characteristics, color fastness, surface zeta potential, wettability, antioxidant activity, and antibacterial effectiveness. All ink formulations exhibited excellent stability throughout an 8-day period. Among the tested substrates, silk demonstrated the deepest color and outstanding ink penetration, showcasing its compatibility with the natural ink systems.

Remarkably, all treatment methods produced fabrics with notable bioactivity. Antioxidant activity exceeded 95%, and antibacterial efficacy surpassed 85% across all samples. Treatments M1 and M3, in particular, displayed improved color fastness and maintained strong antioxidant and antibacterial properties even after ten laundering cycles, signifying excellent durability of the bio-functional attributes. This research showcases the practicality of utilizing curcumin as a natural dye and tannin acid as a functional bio-based mordant to accomplish both aesthetic and functional results. The incorporation of these natural substances within a digital inkjet printing approach not only yields vibrant and stable yellow colors but also provides health-promoting features to the textiles. These outcomes highlight the potential of bio-based inks to revolutionize traditional textile processing, presenting a sustainable and multifunctional alternative to conventional dyeing and finishing practices. This method aligns with global initiatives aimed at minimizing the environmental footprint of textile production while enhancing the value-added functionalities of end products (Zhou et al., 2019).

The growing demand for high-quality digitally printed wool textiles is often hindered by inconsistent print quality due to the natural scales and hydrophobic characteristics of wool fibers. To address this issue, this research presents a sustainable, chemical-free pretreatment approach utilizing a combination of hydrogen peroxide (H_2O_2) and the proteolytic enzyme papain to enhance the surface properties of wool and its capacity for dye absorption. Wool fabrics underwent sequential treatment with a low concentration of H_2O_2, followed by papain application. The oxidative effects of H_2O_2 specifically target the disulfide (–S–S–) bonds in wool keratin, partially disrupting the fiber surface and generating reactive groups, such as aldehydes. These changes create more accessible sites for papain, which subsequently hydrolyzes the peptide bonds on the surface of the fibers. The cooperative effects of these two treatments effectively smooth out the wool scales and add more hydrophilic, reactive sites. This synergistic pretreatment led to significant improvements in wettability and zeta potential, making the wool more hydrophilic and enhancing its attraction to aqueous dye solutions. Analytical techniques, including FTIR and XPS, validated the chemical changes that occurred: an increase in carbonyl content and a decrease in disulfide signals indicated the oxidation and hydrolysis processes, while signals related to nitrogen demonstrated the presence of new amino and peptide features on the surface. After the pretreatment, wool fabrics were printed with reactive dyes using an inkjet method under controlled conditions. The measurements of reflectance revealed a significant enhancement in color strength (K/S value) compared to untreated wool, indicating more effective dye fixation. These improvements are attributed to the greater number of dye-binding sites and enhanced interactions between the fiber and dye resulting from the pretreatment. Importantly, the entire pretreatment process is environmentally safe, producing no harmful byproducts or waste. By using low concentrations of H_2O_2 in combination with a biodegradable enzyme, this method adheres to the principles of green chemistry, reducing excessive chemical usage, lessening wastewater toxicity, and potentially decreasing operational costs in textile processing (Shi et al., 2021).

Creating high-value, multifunctional wool textiles is becoming increasingly essential to satisfy various consumer and industrial needs. This research utilizes chitosan's unique macromolecular properties, including its reactive amino groups, film-forming capability, and natural biocompatibility, to investigate a padding-cross-linking method aimed at producing wool fabrics with outstanding printability, shrink resistance, hydrophilicity, and antibacterial properties. Through a padding process, chitosan was effectively applied to wool fibers via Schiff base reactions involving the fabric's aldehyde or oxidized carbohydrate groups. Consequently, the initially hydrophobic wool fibers demonstrated a significant increase in hydrophilicity, greatly improving their ability to absorb dye and ink, an essential factor for high-definition digital printing purposes. Inkjet printing tests using a reactive dye ink indicated significant improvements: the color strength (measured by the K/S value) rose remarkably from 20.48 (for untreated wool) to 26.6 (for chitosan-treated wool), clearly highlighting the enhanced dye affinity resulting from the chitosan treatment. Another key aspect of this research was shrinking resistance. After complete treatment, the wool samples exhibited an area shrinkage of only 2.53%, a significant advancement compared to the 10.96% shrinkage recorded in untreated fabrics. This improvement positions the padding-cross-linked textiles as promising candidates for high-water, high-temperature textile processing. In addition to physical and color attributes, the functional advantages of chitosan were demonstrated in biological performance evaluations. The treated fabrics showed significant antibacterial effectiveness against both E. coli and S. aureus, due to the cationic amino groups in chitosan, which disrupt bacterial cell membranes. This functionality opens up potential applications in sanitary and medical textile fields (Shi et al., 2022).

The sustainable digital inkjet printing of wool and cotton textiles utilizes natural, plant-based inks, specifically addressing the reduction of the coffee-ring effect—a frequent issue in inkjet-printed designs. The fabrics underwent plasma surface treatment (PST) beforehand, a sustainable technique that improves fabric wettability and surface energy without leaving harmful residues. After this treatment, the fabrics were printed using plant-derived inks in shades of blue, red, yellow, and black. These inks were created from the extracts of bio-indigo leaves (for blue), quebracho red bark (for red), and flame-of-the-forest flowers (for yellow), all of which are abundant in chromophoric phytochemicals like indigotin, tannin, and butein. The pre-treated fabrics showed better print consistency, enhanced pigment diffusion, and greater ink penetration in comparison to untreated samples. A primary goal of the research was to investigate and reduce the coffee-ring effect, which occurs when pigment particles aggregate at the edges of a drying droplet, leading to ring stains and inferior print quality. Importantly, inks sourced from quebracho and flame-of-the-forest extracts showed notably smaller dispersion diameters on PST-treated fabrics, suggesting more uniform ink distribution and less noticeable coffee-ring effects. This enhancement can be attributed to the Marangoni flow phenomenon, which causes inward fluid movement within a drying droplet due to surface tension variations, counteracting the outward capillary flow that typically results in ring formation. These results are particularly significant for potential future uses in digital textile printing and printed photonics, where precise design is essential. The study also delivered a detailed overview of ink synthesis procedures, color matching techniques, and the digital printing workflow. Color performance was evaluated using Data color tools, producing color patch evaluations, percentage reflectance graphs, and color difference metrics. The printed textiles achieved robust color strength (K/S values), with wool and cotton achieving values of 3.6396 and 3.9606, respectively—indicating rich, vivid coloration attained through natural dyes. Additionally, the color fastness characteristics were thoroughly assessed. Wash fastness ratings varied from good to very good (grades of 4–5 on the gray scale), while both dry and wet rub fastness consistently scored between 4 and 5. Light fastness evaluations indicated moderate performance on wool (rating 3) and more favorable stability on cotton (ratings of 4–5), highlighting the practical capabilities of plant-based inks in actual textile applications. This method presents a feasible alternative to traditional synthetic dye systems and aids in the progress of eco-friendly, effective, and high-performance digital printing technologies for both the textile and emerging photonic sectors (Madhukar Thakker & Sun, 2023).

The creation and utilization of plant-derived inks as eco-friendly substitutes for petroleum-based inks in digital inkjet printing on wool and cotton materials were explored. The natural inks were created from plant extracts: bio indigo for the blue hue, quebracho red bark for red, and flame of the forest herb for yellow. To produce black ink, these three primary color extracts were mixed together. All inks underwent a multi-stage process that included milling, filtration, and homogenization to ensure fine particle distribution and to avoid clogging during the inkjet printing process. The final ink formulations were subjected to double filtration using 0.22 μm microfilters to achieve sterile, colloidal ink solutions. The ink compositions were made up of plant extracts, distilled water, and glycerol as a humectant and were optimized for rheological characteristics shown in Table 5.1 such as viscosity, density, pH, surface tension, and conductivity, ensuring compatibility with inkjet print heads. The findings indicated that the plasma surface treatment enhanced ink absorption, color strength (K/S values), and print clarity due to increased fabric hydrophilicity and the presence of functional polar groups (e.g., –OH, –COOH) that interact effectively with phytochemical chromophores such as indigotin, tannin, and butein found in the plant-based inks. Furthermore, fastness properties, including wash, rub, and light fastness, were greatly enhanced, particularly when paired with plasma treatment. This research validates that plant-based inks can be successfully developed and used in digital textile printing methods, representing an environmentally friendly and effective alternative to traditional synthetic dyes. These results underscore the promise of herbal inks (Fig. 5.10) in sustainable textile manufacturing, contributing to reduced environmental impacts and safer production practices (Thakker et al., 2022).

The process of printing on wool fabrics using water-based inkjet technology involves gardenia yellow, a dye derived naturally from Gardenia jasminoides. The ink was created by milling, filtering, and homogenizing the aqueous extract, resulting in a stable colloidal solution that is finely dispersed. Comprehensive light scattering analysis demonstrated the ink's exceptional stability, indicating no particle aggregation over extended storage periods. To improve droplet formation and print quality, the surface tension and viscosity of the ink were carefully adjusted by adding biocompatible polyol solvents like glycerol and ethylene glycol. This modification led to the creation of well-defined droplets that are compatible with piezoelectric inkjet nozzles. Additionally, a sodium polyacrylate sizing agent was added to enhance color

Table 5.1 Physical properties of herbal inks (Thakker et al., 2022)

Herbal ink color	Relative density	Viscosity (cP)	Surface tension	Conductivity (mS/cm)	pH
Distilled Water	1	4.07	74	0.0	6.34
Bio Indigo, C (Cyan)	1.06	7.6	58	09.6	7.69
Quebracho Red M (Magneta)	1.06	8.16	54	02.5	5.21
Sacred Tree, Y (Yellow)	1.06	9.66	49	04.2	5.47
Bio indigo + Quebracho Red + Sacred Tree K (Black)	1.06	9.46	59	10.3	5.88

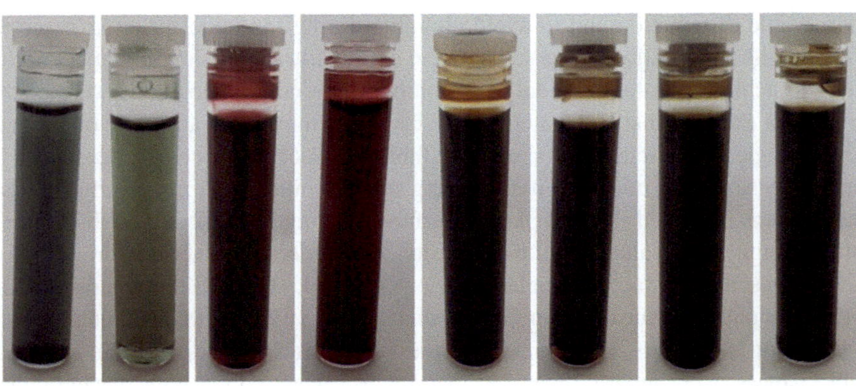

Fig. 5.10 Double-filtered plant-based inks from left to right: cyan, light cyan, magenta, light magenta, yellow, black, light black, and light black inks (Thakker et al., 2022)

depth and outline sharpness. In-depth assessments showed that the sizing played a crucial role: improved color strength (K/S) and sharper design outlines were realized due to increased dye retention along the ink boundaries. Furthermore, applying a tannic acid post-treatment to fix the printed ink greatly enhanced color fastness, resulting in wash-resistance ratings appropriate for practical textile applications. Besides enhancing color, the gardenia yellow also provided functional advantages: the treated wool fabrics demonstrated antibacterial efficacy greater than 95% against Staphylococcus aureus and Escherichia coli, which can be attributed to the bioactive components found in the dye extract. Notably, the entire printing process, including ink formulation, fabric sizing, and color fixation utilizes non-toxic, environmentally friendly chemicals and does not produce harmful waste. This study not only showcases the viability of using gardenia-based beeswax-free inks in digital textile printing but also emphasizes a scalable, eco-friendly approach to create vibrant, durable, and functional wool textiles with a minimal environmental impact (Wang et al., 2023).

Curcuminoids, which are naturally occurring compounds derived from Curcuma longa, are appreciated not only for their bright yellow color but also for their fluorescent characteristics and sensitivity to pH changes. This study extracted a curcuminoid-rich dye (CurD), which contains 38.34 $\pm$ 0.66% w/w total curcuminoids, using microwave-assisted extraction and formulated it into solvent-based (CurD Si) and water-based (CurD Wi) screen-printing inks. At a concentration of 0.1% w/v, the CurD solution showed a peak fluorescence emission at 526 nm. When printed with both ink types on clear polyvinyl chloride (PVC) and paper surfaces, CurD Si and CurD Wi exhibited similar viscosity and drying times, demonstrating uniform printability across different binder systems. Under standard illuminants D50 and D65, the printed inks appeared yellow, but emitted a green fluorescence under UVA (365 nm), highlighting their potential use in anti-counterfeiting and decorative applications. Importantly, the total color difference (ΔE), rather than fluorescence intensity, increased with higher CurD concentrations, indicating more significant observable color variations. A significant discovery was that the CurD Wi ink containing 1% w/w CurD was able to create latent graphics on paper labels. These graphics were invisible under normal lighting but displayed strong green fluorescence when exposed to UV light. Furthermore, these prints displayed dynamic responsiveness to alkaline conditions—changing color from yellow to orange-brown—achieving ΔE values of 15.02 in solution and 26.62 in vapor tests, which indicates effective pH sensitivity. In contrast, while CurD Si inks were also fluorescent, they did not exhibit any noticeable color change when exposed to alkaline conditions (Khankaew & Panichayupakaranant, 2023).

A natural yellow ink was created from marigold petals and applied through inkjet printing to various natural textile substrates, including wool, Silk, cotton, and linen. The dye extraction method involved boiling marigold petals in distilled water at 100 °C for two hours, resulting in a plant-based colorant in powdered form. This powdered dye was then blended into a complex ink formulation that included polyethylene glycol (PEG) solutions, alongside several additives such as stabilizers, humectants, anti-foaming agents, thickeners, and acidity regulators, to optimize the ink's properties for compatibility with inkjet printing. Among the fabrics tested, silk showed the most significant color yield, followed by cotton, indicating enhanced dye absorption and fixation on protein-based fibers. The printed samples underwent evaluation for color fastness properties, with findings revealing wash, rub, and light fastness ratings ranging from 4 to 4–5 across most substrates, suggesting reasonably durable performance that is suitable for practical use. However, the study does highlight specific limitations concerning environmental sustainability. The extraction method necessitated extended high-temperature treatment, resulting in elevated energy consumption, and several components utilized in the ink formulation, such as PEG and other synthetic auxiliaries, are derived from petroleum, thus restricting the eco-friendliness of the method (Thakker & Sun, 2024).

Strategies for the sustainable coloring and functionalization of textiles made from protein using natural, plant-based inks in digital inkjet printing have been explored. Formulations based on curcumin, enhanced with tannic acid and environmentally safe surfactants, exhibited strong color intensity and lasting bioactivity, achieving antioxidant and antibacterial effectiveness of over 85% even after several wash cycles. The application of enzymatic and plasma surface treatments further enhanced the printability and dye absorption of wool. Moreover, inks derived from plants such as marigold, gardenia, and bio-indigo provided outstanding color quality, durability, and additional functional characteristics. These advancements underscore a practical transition toward bio-based, multifunctional, and eco-friendly technologies in textile printing.

5.5 Economic and Commercial Viability

The shift from synthetic dyes to natural, plant-derived inks in digital textile printing is gaining recognition for its environmental advantages as well as its economic and commercial viability. Incorporating inks rich in curcumin and other phytochemicals into production processes could greatly decrease costs related to environmental compliance, enhance brand reputation, and satisfy the increasing consumer interest in sustainable and functional textiles.

5.5.1 Raw Material Cost and Availability

Natural colorants such as curcumin (derived from turmeric), tannin acid (extracted from plant bark or fruit), gardenia yellow, marigold petals, bio-indigo, and flame-of-the-forest flowers are commonly found in tropical and subtropical areas, frequently as byproducts of agricultural or herbal industries. For example, curcumin can be obtained from the waste produced during turmeric processing, making it both an economical and environmentally friendly option. The capability to source raw materials locally not only decreases reliance on imported synthetic dyes but also promotes rural bio-economies and enhances value in agro-industrial supply chains.

5.5.2 Manufacturing and Operational Economics

Formulating ink with natural dyes includes processes such as milling, filtration, homogenization, and adjustments for pH and viscosity, which use significantly less energy compared to the high-temperature and solvent-intensive methods associated with synthetic dye manufacture. Employing aqueous-based systems further minimizes the reliance on organic solvents and decreases the expenses related to waste treatment. Energy-efficient pretreatment techniques like enzyme-assisted (papain), hydrogen peroxide oxidation, or chitosan padding are also biodegradable and can be easily scaled using current finishing lines. Digital inkjet printing enables precise, contactless application with minimal waste and high resolution. In contrast to screen printing, which necessitates costly screens and large quantities of water and dyes, inkjet printing supports on-demand manufacturing and short production runs, thereby lowering inventory expenses and enhancing responsiveness to fashion trends.

5.5.3 Functional Value Addition

In addition to their visual appeal, inks made from curcumin and tannins offer antioxidant, antibacterial, and anti-inflammatory benefits to fabrics, resulting in high-value products aimed at specific markets like wellness clothing, hygienic home textiles, hospital linens, and baby apparel. Functional textiles can command a higher price in the market, yielding a better return on investment (ROI) than traditional textiles. The incorporation of these additional features also minimizes the requirement for separate chemical finishing processes, which are usually costly and harmful to the environment. For instance, chitosan crosslinking not only enhances color richness and wash durability but also introduces antibacterial qualities, effectively streamlining multiple processes into one sustainable step.

5.5.4 Market Trends and Consumer Demand

There is a rising global demand for sustainable, ethically sourced, and multifunctional textiles. Studies indicate that the eco-textile market is escalating at a compound annual growth rate (CAGR) of over 11%, with a predicted worth of more than USD 110 billion by 2030. Companies like Patagonia, Stella McCartney, and Eileen Fisher exemplify businesses that have centered their models around sustainability, shaping consumer expectations and behavior. Digital printing with plant-based inks aligns seamlessly with these trends. Its capacity to deliver vibrant, customizable designs while minimizing environmental impact appeals to both the fast-fashion and luxury markets looking for innovation with integrity. Additionally, the growing prevalence of green certifications (such as GOTS, OEKO-TEX, Cradle-to-Cradle) strengthens the market's acceptance and credibility of naturally dyed products.

5.5.5 *Regulatory and Policy Support*

Governments around the globe are implementing more stringent regulations regarding synthetic dyes and wastewater (such as ZDHC, REACH, and BIS standards in India), which is compelling industries to transition to cleaner technologies. The use of natural inks can greatly lower compliance expenses, environmental fees, and the challenges associated with wastewater treatment, particularly in areas facing water shortages. Furthermore, policy incentives aimed at promoting sustainable production—like green textile initiatives, carbon credits, and subsidies for micro, small, and medium enterprises—can enhance the financial attractiveness of adopting plant-based printing methods.

5.5.6 *Scalability and Industrial Integration*

The ability to scale these natural ink formulations is shown by their suitability for existing piezoelectric inkjet systems. Several of the ink formulations, such as those based on curcumin, gardenia yellow, and marigold, fulfill critical requirements regarding viscosity, surface tension, and conductivity, allowing for smooth application without the risk of clogging or damaging the print head. As microfiltration, stabilizer innovations, and shelf-life enhancement continue to progress, the stability and consistency of these inks are increasingly aligning with industry standards.

5.6 Challenges in Implementation

While sustainable digital inkjet printing using plant-based inks and eco-friendly treatments shows promising potential, several major hurdles impede its widespread industrial adoption. These challenges include technical, economic, logistical, and regulatory issues. Addressing these concerns is essential to transition these innovations from research settings to commercial production.

5.6.1 Ink Formulation and Stability

A significant challenge is creating plant-based inks that possess the necessary physicochemical characteristics suitable for inkjet printing. Natural dyes like curcumin, marigold, and gardenia frequently struggle with:

- Poor solubility and lack of dispersion stability can result in sedimentation or blockage of the print heads.
- Variability in particle size distribution necessitates thorough milling and filtration (for instance, <0.22 μm microfiltration).
- The dye's high sensitivity to pH, temperature, and light can lead to degradation, subsequently shortening its shelf-life.
- Differences in plant materials, extraction techniques, or seasonal fluctuations lead to batch-to-batch variability, affecting both reproducibility and print quality.
- Creating a stable ink formulation with regulated viscosity (6–10 cP), surface tension (45–60 mN/m), and conductivity is challenging and often involves using synthetic additives like stabilizers and humectants, which may compromise the ink's environmentally friendly attributes.

5.6.2 Limited Color Gamut and Fastness Properties

Natural dyes usually yield a limited spectrum of colors, making it challenging to achieve rich blues, deep blacks, and vibrant fluorescent shades when compared to synthetic alternatives. This limitation affects their use in fashion or industrial design sectors that demand a wide variety of bright and diverse colors. Additionally, although advancements like chitosan cross-linking and tannin post-treatment enhance color retention, natural inks often fail to match the performance of synthetic dyes:

- The lightfastness of plant-based inks is especially important on materials like wool and silk.
- Rub fastness is crucial, particularly for darker colors.
- Wash fastness is an issue unless the inks are treated with mordants or undergo further processing.

5.6.3 Substrate Compatibility and Pretreatment Complexity

Fibers derived from proteins, such as wool and silk, possess intricate surface chemistries that include hydrophobic epicuticle layers, which obstruct ink adhesion and dye absorption. While environmentally friendly pretreatments like enzymatic (papain), oxidative (H_2O_2), plasma surface treatment (PST), and chitosan padding have indicated potential, they also come with certain challenges:

- Achieving process standardization is challenging due to variations in treatment efficiency that depend on the fiber source, weaving technique, and fabric density.
- Utilizing multiple pretreatment stages extends both processing time and labor demands.
- Enzyme treatments necessitate careful regulation of pH, temperature, and enzyme activity, complicating consistency at a large scale.
- The financial investment required for plasma or UV treatment systems might be too high for small-scale operations or micro, small, and medium enterprises (MSMEs).

5.6.4 Scalability and Industrial Adaptation

Even though there are encouraging results at the laboratory level, the implementation on an industrial scale remains restricted because of:

- Absence of commercially available natural ink formulations that have a specified shelf life and proper certifications.
- Issues with compatibility in high-speed industrial printers, which need inks with precisely regulated rheology and evaporation properties.
- Challenges related to maintenance, including the need for frequent cleaning of print heads due to clogging from particles in plant-based residues.
- Elevated costs associated with high-purity dye extracts, particularly those enhanced with bifunctional additives such as tannin or chitosan.

5.6.5 Regulatory, Certification, and Consumer Perception

While bio-based inks support sustainability initiatives, formal regulations and standards for natural inks remain inadequately developed. Issues include:

- No established test protocols for assessing functional properties such as antioxidant and antibacterial effectiveness.
- Lack of globally recognized ecolabels or certifications dedicated to digitally printed bio-inks.
- Insufficient consumer knowledge regarding the functional and ecological advantages of naturally dyed textiles, which may diminish their willingness to pay higher prices.

5.6.6 Supply Chain and Resource Constraints

It can be challenging to achieve a dependable supply of uniform, top-quality plant raw materials because of:

- Variation in the yield and quality of plant pigments throughout different seasons.
- Disorganized supply chains for bio-resources such as marigold, gardenia, and bio-indigo.
- Issues with post-harvest handling and storage that can compromise the quality of colorants.

5.6.7 Market Entry Barriers and Economic Uncertainty

The shift towards sustainable inks and pretreatments encounters several market-related obstacles:

- Significant initial research and development and implementation expenses, particularly for small and medium-sized enterprises.
- Ambiguity in return on investment, attributed to a gradual consumer adoption and an existing limited demand.
- Opposition from conventional dye houses or textile manufacturers that are committed to synthetic processes.
- A shortage of technical expertise and trained staff to effectively implement, maintain, and resolve issues with environmentally friendly digital printing technologies.

5.7 Future Trends and Research Directions

The advancements in natural ink creation, functional textile breakthroughs, eco-friendly processing methods, and digital manufacturing integration are shaping the future of sustainable textile coloration. As the sector shifts away from synthetic dyes derived from petroleum, researchers are increasingly focusing on crafting stable and multi-chromatic natural inks sourced from plants such as curcumin, gardenia, marigold, and indigo. To encourage wider commercial use, it is vital to broaden the color range and enhance the stability of these natural inks. Scientists are investigating techniques like dye blending, co-pigmentation, microencapsulation, and nanoparticle-based carriers to improve dispersion, shelf life, and inkjet compatibility, all while upholding environmental sustainability standards.

At the same time, there is a strong emphasis on developing functional textiles. Anticipated innovations are likely to incorporate bioactive features, including antimicrobial, antioxidant, UV protection, and pH responsiveness, directly into ink formulations. These intelligent textiles could be applied in various settings, from medical and wellness apparel to anti-counterfeiting measures and responsive fashion designs. Biocompatible carriers may be utilized in controlled-release systems, allowing textiles to provide continuous functional advantages over time. Research is also directed toward enhancing eco-friendly pretreatment and finishing processes that optimize printability, color durability, and long-lasting quality. Techniques involving enzyme engineering, advanced oxidative treatments, and natural mordants like chitosan, alginate, or pectin are expected to replace traditional chemical processes.

The digital characteristics of inkjet printing make it a perfect fit for Industry 4.0 frameworks. Upcoming research is likely to incorporate artificial intelligence, machine learning, and real-time sensing technologies into the textile printing process. These advanced systems could autonomously refine ink formulations, printing settings, and fabric-specific configurations, boosting efficiency and minimizing waste. The application of digital twins and predictive modeling will further enable tailored production, reduced inventory, and improved sustainability.

An important area of progress is the creation of environmentally friendly solvents and additives. Natural substitutes for petroleum-derived humectants, surfactants, and thickeners like glycerol, bio-surfactants, and plant-based gums will increasingly be crucial for achieving complete eco-compatibility in ink formulations. These green additives must be developed to ensure they preserve essential rheological characteristics, jetting performance, and long-term storage stability while reducing toxicity and environmental impact.

Standardization and certification will be essential in driving the commercial acceptance of plant-derived inks and treatments. The future will witness the creation of testing protocols and regulatory standards specifically tailored for bio-based digital textile printing. Life Cycle Assessment (LCA) tools will be crucial for quantifying environmental effects at each phase, from dye cultivation to post-consumer disposal, guiding both industrial strategies and consumer decisions. Certifications that align with GOTS, OEKO-TEX, and Cradle-to-Cradle principles will help substantiate sustainability claims and foster market confidence.

5.8 Conclusion

The transformation of textile printing from early manual methods to contemporary digital inkjet technologies signifies a significant change in how fabrics, particularly wool and its blends, are treated both aesthetically and functionally. Digital inkjet printing is recognized as a sustainable and high-resolution option compared to traditional techniques, enabling precise color application, customized designs, and reduced environmental consequences.

Wool boasts remarkable qualities such as thermal insulation and excellent dye absorption, but it also poses specific challenges because of its scaly, hydrophobic surface. Nevertheless, advancements in specialized reactive and acid inks, along with environmentally friendly pretreatment strategies like enzymatic processes, plasma treatment, and green chemicals have greatly enhanced the printability of wool. These methods improve dye adhesion, color intensity, and bifunctional characteristics, including antibacterial and antioxidant properties. Additionally, the integration of natural dyes like curcumin and tannic acid into digital inkjet printing systems showcases the potential to blend aesthetic appeal with sustainability. This strategy not only decreases reliance on petrochemical-based inks but also provides added functionalities that cater to the growing consumer preference for eco-friendly textiles.

As the textile sector shifts towards a circular economy and complies with strict environmental guidelines, employing digital inkjet printing with biodegradable and low-toxicity inks on wool and its blends presents a promising future. Ongoing innovation in ink composition, fiber surface enhancement, and digital technology is crucial to fully realize the potential of sustainable wool printing and fulfill the requirements of a rapidly evolving, design-centric market.

Declaration The authors declare that they have no known competing financial interests or personal relationships that could have appeared to influence the work reported in this chapter. Generative AI and AI-assisted technologies have been used in the writing process to improve the readability and language of the work only. After using this tool, the authors carefully reviewed and revised the content to ensure accuracy and quality. The authors take full responsibility for the final version of this publication.

References

An, F., Fang, K., Liu, X., Li, C., Liang, Y., & Liu, H. (2020). Rheological properties of carboxymethyl hydroxypropyl cellulose and its application in high quality reactive dye inkjet printing on wool fabrics. *International Journal of Biological Macromolecules, 164*, 4173–4182.

An, F., Fang, K., Liu, X., Yang, H., & Qu, G. (2019). Protease and sodium alginate combined treatment of wool fabric for enhancing inkjet printing performance of reactive dyes. *International Journal of Biological Macromolecules.*

Barton, J. (2000). *International dyer* (p. 185).

Bechtold, T., & Mussak, R. (2007). *Handbook of natural colorants.* Wiley & Sons.

Broadbent, P. J., & Rigout, M. L. A. (2013). Wool printing. In *The coloration of wool and other keratin fibres.*

Broadbent, A. D. (2001). Protein fibres. In: *Basic principles of textile coloration.* Bradford: Society of Dyers and Colourists.

Christie, R. M., Mather, R. R., & Wardman, R. H. (2000). *The chemistry of colour application.* (Cross Reference)

Christoe, J. R., Denning, R. J., Evans, D. J., Huson, M. G., Jones, L. N., Lamb, P. R., et al. (2000). *Kirk-Othmer encyclopedia of chemical technology.* An Interscience Publication.

Clark, M. A., Patel, M. R., & Kelly, P. A. (1994). Stable reactive dye inks for digital textile printing. US Patent 5300146. (Cross Reference)

Cristea, M., & Vilarem, G. (2006). Improving dyeing process of natural and synthetic fabrics by using enzymes. *Autex Research Journal, 6*(1), 8–15.

CSIRO. (2006). The chemical and physical structure of Merino wool. CSIRO Textile & Fibre Technology, Australia. Retrieved 14 Oct 2025 from https://coilink.org/20.500.12592/2f1lcm2. COI: 20.500.12592/2f1lcm2

Dawson, T. L., & Hawkyard, C. J. (2000). A new millennium of textile printing. *Review of Progress in Coloration and Related Topics, 30*, 7–20.

Ellen Mac Arthur Foundation. (2021). *The jeans redesign guidelines.*

Eltz, H., & Russ, C. (1993). Reactive dyes with cyanamide-based stabilisation for inkjet printing. DE Patent 4222032. (Cross Reference)

Ethical, R. (2021, August 17). Toxic fashion. *Ethical Rebel Plastics and Pollution Magazine*, (02B), 19–20.

Faisal, S., & Lin, L. (2020) Green synthesis of reactive dye for ink-jet printing. *Coloration Technology, 136*(2), 110–119. ISSN 1472-3581.

Faisal, S, Lin, L., & Clark, M. (2020) Synthesis, stability and printing properties of a novel 2-sulphophenoxy-4-chloro-s-triazine reactive dye for inkjet printing of wool. *Coloration Technology, 136*(2). 153–166. ISSN 1472-3581.

Fralix, M. T. (2001). From mass production to mass customization. *Journal of Textile and Apparel, Technology and Management, 1*, 1–7.

Fryberg, M. (2005). Dyes for ink–jet printing. *Review of Progress in Coloration and Related Topics, 35*, 1–30.

Fu, Z. (2006). Pigmented ink formulation. In H. Ujiie (Ed.), *Digital printing of textiles* (pp. 218–232). (Cross Reference)

Ghoranneviss, M., Shabani, M., & Taheri, M. (2014). Surface modification of wool using atmospheric pressure plasma to improve dyeability with acid dyes. *Surface and Coatings Technology, 245*, 1–6.

Gisler, R. (1989). Trisazo dyes based on fluoropyrimidine reactive groups for inkjet printing. DE Patent 3819689. (Cross Reference)

Glover, B. (2003). The latest technology developments in inkjet printing from ITMA 2003 Textile Inkjet printing-a review of ink jet printing of textiles, including ITMA.

Kanik, M., & Hauser, P. J. (2002). Printing of cationised cotton with reactive dyes. *Coloration Technology, 118*, 300–306.

Khankaew, S., & Panichayupakaranant, P. (2023). Development of multifunctional curcuminoid dye-based inks and applications. *Progress in Organic Coatings, 182*, 107707. ISSN 0300-9440.

Le, H. P. (1998). Progress and trends in ink-jet printing technology. *Journal of Imaging Science and Technology, 42*, 49–62.

Lewis, D. M. (1982a). The dyeing of wool with reactive dyes. *Journal of the Society of Dyers and Colourists, 98*, 165–175.

Lewis, D. M. (1982b). *Textile Research Journal, 52*, 580. (Cross References)

Lewis, D. M. (1989). *Journal of the Society of Dyers and Colourists, 105*, 119. (Cross References)

Lewis, D. M. (2013). Dyeing wool blends. In *The coloration of wool and other keratin fibres.*

Madhukar Thakker, A., & Sun, D. (2023). Assessment of coffee-ring effect on wool and cotton fabrics inkjet printed with herbal inks. *The Journal of the Textile Institute, 115*(8), 1193–1206.

Magdassi, S. (Ed.). (2012). *Chemistry of inkjet inks.* New Jersey: World Scientific Publication Ltd. (Cross Reference)

Maheshwari, P. (2020). Modified yellow reactive dye for improved fixation on protein fibres via digital printing. *Journal of Textile Science & Engineering, 10*(2), 1–8.

Miles, L. W. C. (2003). Traditional methods. In: L.W.C. Miles (Ed.), *Textile printing.* Bradford: Society of Dyers and Colourists.

Moser, L. S. (2003). ITMA 2003 review: Textile printing. *Journal of Textile and Apparel Technology and Management, 3*, 1–15.

Provost, J. (1994). Ink jet printing on textiles. *Surface Coatings International, 77*, 36–41.

Schindler, W. D., & Hauser, P. J. (2004). *Chemical finishing of textiles.* Woodhead Publishing.

Shi, F., Liu, Q., Zhao, H., Fang, K., Xie, R., Song, L., et al. (2021). Eco-friendly pretreatment to the coloration enhancement of reactive dye digital inkjet printing on wool fabrics. *ACS Sustainable Chemistry & Engineering, 9*(30), 10361–10369.

Shi, F., Wang, M., Fang, K., Zhao, Z., Zhao, H., & Chen, W. (2022). Fabrication of chitosan-loaded multifunctional wool fabric for reactive dye digital inkjet printing by Schiff base reaction. *Langmuir, 38*(33), 10081–10088.

Textile inkjet printing acid dye ink and production method thereof, Patent number CN101899786A.

Thakker, A. M., & Sun, D. (2024). Inks for digital printing of textiles. *Journal of the Textile Institute, 115*(12), 2377–2390.

Thakker, A. M., Sun, D., & Bucknall, D. (2022). Inkjet printing of plasma surface–modified wool and cotton fabrics with plant-based inks. *Environmental Science and Pollution Research, 29*, 68357–68375.

The Chemical & Physical Structure of Merino Wool. (2006).

Tyler, D. J. (2005). Textile digital printing technologies. *Textile Progress, 37*, 1–65.

Ujiie, H. (2010). Inkjet textile printing status report 2010. In: *Digital inkjet textile Hangzhou, China.*

Walter. R. (1987). Dyeing of polyester/wool mixtures (Leverkusen, Germany: Bayer). (Cross References)

Wang, M., Shi, F., Zhao, H., Sun, F., Fang, K., Miao, D., et al. (2022). The enhancement of wool reactive dyes ink-jet printing through air plasma pretreatment. *Journal of Cleaner Production, 362*, 132333. ISSN 0959-6526.

Wang, M., Zheng, S., Fang, K., Zhao, Z., Shu, D., Xie, R., & Chen, W. (2023). Green fabrication of inkjet printed antibacterial wool fabric with natural gardenia yellow dye. *Industrial Crops and Products, 206*, 117700. ISSN 0926-6690.

Xiao, H., Yu, Q., Tu, J., Wang, M., Alebeid, O. K., Wang, P., & Lu, M. (2021). *Ink-jet printing of wool/acrylic blend fabrics with cationic reactive dyes fibers and polymers.*

Yang, K., Clark, M., & Lewis, D. M. (2019). *Synthesis of a di-(p-sulphophenoxy)-striazine reactive dye and its application in wool fabric ink-jet printing.* Coloration Technology.

Yuen, C., Kan, C., Jiang, S., Ku, S., Choi, P., & Wong, K. (2010). Optimum condition of ink-jet printing for wool fabric. *Fibers and Polymers, 11*, 229–233.

Zhou, Y., Junchun, Y., Biswas, T. T., Tang, R.-C., & Vincent, A. (2019). Nierstrasz Inkjet printing of curcumin-based ink for coloration and bioactivation of polyamide, silk and wool fabrics. *ACS Sustainable Chemistry & Engineering, 7*(2), 2073–2082.

Chapter 6
Chemistry of Inks and Chemicals Used in Digital Printing

Edward Menezes, Manjiri Paranjape, and Amey Damle

Abstract Digital textile printing represents a transformative advancement in the textile industry, offering remarkable design flexibility, precision, and sustainability. Unlike traditional methods, it employs inkjet technology to apply intricate patterns directly onto fabrics, enabling rapid customization, minimal waste, and reduced environmental impact. The process is fundamentally driven by the underlying chemistry of ink formulation which must meet stringent physicochemical criteria such as viscosity, surface tension, stability, and compatibility with specific firbres and treatments. Textile inks are designed with functional additives including surfactants, dispersants, thickeners, and binders, ensuring optimal print quality, colour vibrancy, and resistance to washing, abrasion, and UV exposure. Moreover, alignment between ink chemistry and substrate pre- and post-treatment is essential to achieve durable and high-performance prints. As the industry shifts toward on-demand production, personalization, and environmentally responsible practices, innovations in sustainable ink systems and greener processing techniques are crucial. This chapter explores the critical role of ink and chemical formulation in digital textile printing and highlights how ongoing advancements in this domain are enabling a more circular, efficient, and responsive textile manufacturing ecosystem.

Keywords Digital textile printing · Inkjet ink formulation · Sustainable printing

6.1 Introduction

Textile printing is one of the most crucial and creative processes in the textile manufacturing industry, playing a central role in the design, functionality, and appeal of fabric products. It involves applying colour or patterns to textiles in specific areas, creating visual designs that enhance the fabric's aesthetic or functional qualities. This process is distinct from dyeing, which uniformly colour the entire fabric.

E. Menezes · M. Paranjape · A. Damle (✉)
Rossari Biotech Limited, Mumbai, India
e-mail: amey.damle@rossari.com

A. Athalye and H. Patil (eds.), *Digital Printing*,
https://doi.org/10.1007/978-981-95-2593-5_6

Textile printing has been practiced for centuries, with block printing emerging as one of the earliest methods. Over time, the industry witnessed the development of screen printing, rotary printing, and engraved roller printing. These techniques are effective for large-scale, repeat-pattern production however, each of these methods had its limitations in terms of setup time, production flexibility, and the ability to handle small batch runs or highly intricate designs (Carden, 2016; Ujiie, 2006).

In the modern textile industry, digital textile printing has revolutionized the printing process. Unlike traditional methods, digital printing uses inkjet technology to print designs directly onto fabrics with high precision and minimal waste. This not only supports customization and on-demand production but also reduces the environmental impact associated with water, chemicals, and energy consumption. Digital textile printing has become a key process for manufacturers and designers, enabling them to rapidly prototype new patterns, create small runs of custom or niche designs, and quickly respond to market trends. It has been particularly influential in sectors where personalization, fast fashion, and short-run production are increasingly in demand (Gulrajani, 2013; Kan & Yuen, 2012).

However, while inkjet printing may seem straightforward, formulating the inks used in digital textile printing requires meeting specific physicochemical standards, such as storage stability, adhesion, and surface tension (Gao et al., 2021). These essential properties make the chemistry and formulation of inkjet inks quite unique. In the case of textile printing, the ink and the pre-treatment of the fabric are critical to achieving high-quality, durable prints. Unlike traditional printing on paper or other substrates, textiles are subjected to regular washing with various detergents, meaning the ink must be resistant to fading, rubbing, and the harmful effects of sunlight (Ujiie, 2006). These challenges make ink formulation for digital textile printing a complex process, requiring a deep understanding of the right ingredients and techniques to ensure optimal performance. Addressing these considerations is key to selecting the right components for a successful inkjet ink formulation that can meet the demands of modern textile printing.

6.2 Classification of Inkjet Printing Inks

Inkjet printing technology works by precisely placing tiny droplets of ink onto paper or other surfaces. The quality and performance of the print largely depend on how the ink is formulated. Inkjet inks are divided into two main components: Bases and Colourants see Fig. 6.1 (Le, 1998).

The base is the liquid medium that delivers the colourant to the printing surface. It plays a key role in determining the ink's physical properties, how it behaves in the printer, and how it interacts with the material being printed on.

The colourant is the component that gives the ink its colour. Colourants can be either dyes (which are soluble in the base) or pigments (which are particles dispersed in the base) (Tawiah et al., 2016).

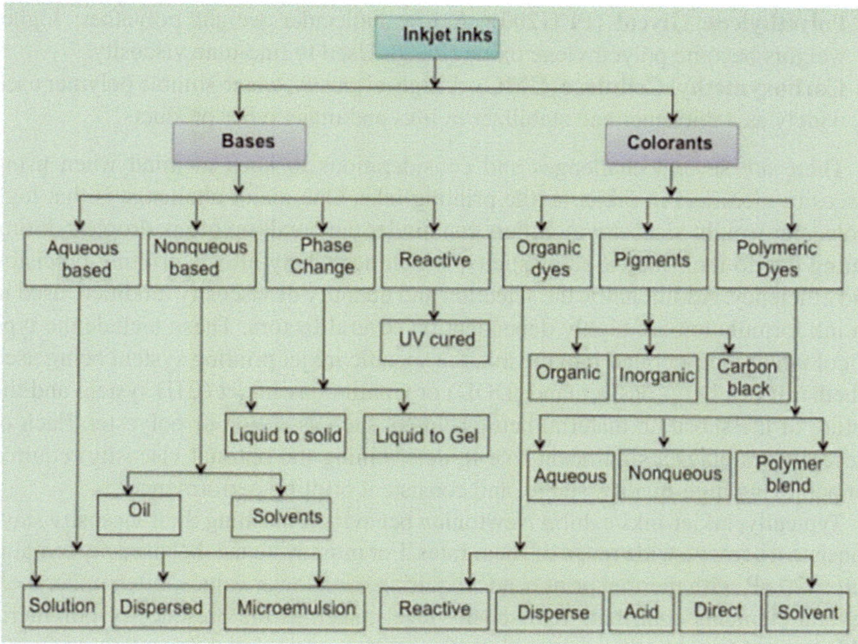

Fig. 6.1 Classification of inkjet inks (Tawiah et al., 2016)

Each of these components has its own subcategories based on their chemical nature and usage. Understanding this classification helps in choosing the right type of ink for specific printing technologies, ensuring compatibility, optimal performance, and desired results.

6.3 Physiochemical Aspects of Inkjet Inks

6.3.1 Viscosity

Viscosity is a crucial parameter in inkjet printing that directly impacts droplet formation, flow behaviour, and print quality. It must be precisely controlled to ensure consistent jetting neither too high to cause nozzle blockage nor too low to cause ink leakage or droplet deflection due to air currents.

Viscosity modifiers stabilize ink behaviour by adjusting fluid flow properties. Common modifiers include.

- **Glycerol**: A clear, viscous alcohol used for its nontoxicity and stability; produced naturally (from soap making) or synthetically (from biodiesel by-product).

- **Polyethylene Glycol (PEG200)**: A low molecular weight polyether; higher weights become polyethylene oxide (PEO). Used to fine-tune viscosity.
- **Carboxymethyl Cellulose (CMC)**: A high-viscosity, water-soluble polymer used widely as a thickener and stabilizer in inks and many other products.

There are several challenges and considerations to keep in mind when using viscosity modifiers in inkjet textile printing inks. One major challenge is that high molecular weight viscosity modifiers can hinder the break-up of ink droplets during jetting due to their increased elasticity, which negatively affects printing precision and efficiency. Additionally, the selection and quantity of viscosity modifiers used in an ink formulation are highly dependent on several factors. These include the type of colourant incorporated into the ink, the specific inkjet printing system being used whether it is a Drop-on-Demand (DOD) or Continuous Inkjet (CIJ) system and the nature of the substrate material being printed, such as cotton or polyester. Each of these factors plays a significant role in determining the optimal viscosity required for achieving high-quality, stable, and consistent printing performance.

Typically, inkjet-inks exhibit Newtonian behaviour, meaning their viscosity stays consistent across a wide range of shear rates. For most systems, the viscosity remains below 20 cP, with thermal print heads requiring even lower values, often under 3 cP. Several elements contribute to viscosity levels, such as the presence of polymeric additives, surfactants, the composition of solvents, and potential flocculation of particles. Over time, storage-related changes like particle aggregation or unintended polymerization in UV-based inks can increase viscosity and degrade print performance. Therefore, tracking viscosity changes during storage is essential to anticipate and mitigate such problems (Le, 1998; Tawiah et al., 2016; Tyler, 2005).

6.3.2 Surface Tension

Surface tension plays a pivotal role in the effectiveness of inkjet printing for textiles, as it directly impacts how ink droplets form and interact with the fabric surface. Ideal ink formulations maintain surface tension between 25–60 dynes/cm to ensure smooth flow through the printer's nozzles and proper spreading on the textile substrate (Magdassi, 2009; Ujiie, 2006). This balance can be achieved by altering solvent compositions or incorporating surfactants agents that lower surface tension and enhance wetting. For example, adding propanol to water can significantly reduce surface tension, depending on the ratio used (Ali, 2008; Le, 1998). Surfactants, which possess both hydrophilic and hydrophobic components, are commonly employed in concentrations under 1% to improve ink performance and dye dispersion in fabric treatment. These compounds fall into three main categories: fluorochemicals, silicones, and hydrocarbons (Mallinson, 1999). Fluorinated surfactants are the most potent, while hydrocarbons are milder and more environmentally friendly. Although silicones are cost-effective, their airborne nature can lead to substrate contamination and coating defects like "fish-eyes." Once surfactants are integrated into inks or coatings, they often permanently alter the surface energy, making further applications more difficult. To prevent such issues, it's advisable to use the lowest effective dose of the least aggressive surfactant preferably hydrocarbons and ensure substrates are thoroughly cleaned before application. Reactive surfactants, which chemically bond with the ink or coating during curing, also offer a practical solution by minimizing residual surface activity in the final product (Flick, 1993; Nakagawa et al., 1963; Mallinson, 1999).

6.3.3 Conductivity

Ink conductivity is a crucial parameter that directly affects printing performance, particularly in digital inkjet systems. This property is largely influenced by the ink's salt content, which must be carefully managed. Exceeding a salt concentration of 1–1.5% can impair printing quality, often leading to nozzle blockages and inconsistent jetting behaviour (Dawson, 2000). Moreover, for optimal formulation, only highly purified dyes should be used to avoid introducing unwanted conductive impurities. In certain printing systems, especially those with ink recirculation, sensors rely on conductivity signals to monitor and regulate ink flow. This becomes even more critical in continuous inkjet (CIJ) technology, where droplets are directed using an electric field. In such systems, conductivity must be fine-tuned using charge control agents like electrolytes and ionic surfactants, which must be fully soluble in the ink. Precise control is essential, as even minor fluctuations caused by factors such as salt crystallization, interactions with other ingredients, or contact with container surfaces can disrupt performance during storage and application (Bao and Song, 2016; Gregory, 2012).

6.3.4 pH

Maintaining proper pH is vital in water-based ink formulations, as it directly impacts the solubility of components and the overall stability of dispersed pigments. For instance, polymeric binders like acrylic resins can become insoluble under acidic conditions, leading to formulation instability (Campbell, 2008). The stability of pigment dispersions is closely tied to the zeta potential of particles; higher zeta potential values indicate stronger repulsion between particles, reducing the risk of aggregation. Stabilization is often achieved through charged polymers adsorbed onto particle surfaces, whose effectiveness depends heavily on the system's pH. To safeguard against fluctuations caused by changes in water quality or component interactions, buffers are frequently incorporated into ink formulations. To protect sensitive printer components, such as nozzles, from corrosion, ink pH is typically maintained within a mildly alkaline range of 7–10 (Ali, 2008; Magdassi, 2009).

6.3.5 Foaming and Defoamer

One of the critical challenges in inkjet printing is the formation of bubbles and foam within the ink, which can significantly compromise performance (Magdassi, 2009; Karl, 2013). Foaming is particularly common in formulations containing surfactants and polymers, as these ingredients tend to stabilize air pockets. To combat this issue, defoamers are often incorporated into the ink. These compounds work by locally disrupting surface tension, causing rapid thinning and collapse of foam structures examples include amyl alcohol for surface tension reduction and tributyl phosphate for enhancing liquid drainage by lowering surface viscosity (Roberts et al., 1977). When dealing with persistent or severe foaming, it is essential to choose defoamers that remain stable and uniformly dispersed during storage, using the lowest effective concentration to maintain formulation integrity (Karl, 2013). Antifoaming agents are commonly derived from mineral oils, while alternative active ingredients may include silicones, high molecular weight alcohols, fatty acid derivatives, phosphoric acid esters like tributyl phosphate, fluorinated compounds, or combinations of these components. Selecting the right defoaming strategy ensures consistent ink performance and prevents printing defects caused by trapped air or foam buildup (Kaimouz et al., 2010).

6.3.6 Storage and Stability

As inkjet printing becomes more widely used in production, ensuring stable and reliable performance is essential. Several key factors can negatively impact print quality, leading to issues such as white lines, colour inconsistencies, and stains. One major concern is the compatibility of the ink with the printhead (Ali, 2008; Kaimouz et al., 2010). High-speed printers like the Tx series eject extremely small ink droplets at rapid frequencies, and any particles in the ink that are too large can clog the nozzles. Additionally, if the ink's physical properties do not match the speed and motion of the printhead, misting or incomplete droplet ejection may occur (Ujiie, 2006). To avoid such issues, ink formulations must be carefully engineered for long-term stability. Essential properties such as pH, viscosity, surface tension, and particle size distribution must remain unchanged throughout the ink's shelf life, which typically spans up to two years at room temperature, depending on the type of ink (Ali, 2008; Magdassi, 2009). Any change in these parameters can result in performance degradation. Common causes of instability include component interactions, polymerization (especially in UV-curable inks), precipitation due to temperature shifts, and adsorption of surfactants onto container walls, which may alter the ink's surface tension or increase viscosity (Guan et al., 2014; Li & Tincher, 1999). For pigment-based inks, maintaining a stable dispersion is especially critical, as pigment particles tend to aggregate over time if not properly stabilized. Achieving long-term consistency in both ink formulation and printhead compatibility is therefore key to ensuring flawless operation and high-quality results in industrial textile inkjet printing (Ujiie, 2006).

6.4 Chemical Composition of Digital Printing Ink

Achieving targeted performance and exceptional print quality hinges on the deliberate selection and precise blending of ink ingredients. This is a complex process where each component uniquely influences the final product. The core ingredients typically include colourants either pigments or dyes used in concentrations ranging from 0.3 to 15%, depending on the application and intensity required. Humectants such as glycerol, ethylene glycol, or diethylene glycol are included at levels around 20–30% to maintain ink fluidity and prevent premature drying in the printhead nozzles. Small quantities of surfactants (1–2%) enhance substrate wetting by reducing surface tension, while pH stabilizers (~0.5%) and defoamers (~0.1%) ensure chemical consistency and minimize air bubble formation during printing. The bulk of the formulation typically 30–80% is deionized water, serving as the main carrier for all ingredients in water-based inks (Daplyn & Lin, 2003; Fu et al., 2011, 2012; Guan et al., 2014; Hou et al., 2019; Kosolia et al., 2011).

The overall formulation is grounded in key chemical principles such as solubility, molecular interaction, and compatibility between components see Table 6.1. These interactions dictate the ink's stability, performance, and adaptability to various printing systems. Thus, ink formulation is not merely a technical process but a carefully calibrated science, essential for delivering consistent, high-quality printing results.

Table 6.1 Formulations of digital textile printing inks (reactive, acid, pigment, and disperse) showing typical composition ranges (Ali, 2008)

Component	Reactive ink	Acid ink	Pigment ink	Disperse ink
Colourant (dye/pigment)	5–15%	5–10%	3–5%	3–5%
Solvent	20–40%	10–30%	20–40%	20–40%
Binder	–	–	5–10%	–
Additives	–	1–5%	–	–
Surfactants	0.1–1%	0.1–1%	0.1–1%	0.1–1%
Water	40–70%	60–90%	40–60%	40–60%
pH regulator	Adjust to 5–8	–	–	Adjust to 5–7

6.4.1 Inks

Inks are complex fluid formulations designed to deliver colourants onto a surface with precision, durability, and aesthetic quality. In inkjet printing, one of the key challenges lies in balancing ink flow characteristics ensuring smooth jetting through microscopic nozzles while maintaining strong adhesion and vibrant colour on a wide variety of substrates. Examples of commonly used dyes and pigments in inkjet printing see Figs. 6.2 and 6.3 respectively.

At the core of most ink formulations is the liquid medium, which can be either **aqueous** (water-based) or **non-aqueous** (solvent- or oil-based). This classification greatly influences the ink's drying method, environmental impact, print durability, and substrate compatibility.

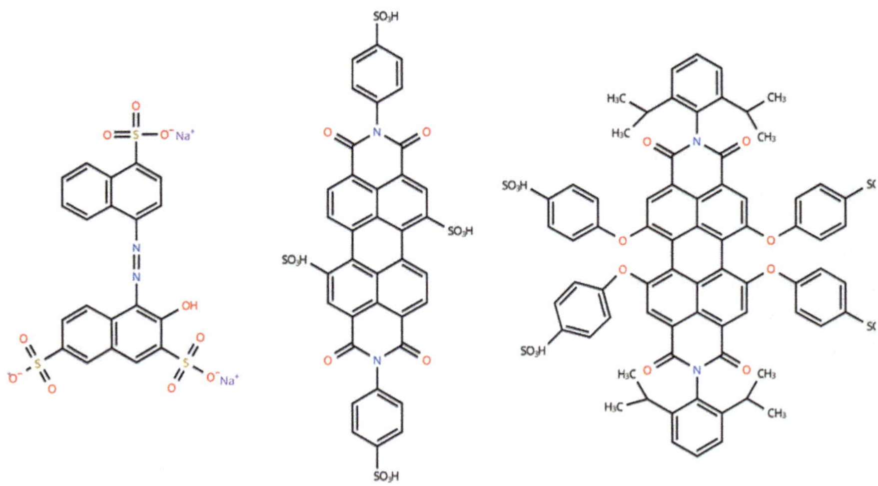

Fig. 6.2 Examples of dyes used in inkjet printing (Choi et al., 2019)

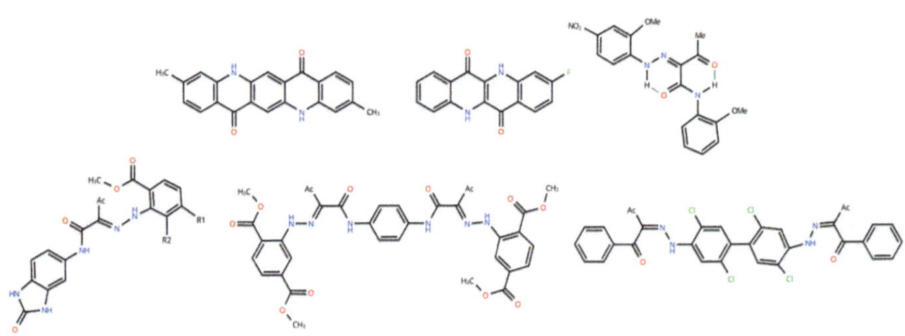

Fig. 6.3 Examples of pigments used in inkjet printing (Magdassi, 2009)

6.4.1.1 Aqueous Inks

Aqueous inks, utilizing water as their main solvent, are versatile and employed across a wide range of printing applications. They can incorporate pigments, dyes, or combinations thereof as colourants. Thermal inkjet printers rely on water for their ink ejection mechanism, making aqueous inks their primary choice. While aqueous inks offer a broad and vibrant colour range, they are not inherently waterproof and require coating or lamination for water resistance. Dye-based aqueous inks are the most affordable but are susceptible to fading when exposed to light or ozone. Pigment-based aqueous inks offer superior durability and UV resistance, often marketed as archival quality, though at a higher cost. While some professional wide-format printers use aqueous inks, most utilize a wider variety of inks, typically requiring piezo inkjet heads and more intensive maintenance (Tawiah et al., 2016).

6.4.1.2 Non-aqueous Inks

Solvent inks utilize volatile organic compounds (VOCs) as their primary solvent, employing pigments for superior fade resistance. These inks are cost-effective and enable printing on uncoated vinyl substrates, commonly used for vehicle graphics, posters, and stickers. However, they produce solvent vapours and require proper disposal of spent solvent. Unlike aqueous inks, solvent-based prints are typically waterproof and UV-resistant, suitable for outdoor applications without additional coatings.

Many high-speed solvent printers incorporate specialized curing systems, utilizing heaters and blowers to dry the ink. The substrate is often preheated before ink application and heated again afterward. Hard solvent inks offer the highest durability without overcoating but necessitate dedicated ventilation to mitigate toxic fume exposure.

Mild or eco-solvent inks offer a compromise, being less hazardous than hard solvents but still requiring some ventilation. They have gained significant popularity due to improved colour quality and durability at a lower cost, making them suitable for use in environments with less stringent ventilation requirements than hard solvent inks (Tawiah et al., 2016).

6.4.1.3 UV-Curable Inks

UV-curable inks consist primarily of acrylic monomers and photo initiators. After printing, intense UV light triggers a chemical reaction, causing the ink components to crosslink and solidify. This curing process, typically achieved with mercury-vapor lamps or UV LEDs, allows for rapid hardening even on heat-sensitive materials due to short, high-intensity exposure. Unlike solvent inks, UV inks cure through a chemical reaction, not evaporation, resulting in near-100% volume utilization.

This rapid curing process leads to immediate drying and a fully hardened image within seconds, enabling high-speed printing. The absence of solvents penetrating the substrate ensures high-resolution prints. UV-curable inks offer benefits like instant drying, compatibility with diverse uncoated surfaces, and high durability.

However, they are expensive, require specialized curing chambers, and result in a noticeable relief on the substrate due to the high volume of cured ink. While advancements are ongoing, the cured ink's rigidity can cause cracking on flexible substrates. Consequently, UV-curable inks are commonly used in large flatbed printers designed for rigid materials like plastic, wood, and metal, where flexibility is not a concern (Le, 1998; Tawiah et al., 2016).

6.4.1.4 Sublimation Inks

Sublimation inks utilize dyes that, historically believed to transition directly from solid to gas, are used for printing on polyester-rich textiles. During heating, these dyes sublimate and permeate the fibres, producing vibrant, durable images. This technology is popular for decorating apparel, signage, posters, and various novelty items like phone cases, plaques, and mugs designed for sublimation.

Sublimation inks employ dispersed dyes, which are organic colourants insoluble in water and other common solvents. When used with hydrophobic synthetic polymers, these dyes disperse evenly throughout the material at a molecular level, rather than simply coating the surface. This creates a vibrant colour impression without the need for gloss coatings or other surface treatments. Light is refracted from within the polymer matrix and the polymer-air interface, resulting in a rich, deep colour (Tyler, 2005; Xu, 2017).

6.4.1.5 Solid Inks

Solid inks, also known as hot melt inks, are primarily composed of paraffin-based materials. They function by melting to a liquid state for printing and then solidifying upon contact with a cooled substrate. In graphic printing, they're often used for masking. Unlike traditional liquid inks or toner powders, solid ink technology uses solid ink sticks, crayons, pearls, or granules. These solid ink forms are stored in a hopper and then either transported to the print head by a worm gear or melted on demand. Once melted, the ink is applied to the substrate, resembling offset or conventional printing processes.

6.4.1.6 Metal Inks

Metal nanoparticle inks, also called functional inks, are essential for printed electronics manufacturing, enabling the additive creation of electronic circuits, electrodes, and connectors through their electrical conductivity, semi-conductivity, electroluminescence, and insulating properties; these inks consist of metal nanoparticles like silver, copper, or gold dispersed in solvents like water, alcohols, or tetradecane, with water offering environmental benefits and nonpolar solvents allowing for higher metal concentrations; these inks utilize nanoparticles with at least one dimension under 100 nm, produced either by breaking down larger particles (top-down) or growing them from smaller units (bottom-up), allowing for high-capacity, continuous printing required in industrial applications (Kang et al., 2022; Tyler, 2005).

6.4.2 Binders

A key factor in achieving high-quality results in digital textile printing particularly pigment-based inkjet printing is the use of binders. These film-forming agents play a vital role in anchoring pigments onto fabric surfaces, enhancing fastness to washing and rubbing, and influencing the overall tactile properties of the printed textile. The most widely employed binders in this domain include styrene-butadiene (SB), styrene-acrylate (SA), and vinyl acetate-acrylate (VAc-A) copolymers, each offering distinct advantages rooted in their molecular composition and interaction with different textile firbres (Kutanaee, 2011; Stankovská et al., 2014; Zhang et al., 2015).

Styrene-butadiene binders are synthesized through copolymerization of styrene and butadiene, resulting in a polymer matrix that balances flexibility with structural integrity. Their excellent film-forming ability, combined with good wash and colour fastness, makes them particularly suitable for pigment printing on cotton and polyester fabrics (Song et al., 2020). Styrene-acrylate binders, incorporating various acrylic monomers, offer superior water resistance, thermal stability, and fabric compatibility attributes that make them highly effective in inkjet applications where print durability is essential (Abd Abdel-Wakil et al., 2020; El-Ghaffar et al., 2014; Song et al., 2020). Vinyl acetate-acrylate binders, formed from vinyl acetate and acrylic monomers, are known for their eco-friendly, water-based composition. These binders deliver a soft hand, strong adhesion, and compatibility with low-VOC ink systems (Wang, 2003; Zhang et al., 2015).

The functionality of these binders is determined by their monomer mix styrene imparts rigidity, butadiene and vinyl acetate add elasticity, and acrylic acid enhances wash resistance and adhesion (El-Ghaffar et al., 2014). These materials are integrated into digital printing through techniques like inkjet, screen, and pad printing, ensuring uniform pigment fixation and durability. Additionally, the shift toward water-based, formaldehyde-free, and APEO-free systems has supported the development of sustainable binders that minimize environmental impact (Robert, 2019; Kozake et al., 2021; Ozcan et al., 2024). Combined with inkjet printing's precision and low waste generation, these innovations support a more responsible and efficient approach to textile colouration (King, 2013; Song et al., 2020).

6.4.3 Surfactants

Surfactants are crucial additives in digital textile printing inks, significantly improving performance by modifying surface tension and enhancing ink interaction with fabric. They ensure uniform printing, prevent nozzle clogging, and boost colour vibrancy, particularly in water-based inks, by improving wetting, spreading, and penetration. Surfactants, categorized by their ionic nature and chemical structure (non-ionic, anionic, cationic, amphoteric, and silicone-based), offer diverse properties tailored to specific ink and fabric requirements. They reduce surface tension for optimal wetting, improve ink spread for even coverage, enhance penetration for better colour yield, stabilize pigment dispersions, control foam, and improve ink-fabric compatibility. Non-ionic surfactants, widely used for their versatility and stabilizing properties, are compatible with various systems. Anionic surfactants are effective at low concentrations for hydrophobic fabrics, while cationic surfactants enhance adhesion to negatively charged fibres. Amphoteric surfactants offer excellent wetting and foam control, suitable for eco-friendly formulations. Silicone-based surfactants, though more expensive, provide superior ink spread and penetration for high-quality prints (Magdassi, 2009; Tawiah et al., 2016; Tyler, 2005).

6.4.4 Dispersants

Dispersants are essential in pigment-based digital textile printing inks, preventing pigment clumping and ensuring uniform colour distribution. By stabilizing pigment particles, they enhance colour strength, ink stability, and print quality, preventing nozzle clogging and maintaining smooth ink flow. Categorized by chemical structure and mechanism (anionic, non-ionic, cationic, polymeric, and silicone-based), each dispersant type offers unique benefits tailored to specific ink formulations and printing conditions. They stabilize pigments through electrostatic, steric, or electrostatic forces, creating repulsions or barriers between particles. The selection of a dispersant depends on pigment type, ink formulation, fabric, printing system, environmental impact, and cost, requiring careful consideration to avoid compatibility issues or performance reductions. Used throughout ink manufacturing, storage, and printing, current trends focus on eco-friendly and nanotechnology-enhanced dispersants. Choosing the appropriate dispersant is critical for high-quality prints and depends on the specific requirements of the printing system, ink, and fabric, with ongoing developments continuously improving performance and sustainability (Magdassi, 2009).

6.4.5 Thickeners

In digital textile printing, especially in pigment-based systems, thickeners play a pivotal role in controlling ink rheology, ensuring sharp print definition, stability, and minimal bleeding. Among the various thickener types, acrylic and polyurethane-based thickener have emerged as high-performance solutions due to their versatile chemistry and superior application properties (Zhou et al., 2016).

Acrylic thickeners are synthetic, water-soluble polymers primarily based on acrylic acid and its copolymers, including methacrylic acid, acrylamide, and other functional monomers. Their effectiveness is derived from two primary thickening mechanisms: the polyelectrolyte swelling effect where carboxylic acid groups ionize under alkaline conditions causing polymer chain repulsion and viscosity build-up, and associative thickening, where hydrophobic moieties interact to form structured networks (US Patent US6696508B2, 2004). Acrylic thickeners are available in different forms such as ASE (alkali-swellable emulsions), HASE (hydrophobically modified ASE), and HEUR (hydrophobically modified urethanes), each tailored for specific digital printing applications. For example, ASE types are often used in reactive dye pre-treatment, while HASE and HEUR variants are preferred in pigment printing due to their superior control over ink penetration and dot definition.

Copolymers of acrylic acid with monomers like acrylamide or hydroxyethyl acrylate offer significant functional advantages. They enable shear-thinning behavior essential for smooth inkjet jetting, enhance salt tolerance critical in pigment systems, and provide tunable pH responsiveness (Karga et al., 2022). The incorporation of crosslinkers such as divinylbenzene or TMPTA further improves thickening efficiency and structural integrity, forming a gel-like matrix that resists phase separation and syneresis during ink storage and application (Ragab et al., 2021).

On the other hand, polyurethane-based (PU) thickeners function primarily through associative thickening mechanisms. These non-ionic polymers consist of a polyurethane backbone with both hydrophilic polyether segments and hydrophobic groups. In aqueous systems, the hydrophobic segments form reversible physical networks, offering excellent viscosity control and shear-thinning behavior. Unlike ASE or HASE thickeners, PU variants do not rely on pH changes, making them suitable across a broad pH range (4–10). Their non-ionic nature ensures compatibility with various formulation components including surfactants, dyes, and binders (Grant, 2001; Yang & Zhou, 2023).

PU thickeners are particularly favored in pigment printing pre-treatment and ink formulations for their high thickening efficiency at low dosages, superior clarity, and excellent jetting performance. They also contribute to better print sharpness, pigment fixation, and fabric softness. However, they are sensitive to surfactants and generally more costly than acrylic counterparts (Kasymova et al., 2024; Meng et al., 2020; Zhou et al., 2017).

Overall, both acrylic and PU-based thickeners offer advanced rheological solutions for digital textile printing. While acrylic thickeners provide pH-responsive viscosity and high formulation versatility, PU thickeners excel in providing stable, efficient, and shear-responsive flow properties. Selection depends on the specific requirements of the ink system, fabric type, and performance targets.

6.4.6 *Other*

Defoamers are critical for maintaining print quality in digital textile inks by preventing foam formation during manufacturing, storage, and printing. Foam can lead to uneven prints and nozzle clogging, which defoamers counteract by destabilizing bubbles through spreading, bridging, and dewetting. The choice of defoamer—silicone, mineral oil, polyether, alcohol, or organic-based—depends on ink compatibility, foam severity, printing system, environmental impact, and cost. Careful consideration is needed to avoid issues like residue formation and concentration control, ensuring optimal print performance.

Biocides, buffers, and dye-fixing additives are essential components in digital textile printing inks, each serving distinct purposes. Biocides, like isothiazolinones and silver-based compounds, prevent microbial growth and extend ink shelf life by disrupting cell functions, but require careful handling due to potential toxicity. Buffers, such as sodium carbonate and citric acid, maintain the ink's optimal pH, crucial for dye fixation and printhead protection. Dye-fixing additives, including cationic fixatives and urea derivatives, enhance print durability by strengthening dye-fibre bonds, improving wash and rub fastness, though overuse can affect fabric feel and colour. These additives collectively ensure ink stability, print quality, and fabric longevity and performance (Tawiah et al., 2016; Tyler, 2000; Ujiie, 2006).

6.5 Mechanism of Ink-Fibre Interaction

In today's digital textile printing landscape, four primary types of dyestuffs dominate commercial use. Each category is engineered to interact with specific fibre types, with distinct colour performance, application methods, and fixation requirements. As Table 6.2 and Fig. 6.4 shows insight into the chemical behaviour and functional outcomes of each ink type.

Fixation is a crucial component of the digital printing workflow. Once a design is digitally applied to fabric, a finishing process is necessary to ensure that the ink bonds permanently to the substrate. The fixation method varies depending on the ink formulation, influencing not only durability but also the final colour brilliance and fastness properties of the print.

One of the key advantages of digital printing over conventional methods is the relatively "dry" appearance of the fabric post-print. This is due to the precise, low-volume application of ink. However, for optimal performance, additional drying is often required especially to stabilize the ink and ensure proper fixation. This is typically achieved through thermal processes, and many digital printing systems are

Table 6.2 Textile digital inkjet printing inks-fibres interaction techniques (Tawiah et al., 2016)

Colourant	Fibre type	Colour fibre-interaction	Colouristic properties	Fixation
Pigment	All fibres	No interaction—complex surface polymer (binder) bonding mechanism	Good washing fastness Excellent light fastness, good rubbing fastness depending on binder content	Oven curing at 160–180 °C for 30–90 s
Reactive dye	Cotto, silk and wool, linen	Covalent fibre bonding	Bright colours, excellent washing fastness, excellent rubbing fastness poor light fastness	Steaming for 90–120 °C for 8–30 min depending on steamer type, washing and drying
Dispers dye	Polyester	Hydrophobic-solid state mechanism	Excellent light, washing and rubbing fastness, bright colours	Transfer press or "thermosoled" or oven cured depending on type of disperse ink
Acid dye	Nylon, silk, wool, leather	Electrostatic and hydrogen bonding with fibre	Bright colours, excellent light fastness, good washing and rubbing fastness	Steaming for 20–60 min depending on steamer type at 20–120 °C depending on shade and fibre type, washing and drying (except leathers)

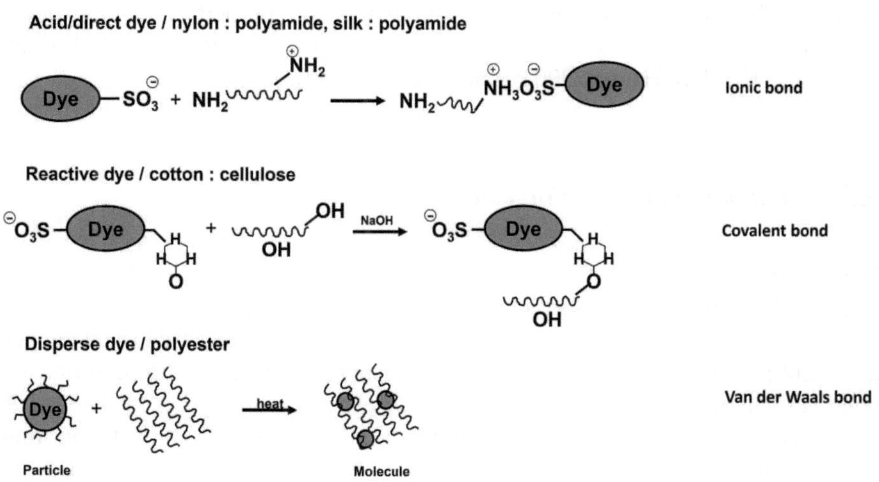

Fig. 6.4 Chemistry of dye-fibre interaction (Yoshiki, 2005)

equipped with supplementary heating elements, particularly when fabrics are rolled for storage or further processing.

In specific scenarios, such as fabric printing for catalogue photography or sampling purposes, full fixation may not be necessary. However, for production textiles intended for end-use, complete fixation and subsequent washing are essential. These steps not only lock in the dyes for lasting performance but also enhance the overall vibrancy and clarity of the print key selling points in competitive textile markets.

For brands and manufacturers, understanding the link between ink chemistry, fixation processes, and final fabric performance is vital. It ensures production consistency, optimizes quality control, and supports strong market positioning through superior textile aesthetics and durability (Cie, 2015; Daplyn & Lin, 2003).

6.5.1 Reactive Dyes—Cellulosic Fibres

Reactive dyes chemically react with cellulosic fibres through a covalent bond formation between dye molecules containing reactive chlorine groups and the hydroxyl groups in cellulose. This process, typically performed under alkaline and heated conditions, forms a durable bond while releasing hydrochloric acid, which is neutralized by the alkali.

In digital applications, pre-treatment with alkali is necessary. Fixation usually occurs via steaming, followed by washing to remove unfixed dye and enhance colour fastness. Although modern reactive dyes show improved properties, ongoing development continues to focus on enhancing black intensity, lightfastness, and bleach resistance (Kanik & Hauser, 2003; Knoblich et al., 2003).

6.5.2 Disperse Dyes—Polyester

Due to polyester's hydrophobic nature and tightly packed molecular structure, disperse dyes are engineered as finely milled, sparingly water-soluble particles that physically embed into the fibre matrix. Dyeing is conducted under elevated temperature and pressure to swell the polyester, allowing dye molecules to penetrate and bond via Van der Waals and dipole forces.

Disperse dyes are unique in their sublimation capability ideal for transfer printing. Prints are made on paper, then transferred to polyester fabric under heat, allowing gaseous dye molecules to migrate and bond effectively with the fibre.

6.5.3 Acid Dyes—Nylon and Protein Fibres

Acid dyes bond ionically with positively charged amino groups in protein-based fibres (like wool and silk) and nylon. These dyes are water-soluble and form strong ionic and secondary bonds (e.g., hydrogen bonding) under acidic conditions.

Fixation is efficient with minimal residual dye, though care is needed as alkaline conditions can reverse bonding. Crosslinkers can also enable acid dyes to bind with cotton, enhancing versatility (Chen et al., 2002; Yang & Li, 2003).

6.5.4 Pigments—All Fibres

Pigment inks differ from dyes in that they use insoluble colourant particles suspended in a fluid medium. Fixation is achieved through heat-activated resin binders that form a polymer matrix, locking the pigment onto the textile surface.

Although early pigment inks suffered from poor colour intensity and gamut, newer formulations particularly those incorporating advanced polymeric dispersing agents offer improved fastness and vibrancy. They are especially beneficial for printing whites and metallics, areas where traditional dyes fall short.

However, resin binders can alter fabric feel ("handle") and require careful formulation to minimize negative effects. Carrier options include water, oil, and solvent-based systems, each with advantages and limitations depending on viscosity, drying time, and compatibility with substrates (Hees et al., 2002, 2003, 2004).

6.5.5 Oil-Based and Phase-Change Inks

Oil-based inks use carriers such as glycols and hydrocarbons, offering potential in high-speed industrial printing systems. Although promising, challenges remain in achieving dye performance equal to water-based counterparts particularly for sublimation dyes.

Phase-change inks, typically wax-based, transition from liquid to solid during the printing process. While they offer high print fidelity and environmental cleanliness, their adhesion and durability on textiles are insufficient for most commercial applications (Le, 1998).

6.5.6 UV-Curable Inks

UV-curable inks represent a transformative technology in ink chemistry. These inks consist of monomers, oligomers, and photoinitiators, along with pigments and wetting agents. Upon UV exposure, photoinitiators trigger a rapid polymerization reaction, resulting in a solid, durable film that encapsulates the colourants.

Key advantages include:

- Instantaneous drying
- Resistance to chemical and physical wear
- Minimal VOC emissions
- Compatibility with a wide range of substrates.

However, UV-curable inks have drawbacks: higher costs, thicker ink films, potential changes to fabric texture, and health concerns related to skin contact or inhalation of misted particles. Adoption of digital printing technologies in the textile industry remains limited but is experiencing growth, particularly in the domain of household textiles. This expansion is driven by technological advancements such as partial ultraviolet (UV) curing of ink droplets by "freezing" technique, which enhances droplet placement accuracy and expands the achievable colour gamut (Tyler, 2005).

6.6 Printing Process

Achieving vibrant, durable, and commercially viable digital pigment textile prints hinges on developing a "best recipe" those functions as a finely tuned, synergistic system. This isn't merely a collection of ingredients; rather, it's a precisely formulated solution where each component pigment, alkali, thickener, and binder must work in concert. The thesis of successful digital pigment printing, therefore, lies in the meticulous selection of high-quality raw materials, precise formulation, and a thorough understanding of the desired end-use properties, ensuring that every element contributes optimally to the final print's aesthetic and functional integrity.

Guideline for printing

Recommended Recipe for Viscose/Cotton/Linen Fabric

Thickener—5.0–7.0%

Sodium-bi-carbonate—3.0%

Soda Ash—1.0%

Urea—10–15%

Resist salt—1.0–2.0%.

Recommended Recipe for Polyester Fabric

Thickener—5.0–7.0%

Buffer pH 4–4.5—0.75–1.0%

Loop accelerator—0.25%.

Step 1—Pre Treatment:

Apply coating chemical to the fabric before printing. This is done on padding machine. Kindy see that the drying temperature should just dry the fabric for better results.

Step 2—Ready Your Design:

Choose the profile according to the fabric, and RIP the design to get it ready to print.

Step 3—Printing:

Your fabric is placed flat on the sticky printer belt. Coated fabric is then fed into the direct digital printer for printing. The fabric then gets printed.

Step 4—Drying:

Your printed fabric gets dried by the heater attached with the digital printer.

Step 5—Post Treatment:

The printed fabric is than done the steaming, washing, drying, and finishing.

Note—Steaming process (for fixation of dyes)—Temperature and Time varies with the quality of fabric.

For Viscose/Cotton/Linen Fabric—102–105 °C for 10–15 min.

For Polyester Fabric—160–165 °C for 8 min.

6.6.1 Pre and Post Treatment

To ensure the production of high-quality and long-lasting digital textile prints, both pretreatment and post-treatment are indispensable. Pretreatment optimizes the fabric's receptiveness to ink, promoting strong adhesion, vibrant colours, and sharp details. This preparatory stage, often involving chemical applications, is vital for preventing colour bleeding and enhancing print durability. Conversely, post-treatment, such as heat setting or steaming, securely fixes the ink to the fabric fibres, maximizing wash and abrasion resistance. These processes collectively transform unprocessed fabric into a visually appealing and durable printed textile, ultimately leading to customer satisfaction and extending the textile's usable life (Cahill, 1998; Ujiie, 2006).

6.6.1.1 Pretreatment

Pre-treatment is a vital step in digital textile printing, meticulously preparing fabric surfaces to enhance ink reception, print vibrancy, and longevity. This process involves applying a tailored chemical solution to improve ink absorption, prevent bleeding, and ensure colour fixation. The pre-treatment solution contains additives like thickeners (sodium alginate, HEC) for viscosity control, alkali agents (sodium carbonate) for reactive dye bonding, urea for moisture retention, surfactants for uniform wetting, anti-migration agents (polyacrylates) for sharp prints, humectants (glycerol) for moisture, and fixation promoters for dye-fibre bonding.

As Table 6.3 shows the process includes cleaning the fabric, applying the pre-treatment via padding, spraying, or foam application, and then drying. Surfactants improve wetting, thickeners and anti-migration agents control ink flow, and alkali agents and fixation promoters optimize dye bonding. Pre-treatment is customized for specific applications: reactive dyes on cellulose need alkali agents and urea, acid dyes on protein fibres require slightly acidic solutions, disperse dyes on synthetics need sublimation-facilitating treatments, and pigment printing uses binders and thickeners. This customization ensures optimal print quality, emphasizing pre-treatment's crucial role in digital textile printing (Gao et al., 2021; Kan & Yuen, 2012; Tyler, 2005).

6.6.1.2 Post Treatment

Post-treatment is essential in digital textile printing, solidifying print quality by ensuring colour permanence and enhancing fabric properties. It includes fixation (steaming, heat setting, or curing), washing, drying, and finishing (softening, antimicrobial treatment). Fixation bonds colourants to fibres, washing removes excess chemicals, drying eliminates moisture, and finishing enhances fabric characteristics. Post-treatment mechanisms vary fixation activates dyes, washing removes unfixed substances, and finishing alters fabric properties. Processes are tailored: reactive and acid dye printing involve steaming and washing, disperse dye printing uses heat setting and washing, and pigment printing utilizes curing and washing. Challenges include high resource consumption and environmental impact. Innovations focus on eco-friendly methods, digital finishing, and wastewater recycling. Post-treatment is crucial for high-quality prints, with advancements improving efficiency and sustainability (Kan & Yuen, 2012; Tyler, 2005).

Here's a concise list of commonly used chemicals in digital textile printing post-treatment,

Common Post-treatment Chemicals and their Roles:

Washing-off Agents (Detergents/Soaping Agents): Used to remove unfixed dyes, excess thickeners, and other auxiliary chemicals from the fabric surface after printing and fixation. This is critical to prevent colour bleeding onto other parts of the fabric or other garments during washing, and to improve wash and rub fastness. Soaping at high temperatures is often used for reactive dyes.

Table 6.3 Pre-treatment of textile digital printing (Tawiah el al., 2016)

No.	Pretreatment chemical used	Type of colourant	Substrate/fabric
1	Alginate ester/seaweed	Reactive dye ink	Cotton
2	Urea/alkali/lyoprint R. G	Reactive dye ink	Cotton/lyocell
3	Atmospheric pressure plasma (APP)	Reactive dye ink	Cotton
4	Plasma treatment using Radio Frequency O_2 Frequency Plasma (RFOP)	Pigment ink	Polyester
5	Enzymatic treatment	Disperse dye ink	Polyester
6	Chitosan	Reactive dye ink	Cotton
7	Gas plasma	Pigment ink	Cotton
8	Foam	Pigment ink	Silk
9	β-cyclodextrin and citric acid	Pigment ink	Polyester
10	Fluoroacrylate copolymer and Guar Gum	Disperse dye ink	Polyester
11	Nano-modification using atmospheric-pressure air/Ar plasma (AAP)	Pigment ink	Polyester
13	Amino compounds	Pigment ink	Silk
14	Ebecryl 2002 polyurethane acrylate (PUA) based on PEG	Pigment ink	Viscose, nylon, wool, cotton, polyester
15	Cationization with 2,3-epoxypropyltrimethylammonium chloride (EPTAC)	Reactive dye ink	Cotton
16	Cationization with dodecyl trimethylammonium bromide (DTAB), tetra methyl ammonium hydroxide (TMAH)	Reactive dye ink	Linen
17	Cationization with 3-chloro-2 hydroxypropyl trimethylammonium chloride	Reactive dye ink	Cotton
18	Tertiary amine cationic polyacrylamide	Reactive dye ink	Cotton
19	Amino-terminated hyper branched polymer ($HBP-NH_2$)	Reactive dye ink	Cotton
20	Cationization with (3-chloro-2-hydroxylpropyl) trimethyl-ammoniumchloride (CHPTAC)	Reactive dye ink	Cotton
21	Pretreatment with polyamidoamine (PAMAM) dendrimer	Reactive dye ink	Cotton
22	BSF LupreJet HD	Disperse dye ink	Polyester

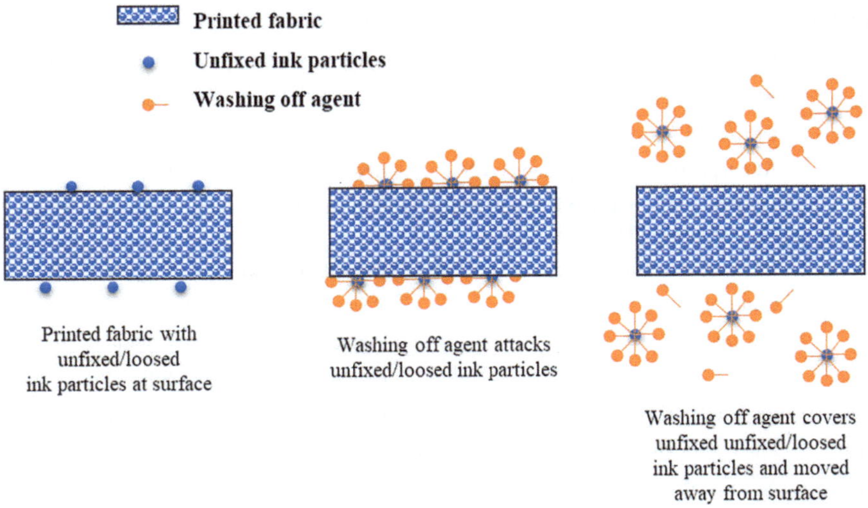

Fig. 6.5 Washing off agent unfixed ink particles interaction

Polyacrylates and Copolymers of Acrylic Acid are among the most common and effective polymeric washing-off agents. These water-soluble polymers feature carboxylic acid functional groups, exemplified by sodium polyacrylate and copolymers of acrylic acid and maleic anhydride. Their primary function stems from their anionic charge, which allows them to efficiently bind to and disperse dye particles within the wash liquor. Additionally, they act as powerful sequestrants, binding to undesirable metal ions in the water or textile.

It works by (shown in Fig. 6.5) promoting dye diffusion from the fibre into the wash bath, where it then forms a protective colloid to effectively disperse and suspend the dye, preventing re-deposition. Furthermore, it also assists to suspend insoluble particles, including those formed by hardness ions (like Ca^{2+}), which ensures a cleaner wash and prevents fabric dullness.

Reducing Agents (e.g., Sodium Hydrosulphite, Thiourea Dioxide): Used for reduction clearing, especially for disperse dyes on polyester, to remove surface dyes and improve wash fastness and rub fastness.

This process typically involves a hot (70–80 °C) aqueous solution containing a strong reducing agent like sodium dithionite ($Na_2S_2O_4$) and an alkali such as sodium hydroxide (NaOH). The chemical action of this solution degrades the dye's chromophoric groups, notably azo (–N = N–) or nitro ($–NO_2$) functionalities, converting them into water-soluble leuco compounds. These colourless, solubilized by-products are then easily rinsed away, preventing any re-deposition of the dye onto the fabric.

Anti redeposition agents:

Their main job is to stop loose or unfixed dye particles from re-sticking to the fabric. This is crucial for keeping printed designs clear and vibrant, preventing "back-staining" or dull colours, especially on multi-coloured or light backgrounds. These agents work by dispersing the detached dye in the wash water, allowing it to be rinsed away completely.

These chemical auxiliaries primarily operate through principles of dispersion and colloidal stabilization. Their fundamental mechanism involves maintaining extraneous dye particles and other process-related impurities in a uniformly suspended state within the aqueous wash liquor. This sustained dispersion is critical for inhibiting the spontaneous re-adsorption or physical deposition of these contaminants onto the firbre surface, a phenomenon that would otherwise compromise colour fastness and introduce undesirable staining or dullness.

Within digital textile printing applications, a diverse array of anti-redeposition agents are strategically employed. Anionic polymers constitute a prominent class due to their highly effective dispersing capabilities. For instance, Sodium Polyacrylate (PAA) functions by adsorbing onto the surface of dye particles, thereby creating an electrostatically charged protective colloidal layer that prevents agglomeration and facilitates their removal. Similarly, Polyvinyl Pyrrolidone (PVP) is recognized for its efficacy, particularly in the washing-off of reactive and direct dyes. Carboxymethyl Cellulose (CMC), another anionic polymer, is widely utilized, especially for its demonstrably high efficacy in preventing redeposition on cellulosic substrates like cotton.

Fixing Agents (e.g., Cationic Fixing Agents, Cross-linking Agents): Improve the wet fastness properties (wash, perspiration, rub fastness) of certain dyes by forming insoluble complexes with the dye or by cross-linking with the firbre.

Dye fixing agents work by either creating an insoluble complex with unfixed dye molecules (for reactive/acid dyes) or by forming a durable, cross-linked polymer network that traps pigment particles (for pigment inks), ultimately anchoring the colourants firmly to the textile fibres and significantly enhancing the overall print quality and fastness.

The choice of fixing agent depends on the dye type: alkalis like sodium carbonate and urea are used for reactive dyes on cellulosic fibres, enabling chemical bonding and better dye penetration. For acid dyes on protein or polyamide fabrics, cationic polymeric compounds form ionic bonds to "lock in" the dye and improve wash fastness. When using pigment inks, binders (often acrylic or polyurethane resins) act as the fixing agent, forming a protective film that mechanically adheres the pigment to the fabric, with crosslinking agents sometimes added for extra durability.

Depth Enhancer:

These are specialty chemical additive designed to enhance the intensity, richness, and depth of colours on printed fabrics. This is particularly important for achieving vibrant, saturated colours, especially dark shades like black, navy, or deep reds, which can sometimes appear dull or less intense without such an agent.

Depth enhancer agents are typically made from materials like acrylic-based resins, which form thin films that help reduce light reflection. Silicone-based emulsions are also common, contributing to colour depth while often giving the fabric a softer feel.

These agents work by optimizing how light interacts with the fabric's surface. By forming a low refractive index film or creating a slightly textured surface, they cut down on superficial light reflection. This allows more light to be absorbed by the dyes within the fibres, making colours appear deeper, richer, and more vibrant.

Softening Agents: Applied after washing to impart a soft and pleasant hand feel to the printed fabric, especially important if the printing process or binder use has resulted in a stiffer hand.

pH Adjusters (Acids/Alkalies): Used to neutralize the fabric after alkaline treatments (e.g., reactive printing) or to adjust the pH for optimal performance of subsequent finishing agents.

Also steaming and curing are crucial, distinct steps that trigger the chemical reactions needed to permanently fix dyes to fabrics.

- **Steaming** (used for reactive and acid dyes) provides the essential heat and moisture for the dyes to react with the firbres.
- **Dry heat** (used for disperse and pigment inks) serves a different purpose: it helps disperse dyes sublimate (turn directly from solid to gas) onto the fabric or cures the binder in pigment inks, making them permanent.

The selection and precise formulation of these chemicals are critical for achieving high-quality digital prints with desired colour vibrancy, sharpness, durability, and desired fabric hand feel, while also considering environmental impact and cost-effectiveness.

6.7 Trouble Shooting

Troubleshooting in digital textile printing isn't merely beneficial; it's absolutely essential. The complex dance between sophisticated machinery, delicate ink chemistries, and diverse fabric types means issues are bound to arise at any stage, from pre-treatment to post-treatment. Without a methodical way to spot and fix these glitches, businesses risk significant material waste, crippling production delays, dissatisfied customers, and ultimately, severe financial hits.

At its core, troubleshooting in digital textile printing is far more than just fixing immediate problems see Table 6.4; it's a continuous drive for process optimization. It requires a systematic approach: identifying symptoms, hypothesizing potential causes, rigorously testing solutions, and carefully documenting everything learned. This ongoing cycle is what builds a printing workflow that is not only robust and reliable but also consistently delivers high-quality results.

Table 6.4 Digital textile printing faults and remedies

Faults of digital textile printing with remedies		
	Causes	Remedies
Colour	Staining In case of dark colour printing, colour staining is appeared in the unprinted area. Fabric should not come into the contact with each other during loop formation in steaming. Coating recipe should be suitable. Clearly remove the excess colour	Fabric should not come into the contact with each other during loop formation in steaming. Coating recipe should be suitable. Clearly remove the excess colour
White spot	Due to uneven singeing	Remove protruding fibres from the fabric surface properly before coating. Chemical tank should be clean
Low colour yield	Uneven singeing and mercerizing. Unsuitable coating chemical	Singeing and mercerizing process should be done properly before coating and also using suitable coating chemical
Shade variation	Most critical problem we face in digital textile printing is shade variation. Printing the design perfectly with correct shade but, after finishing we get different shade. Due to improper chemical coating and finishing parameters shade variation appears	Recipe of chemical coating should be perfect and maintain the viscosity and pH of coating solution. During steaming, washing and finishing, all the parameters should be proper

6.8 Sustainability Aspect

Digital printing significantly reduces water consumption in textile printing for both cotton and polyester compared to traditional methods. For cotton, traditional printing can use 50–60+ liters of water per meter due to extensive dyeing and washing, while digital printing lowers this to 5–20 L per meter. Similarly, while traditional polyester printing involves water for cooling and washing, digital techniques like sublimation or direct disperse printing drastically minimize water usage. Overall, digital printing's precise ink application and fewer processing steps make it a more environmentally friendly option for decorating both natural and synthetic fabrics due to its substantially lower water footprint.

The textile industry is rapidly embracing sustainable digital printing inks as a crucial step toward environmental responsibility. While digital textile printing already offers significant ecological benefits over traditional methods by reducing water, energy, and chemical use, the inks themselves are pivotal in further minimizing the environmental impact.

Eco-friendly inks for digital textile printing are increasingly developed from natural sources such as plant extracts, fruit residues, and microbial pigments, offering a sustainable alternative to petroleum-based inks. Notably, herbal inks derived from materials like bio-indigo, quebracho red, and the flame of the forest have shown promising results when applied to wool and cotton fabrics, exhibiting acceptable rheological properties and moderate fastness levels (Madhukar Thakker et al., 2024). The incorporation of natural indigo inks in digital printing have also confirmed their compatibility with cotton fabrics, achieving vibrant colouration and acceptable durability (Lee et al., 2023). Similarly, bio-pigments from fungi (e.g., Draconin red) and plant-based flavins have been successfully used in inkjet formulations, demonstrating added antibacterial and UV protection functionalities (Iyer et al., 2020; He et al., 2022). Bio-based dye such as gardenia blue in combination with polysaccharide-modified silk fabrics, exhibited enhanced print sharpness and wash fastness (Liang et al., 2024). Collectively, these green formulations underscore the growing potential of bio-based and water-soluble inks in replacing petrochemical-based systems while maintaining print quality and expanding functionality in both fashion and technical textiles.

Challenges remain in matching the colour intensity, lightfastness, and wash resistance of synthetic dyes. However, ongoing technological advancements and refined ink formulations are consistently improving the performance of these inks. Given their numerous benefits, they are poised to become a key innovation for a more sustainable and circular textile industry.

6.9 Conclusion

Digital textile printing has transformed the landscape of textile design and manufacturing by offering unmatched flexibility, precision, and sustainability. At the core of this transformation lies the intricate chemistry of ink formulation and fixation processes which are fundamental factors that determine print quality, durability, and environmental compatibility. The development of specialized inks, tailored for different fibres and printing technologies, requires a sophisticated understanding of physical parameters such as viscosity, surface tension, pH, and conductivity, as well as the functional role of additives like surfactants, dispersants, thickeners, binders, defoamers etc. Also the alignment between ink chemistry and textile pretreatment and post-treatment processes, which ensure optimal adhesion, colour vibrancy, and fastness properties.

As consumer demand shifts toward personalization, short production cycles, and sustainability, digital printing stands out as a technology capable of addressing these evolving needs. Continuous innovation in ink formulation, fabric preparation, and eco-friendly processing methods will further cement digital textile printing's role as the cornerstone of modern, responsible textile manufacturing.

Declaration The authors declare that they have no known competing financial interests or personal relationships that could have appeared to influence the work reported in this chapter. Generative AI and AI-assisted technologies have been used in the writing process to improve the readability and language of the work only. After using this tool, the authors carefully reviewed and revised the content to ensure accuracy and quality. The authors take full responsibility for the final version of this publication.

References

Abd El-Ghaffar, M. A., A. Kantouch, F., M. Mahmoud, Z., Haggag, K., I. Hashem, A., & A. Ramadan, A. (2014). Micro-emulsion co-polymerisation of butyl acrylate with acrylic acid as binder for textile pigment printing. *Pigment & Resin Technology, 43*(2), 84–91.

Abdel-Wakil, W., Fahmy, A., Kamoun, E. A., Hassan, W., Abdelhai, F., & Salama, T. M. (2020). A new route for synthesis of polyurethanevinyl acetate acrylate emulsions as binders for pigment printing of cotton fabrics. *Egyptian Journal of Chemistry, 63*(3), 1063–1073.

Ali, M. K. (2008, January). Chemistry of textile inkjet inks. In *NIP & Digital Fabrication Conference* (Vol. 24, pp. 542–545). Society of Imaging Science and Technology.

Bao, F. B., & Song, Y. (2016). *Advanced nanoscale materials for ink manufacture* (p. 67e85). Printing on Polymers, Elsevier Science.

Cahill, V. (1998, January). Digital printing of textiles, obstacles & overview. In *NIP & Digital Fabrication Conference* (Vol. 14, pp. 257–260). Society of Imaging Science and Technology.

Campbell, J. R. (2008). 11 Digital printing of textiles for improved apparel production JR CAMPBELL, Glasgow School of Art, UK. *Advances in Apparel Production*, 222.

Carden, S. (2016). *Digital textile printing*. Bloomsbury Publishing.

Chen, W., Wang, G., & Bai, Y. (2002). Best for wool fabric printing-digital ink-jet. *Textile Asia, 33*(12), 37–39.

Choi, S., Cho, K. H., Namgoong, J. W., Kim, J. Y., Yoo, E. S., Lee, W., et al. (2019). The synthesis and characterisation of the perylene acid dye inks for digital textile printing. *Dyes and Pigments, 163*, 381–392.

Christie, R. (2014). *Colour chemistry*. Royal society of chemistry.

Cie, C. (2015). *Ink jet textile printing*. Elsevier.

Daplyn, S., & Lin, L. (2003). Evaluation of pigmented ink formulations for jet printing onto textile fabrics. *Pigment & Resin Technology, 32*(5), 307–318.

Dawson, T. L. (2000). Ink-jet printing of textiles under the microscope. *Colouration Technology, 116*(2), 52–59.

Fink, J. (2013). *Hydraulic fracturing chemicals and fluids technology*. Gulf Professional Publishing.

Flick, E. W. (1993). Air products and chemicals, Inc. In: *Industrial surfactants* (pp. 2–25). Oxford: William Andrew Publishing.

Fu, S., Zhang, G., Du, C., & Tian, A. (2011). Preparation of encapsulated disperse dye dispersion for polyester inkjet printing ink. *Journal of Applied Polymer Science, 121*(3), 1616–1621.

Fu, S., Du, C., Zhang, M., Tian, A., & Zhang, X. (2012). Preparation and properties of polymer-encapsulated phthalocyanine blue pigment via emulsion polymerization. *Progress in Organic Coatings, 73*(2–3), 149–154.

Gao, C., Zhang, Y., Mia, S., Xing, T., & Chen, G. (2021). Development of inkjet printing ink based on component solubility parameters and its properties. *Colloids and Surfaces a: Physicochemical and Engineering Aspects, 609*, Article 125676.

Grant, A. (2001, January). Pigmented jet-inks for textile applications. In *NIP & Digital Fabrication Conference* (Vol. 17, pp. 431–434). Society of Imaging Science and Technology.

Gregory, P. (2012). *High-technology applications of organic colourants*. Springer Science & Business Media.

Guan, Y., Tawiah, B., Zhang, L., Du, C., & Fu, S. (2014). Preparation of UV-cured pigment/latex dispersion for textile inkjet printing. *Colloids and Surfaces a: Physicochemical and Engineering Aspects, 462*, 90–98.

Gulrajani, M. L., (Ed.). (2013). *Advances in the dyeing and finishing of technical textiles*. Elsevier.

He, Y., Cao, Y., Hwang, H. J., Gutierrez, S. M. V., Li, S., Chen, H. L., et al. (2021). Inkjet printing and in-situ crystallization of biopigments for eco-friendly and energy-efficient fabric colouration. *International Journal of Precision Engineering and Manufacturing-Green Technology*, 1–13.

Hees, U., Freche, M., Kluge, M., Provost, J., & Weiser, J. (2004). Textile ink jet printing with pigment inks. *Textile Ink Jet Printing-A Review of Ink Jet Printing of Textiles, including ITMA, 2003*, 57–63.

Hees, U., Freche, M., Kluge, M., Provost, J., & Weiser, J. (2002, January). Developments in textile ink jet printing with pigment inks. In *NIP & Digital Fabrication Conference* (Vol. 18, pp. 242–245). Society of Imaging Science and Technology.

Hees, U., Freche, M., Kluge, M., Provost, J., & Weiser, J. (2003, January). Textile ink jet printing with low viscosity pigment inks. In *NIP & Digital Fabrication Conference* (Vol. 19, pp. 626–629). Society of Imaging Science and Technology.

Hou, X., Chen, G., Xing, T., & Wei, Z. (2019). Reactive ink formulated with various alcohols for improved properties and printing quality onto cotton fabrics. *Journal of Engineered Firbres and Fabrics, 14*, 1558925019849242.

Iyer, S. N., Behary, N., Guan, J., Orhan, M., & Nierstrasz, V. (2020). Colour-changing intensified light-emitting multifunctional textiles via digital printing of biobased flavin. *RSC Advances, 10*(69), 42512–42528.

Kaimouz, A. W., Wardman, R. H., & Christie, R. M. (2010). The inkjet printing process for Lyocell and cotton fibres. Part 1: The significance of pre-treatment chemicals and their relationship with colour strength, absorbed dye fixation and ink penetration. *Dyes and Pigments, 84*(1), 79–87.

Kan, C. W., & Yuen, C. W. M. (2012). Digital ink-jet printing on textiles. *Research Journal of Textile and Apparel, 16*(2), 1–24.

Kang, D. J., Gonzaléz-García, L., & Kraus, T. (2022). Soft electronics by inkjet printing metal inks on porous substrates. *Flexible and Printed Electronics, 7*(3), Article 033001.

Kanik, M., & Hauser, P. J. (2003). Ink-jet printing of cationised cotton using reactive inks. *Colouration Technology, 119*(4), 230–234.

Karga, M. E., Kargaki, M. E., Iatrou, H., & Tsitsilianis, C. (2022). PH-responsive, thermo-resistant poly (acrylic acid)-g-poly (boc-L-Lysine) hydrogel with shear-induced injectability. *Gels, 8*(12), 817.

Kasymova, G. A., Badanov, K. I., Badanov, I. K., Makhanbetaliyeva, K. T., Badanova, R. B., Togatayev, T. T., et al. (2024). New dispersions for pigment printing binders in the manufacture of light industry goods. *The Journal of the Textile Institute, 115*(10), 2020–2030.

King, K. M. (2013). Inkjet printing of technical textiles. In *Advances in the dyeing and finishing of technical textiles* (pp. 236–257). Woodhead Publishing.

Knoblich, J., Lucks, G., & Grote, U. (2003). Transfer printing with reactive dyes. *Melliand Textilberichte International Textile Reports, 84*(11/12), 994–994.

Kosolia, C. T., Tsatsaroni, E. G., & Nikolaidis, N. F. (2011). Disperse ink-jet inks: Properties and application to polyester fibre. *Colouration Technology, 127*(6), 357–364.

Kozake, K., Egawa, T., Kunii, S., Kawaguchi, H., Okada, T., Sakata, Y., et al. (2021). Environmental impact assessment of flexible package printing with the "LUNAJET®" aqueous inkjet ink using nanodispersion technology. *Sustainability, 13*(17), 9851.

Kutanaee, H. N. (2011). Perpetration of styrene–butadiene acrylate anti bacterial binders and their application on cotton/polyester printing. *African Journal of Microbiology Research, 5*, 1842–1849.

Le, H. P. (1998). Progress and trends in ink-jet printing technology. *Journal of Imaging Science and Technology, 42*(1), 49–62.

Lee, W., Sung, E., Moon, J., Ahn, I., Yoon, K., Park, Y., & Kim, J. (2023). Performance evaluation of ink and digital textile printing fabric using natural indigo. *Firbres and Polymers, 24*(4), 1309–1319.

Li, X., & Tincher, W. C. (1999). New colourant system for ink-jet printing on textiles. *Textile Chemist & Colourist & American Dyestuff Reporter, 1*(3).

Liang, Y., Wang, N., Li, Q., & Jiang, H. (2024). Surface modification of silk fabric by polysaccharide derivatives towards high-quality printing performance using bio-based gardenia blue ink. *Materials, 17*(14), 3611.

Madhukar Thakker, A., & Sun, D. (2024). Inks for digital printing of textiles. *The Journal of the Textile Institute, 115*(12), 2377–2390.

Magdassi, S., (Ed.). (2009). *The chemistry of inkjet inks*. World Scientific.

Mallinson, P. (1999). Application of surfactants in the textile industry. In *Industrial applications of surfactants IV* (pp. 272–278). Woodhead Publishing.

Meng, Y., Sun, G., Huang, Y., Wang, C., Wang, L., Fan, Q., & Shao, J. (2020). Design and preparation of tertiary amine modified polyurethane acrylate oligomer with co-initiation/polymerization bifunctions for photo-curable inkjet printing of textiles. *Macromolecular Materials and Engineering, 305*(7), 2000178.

Nakagawa, T. O. S. H. I. O., & Shinoda, K. Ō. Z. Ō. (1963). Physiochemical studies in aqueous solutions of nonionic surface active agents. In *Colloidal surfactants. Some physicochemical properties* (pp. 97–178). Academic Press New York.

Ozcan, A., Coskun, S., & Kandirmaz, E. A. (2024). Environmentally friendly water-based inkjet ink production and determining printability. In *Proceedings-The Twelfth International Symposium GRID.*

Ragab, M. M., Hassabo, A. G., & Othman, H. (2021). Synthetic thickener in textile printing. *Journal of Textiles, Colouration and Polymer Science, 18*(1), 65–74.

Robert, T. (2019). *Printing inks from renewable resources.*

Roberts, K., Axberg, C., & Österlund, R. (1977). The effect of spontaneous emulsification of defoamer on foam prevention. *Journal of Colloid and Interface Science, 62*(2), 264–271.

Song, Y., Fang, K., Bukhari, M. N., Ren, Y., Zhang, K., & Tang, Z. (2020). Green and efficient inkjet printing of cotton fabrics using reactive dye@ copolymer nanospheres. *ACS Applied Materials & Interfaces, 12*(40), 45281–45295.

Stankovská, M., Gigac, J., Letko, M., & Opálená, E. (2014). The effect of surface sizing on paper wettability and on properties of inkjet prints. *Wood Research, 59*(1), 67–76.

Tawiah, B. E. N. J. A. M. I. N., Howard, E. K., & Asinyo, B. K. (2016). The chemistry of inkjet inks for digital textile printing-review. *BEST, 4*, 61–78.

Tyler, D. J. (2005). Textile digital printing technologies. *Textile Progress, 37*(4), 1–65.

Ujiie, H., (Ed.). (2006). *Digital printing of textiles.* Woodhead Publishing.

United States Patent and Trademark Office. (2004). Thickener for textile printing which is a polymer dispersion and has improved biodegradability. US6696508B2.

Wang, J. (2003, January). Soft-solvent based inkjet inks for uncoated Vinyl printings. In *NIP & Digital Fabrication Conference* (Vol. 19, pp. 265–267). Society of Imaging Science and Technology.

Xu, M. (2017). Dye sublimation inkjet inks and applications. In *Handbook of industrial inkjet printing: A full system approach* (pp. 179–194).

Yang, Y., & Li, S. (2003). Cotton fabric inkjet printing with acid dyes. *Textile Research Journal, 73*(9), 809–814.

Yang, W., & Zhou, X. (2023). Synthesis of salt-resistant hyperbranched waterborne polyurethane associative thickener and its application in textile printing. *The Journal of the Textile Institute, 114*(1), 32–42.

Yoshiki A. (2005). *Fixing of dye on various textile and inkjet printing.* Technical Brochure, Nippon Kayaku Co.

Zhang, Y., Liu, Z., Cao, Y., Li, R. A., & Jing, Y. (2015). Impact of binder composition on inkjet printing paper. *BioResources, 10*(1), 1462–1476.

Zhou, C. E., Zhang, Q., & Kan, C. W. (2016). Rheological properties of thickener for preparing digital printing ink for nylon carpets. *Firbres and Polymers, 17*, 1475–1479.

Zhou, C. E., Zhang, Q., & Kan, C. W. (2017). Some properties of a thickener for preparing inkjet printing ink for nylon carpet. *Colouration Technology, 133*(2), 116–121.

Chapter 7
The Scientific Formulation, Fluid Dynamics and Application Technology of Ink-Jet in Digital Printing

Abhijit Mukherjee and Bijaya Dasgupta

Abstract Digital printing technology has significantly impacted the textile and graphic industries, empowering high-quality, efficient, customizable printing solutions. At the core of this impact is the ink, which plays a crucial role in analyzing and determining the quality, durability, and sustainability of the final output. The ink's chemistry and formulation also determine the jetting characteristics and printing quality. The raw materials in digital ink, including pigments, binders, solvents, and additives, are specifically designed to work with different substrates and printing techniques, establishing the significance of their roles in these industries. This chapter deals with inks, their technology in the digital printing industry, and how it critically influences prints, including a comparative study of different digital printing methods. This study also examines the classifications and uses of different ink-jet technologies, focusing mainly on ink-jet technology because of its dependability and compatibility with a wide range of inks. The various types of inks used in digital printing are discussed briefly, along with how different ink parameters influence their capability in the digital printing industry. Textile printing has been largely influenced by digital printing technology, with good results in print speed, print head technology, and colour gamut. This chapter also provides an overview of the fluid dynamics of ink-jet printing. These consider the wetting phenomena at the nozzle plate, jet generation, the meniscus dynamics, the flow details of the print head and on reaching the substrate side, the droplet impact, merging, wetting of the substrate, followed by droplet evaporation and drying at the end.

Keywords Digital printing · Printing inks · Ink-jet technology · Fluid dynamics · Print head · Jet generation · Droplet impact · Meniscus dynamics

---

A. Mukherjee (✉) · B. Dasgupta
Department of Fashion Technology, National Institute of Fashion Technology, Kolkata, India
e-mail: abhijit.mukherjee@nift.ac.in; drprofabhijit@gmail.com

A. Athalye and H. Patil (eds.), *Digital Printing*,
https://doi.org/10.1007/978-981-95-2593-5_7

7.1 Introduction

The ink in ink-jet printing is sprayed through a jetting device's nozzle in the form of droplets with a volume range of nano to pico-litres; thus, ink-jet printing is also referred to as droplet-based printing. A computer-aided design (CAD) program is used to pre-define the deposition pattern or the structure's design that will be printed. There are many applications of ink-jet printing, such as in the field of bioengineering (Kumar et al., 2021), in the field of 3D printing of microstructures (Wu et al., 2015), and in the field of textile and flexible electronics (Carey et al., 2017).

If we look into history, the first inkjet-like recording device was invented by William Thomson, later Lord Kelvin, in 1858, using electrostatic forces (Wijshoff, 2008). R.G. Sweet of Stanford University showed in the early 1960s that the ink stream could be divided into droplets of consistent size and spacing up to frequencies of 120 kHz by applying a pressure wave pattern to an orifice. The drops could be reliably and selectively charged as they came out of the continuous jet using a controlled jet breakup mechanism. It is possible to charge electrically conducting ink inductively. When travelling through an electric field, the charged drops were redirected in multiple directions, creating an image on the substrate. The gutter caught the uncharged drops and re-circulated them throughout the system. The Ink-jet Oscillograph was the first device to use this printing technique, known as *Continuous Ink-Jet* (CIJ) printing. In drop-on-demand, it is possible to create drops only when an actuation pulse is provided instead of continuously firing drops (Wijshoff, 2010). Henceforth, *Drop-on-Demand* (DOD) printing and *Continuous Ink-Jet* (CIJ) printing are the two groups that classify ink-jet printing technologies.

There are other methods, too, such as needle-based printing (Phung & Kwon, 2019), thermal ink-jet printing, piezoelectric (Liu et al., 2024) and thermal printing, *Electro-Hydro-Dynamic* (EHD) jet printing (Gao & Zhou, 2018), laser-based printing (Gruene et al., 2011), *Aerosol Jet Printing* (AJP) (Salary et al., 2021), surface acoustic waves (SAW) printing, acoustophoretic printing (Foresti et al., 2018), and drop impact printing (Modak et al., 2020).

Each approach has advantages and disadvantages that make it appropriate for different materials, printing resolutions, and applications. Examples of nozzle-based techniques prone to clogging when expanding their use for printing inks include piezoelectric, thermal, needle-based, and acoustophoretic printing. However, the most advanced and dependable technology is piezoelectric ink-jet printing. The restrictor or throttle is a narrow path that connects the ink chamber to the ink cartridge in the mechanical structure of a *Piezoelectric Ink-jet Printhead* (PIP). A membrane made of a piezoelectric substance positioned between two electrodes sits atop the ink chamber. By delivering an electrical pulse to the piezo-driven membrane above the ink chamber, a nozzle delivers ink droplets to the outer substrate. This membrane raises the fluid velocity at the nozzle by applying pressure to the chamber. Droplets form at the nozzle exit due to the fluid's high velocity. *Micro-electro-mechanical systems* (MEMS) are used to fabricate the nozzle, ink chamber, and piezoelectric membrane on wafers. To create a printhead, these structures are joined using a bonding method based on MEMS. The thermal inkjet printhead comprises a resistor, a chamber, and a nozzle. An electrical pulse applied to the resistor heats the fluid, creating a vapour bubble that forces the fluid through the nozzle to create a droplet or a cluster of droplets. According to the organizational structure, piezoelectric microjet devices can be classified into four different models: bend mode, push mode, squeeze mode, and shear mode. Various factors of ink make it compatible and suitable for digital printing; for instance, it depends on factors like viscosity, surface tension, pH, electricity, etc.

The fluid dynamics issues in ink-jet printing are both on the printhead side—i.e., the piezo acoustic actuation and the intricate flow within the printhead, entrained bubbles, the meniscus dynamics, wetting phenomena at the nozzle plate, and jet formation—and on the receiving substrate side—i.e., droplet impact, merging, the wetting of the substrate, droplet evaporation, and drying. These issues are interconnected, as enhancing the processes and materials related to either the printhead or the receiving substrate is insufficient. Since the same ink is disabled and the frequency, velocity, and size of droplets influence both sides, the process needs to be optimized as a whole. One such instance of conflicting printhead side and receiving substrate, respectively, or better still, the paper side demands, is the ink volatility: It would be better if the ink evaporation is avoided near the nozzle to prevent nozzle clogging, but on the paper side, the ink evaporation should be quick to facilitate useful printing and to prevent deformation of the paper (Lohse, 2022).

7.2 Types of Ink-Jet Printing Technology

There are two types of inkjet technology: continuous inkjet and on-demand inkjet technology, as shown in Fig. 7.1.

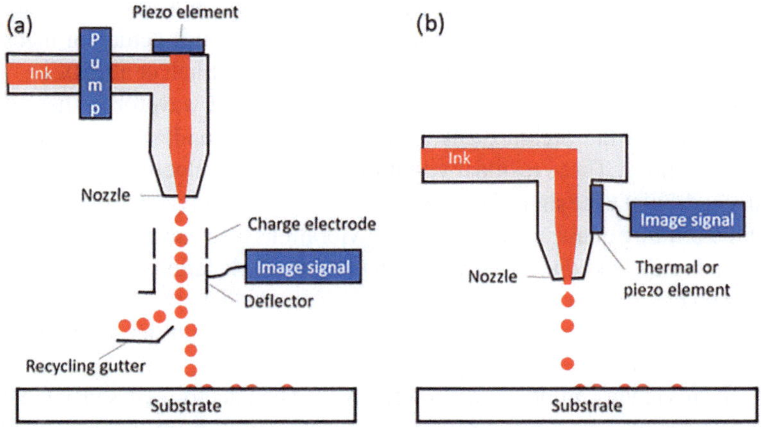

Fig. 7.1 Types of inkjet technology (Adopted from Wiklund et al. (2021))

7.2.1 *Piezoelectric Ink-Jet Printing (PIP)*

Piezoelectric ink-jet printing has two main actuation mechanisms: driving and ejection. Driving entails applying a voltage waveform to the piezoelectric membrane, which deforms and creates pressure, pushing ink into the nozzle for ejection.

7.2.1.1 Actuation Mechanisms and Effect of Voltage Waveform

For the material to be piezoelectric, the actuator device must be poled. Applying a DC voltage across the material is how poling is accomplished. Either the push–pull or pull–push modes can be used to activate the PIP.

- For these modes, the standard voltage waveform, which consists of the rise, dwell, and fall times, is typically utilized.
- The ink chamber expands and produces a negative pressure when the piezoelectric membrane bends upward (z-axis) at the rising edge of the pull–push mode.
- In the meantime, the restrictor and nozzle draw ink toward the ink chamber's centre.
- Throughout the dwell time, the membrane remains bent upward along the z-axis. At the falling edge, it bends toward the ink chamber, which causes the chamber to contract and produce positive pressure.
- The droplet is thus expelled from the nozzle exit.

In contrast to the pull-push mode (Fig. 7.2), the push-pull mode (push mode marked in a blue circle and pull mode marked in a red circle) involves pushing the membrane toward the ink chamber at the rising edge of the voltage waveform and bending it upward in the z-direction at the falling edge. The push-pull actuation mechanism has already been studied, and the change in the voltage waveform's dwell time does not affect the jet pressure or ink droplet velocity. In the pull-push mode, the droplet velocity is high even though the jet pressure and velocity change with dwell time. Researchers have concentrated on the pull-push actuation mechanism because printing applications call for high jet velocities (Shah et al., 2021).

Following the first droplet's ejection using a standard waveform, there are still pressure waves in the pressure chamber. We call these undesired pressure waves residual vibrations. These residual vibrations impact the velocity and volume of the subsequent droplet to be shot. They take several microseconds to decay and ultimately impact the printing quality. These lingering vibrations also cause satellite droplets to form, which may land on a substrate in an undesirable location (Shah et al., 2020).

Based on polarity, the voltage waveform was divided into two categories by pulse type: unipolar (a waveform design method for high-speed ink-jet printing based on self-sensing measurement) and bipolar.

Waveforms having unipolar pulses were used. Piezoelectric self-sensing measurement was used to extract the pressure wave information from the ink chamber by measuring the piezo-current. The unwanted pressure oscillations were suppressed

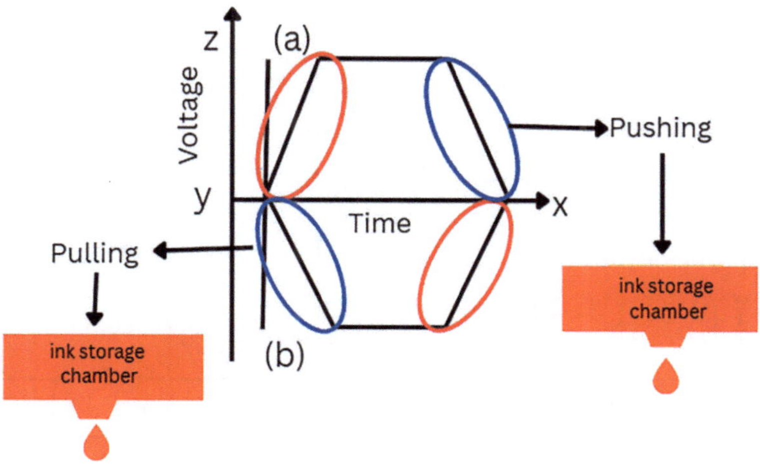

Fig. 7.2 Push–pull mode (self-drawn)

by the second pulse. In order to eliminate the undesirable pressure oscillations, the second waveform magnitude was reduced by half. The unipolar voltage waveform's oscillations have a significant impact on the second waveform's start time. The ideal second pulse start time was 18.8 μs, suppressing unwanted pressure oscillations. If the start time were less or greater than this threshold, the jetting performance would be subpar or worse than the single voltage waveform in the case of a decreased second pulse start time (Kwon & Kim, 2007). The drawback of this waveform type is that it takes longer than the voltage waveform with bipolar pulses to dampen the unwanted pressure oscillations. As a result, the majority of voltage waveforms used in the literature have bipolar pulses. All applications in the ink-jet industry require the effective deposition of various ink types on a substrate. As a result, a mechanism to regulate droplet formation is required. Tuning the voltage waveform is the most straightforward technique. Satellite droplets form after the initial droplet is ejected from the nozzle outlet in a fluid with low viscosity. During the flight to the target location on the substrate, high-velocity satellite droplets can merge to form a main droplet. However, the print quality is diminished by low-velocity satellite droplets because they lack directionality and may land on any random location on the substrate (Shah et al., 2021).

A study (Shah et al., 2021) employed a double waveform with two square pulses (unipolar M-shaped); the actuator successfully ejected the single main droplet without creating satellites. The trajectory of satellite droplets can be enhanced by adjusting the interval of time (t_4) between two pulses. Additionally, they tuned the time interval (t_2) of a unipolar W-shaped waveform to suppress satellite droplets. With ink-jet technology, the printing quality has increased.

Small droplets ejected at higher velocities would be necessary for long-term productivity. By decreasing the pulse amplitude and the nozzle orifice diameter, the droplet size—a dominant parameter—can be reduced. A tiny orifice, however,

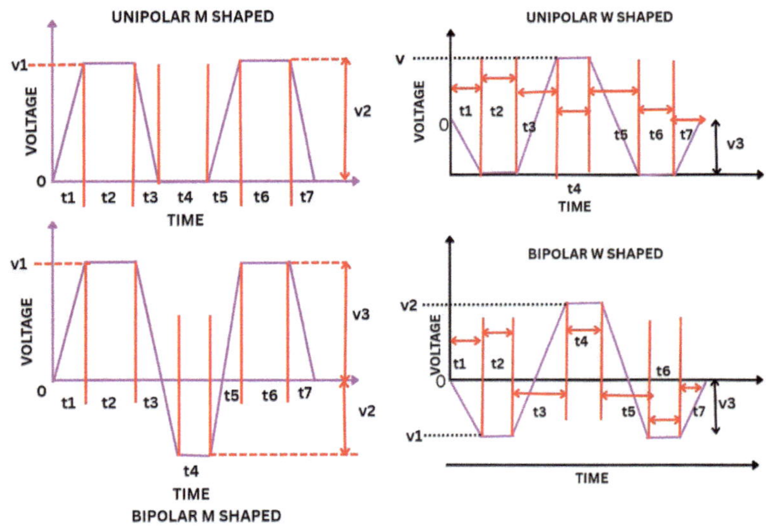

Fig. 7.3 Representation of unipolar M-shaped, unipolar W-shaped, Bipolar M-shaped, bipolar W-shaped waveforms (Adopted and modified from Shah et al. (2021))

can become clogged and necessitate costly manufacturing methods. Additionally, the droplet velocity will decrease as the amplitude decreases. Therefore, it is necessary to investigate alternative techniques for reducing the droplet size. A new waveform consists of successive three square wave voltage pulses, the 1st one being negative, the second being positive, and the third being negative. These pulses have peak amplitudes of $V1 = (-)\ 46V$, $V2 = 56V$, and $V3 = (-)\ 46V$. This leads to a reduction of the droplet volume (A new method for significantly reducing drop radius without reducing nozzle radius in drop-on-demand drop production). The M-shaped and bipolar waveforms can also decrease the droplet volume, but the W-shaped waveform has a more substantial volume reduction effect than the two, as shown in Fig. 7.3 (Shah et al., 2021).

7.2.1.2 Miscellaneous Studies and Recent Research in Piezoelectric Ink-Jet Printheads

The clogging of a nozzle greatly impacts the ink-jet printhead performance in various applications. For example, sieving, bridging, and aggregating particles can stop fluid flow in microfluidic devices (Dressaire & Sauret, 2017). Dirt particles of Silicon (*Si*) or glass stay in the ink channel due to the unusual MEMS (*Micro-Electro-Mechanical Systems*) printhead process. The retention of air bubbles in the tube is caused by cavitation in the bulk ink, resulting from these leftover dirt particles interacting with the oscillating meniscus (Fraters et al., 2019; Shah et al., 2021). Nozzle blockage can be cleared by ink circulation, which can be used to remove the bubbles trapped in the jet and enhance its working reliability (Kwon et al., 2020). Nonaxisymmetric effects (asymmetrical drop formation in different print heads) are another problem with piezoelectric ink-jet printheads, which impair their stability and functionality. Satellite droplets (trailing droplets formed by improper voltage waveform control in the nozzles) have a direct effect on the inkjet printing quality (Wang et al., 2025), as they deviate from the main droplet's trajectory and may not be captured by a nonaxisymmetric droplet. Because satellite droplets force the subsequent droplet to follow the first jetted asymmetrical droplet, high-frequency jets can lessen the effects of nonaxisymmetry. To increase the printhead's stability and performance, it is necessary to investigate the causes and potential fixes. The incorrect alignment of the printhead channel with the nozzle plate, dust particles at the nozzle surface, a wet nozzle plate, a bubble trapped inside the nozzle, and unstable asymmetrical surface modes at the meniscus interface through a Rayleigh–Taylor-like mechanism (The Rayleigh–Taylor-like mechanism, also called Rayleigh–Taylor Instability (*RTI*), is a fundamental macroscopic instability in neutral fluids and plasmas (Zhdankin et al., 2023)) can all result in nonaxisymmetry (Van Der Meulen et al., 2020).

An increased generation of satellite drops results from asymmetrical drop formation. Usually, thin droplet tails break up Rayleigh-Plateau (when the fluid column becomes unstable due to surface tension because of small perturbations) (Barker et al., 2023) to form satellite drops. Because of their disparate velocities, the trailing satellites can readily merge with the following drops in an axisymmetrical scenario. The satellites' likelihood of being caught is decreased in the nonaxisymmetrical scenario, though, if they are steered in a different direction than the main drops (Van Der Meulen et al., 2020).

Important variables influencing print quality and speed are the ink's surface tension and the nozzle's inner wall's wettability (He et al., 2017).

7.2.2 Thermal Ink-Jet Printing (TIJ)

Thermal ink-jet is the most successful method in the market today. Depending on how it is configured, a thermal ink jet can be either a side shooter with an orifice on a side close to the heater or a roof shooter with an orifice on top of the heater. A thermal ink jet consists of an ink chamber with a heater and a nozzle close by Le (1998). Figure 7.4 shows a schematic illustration of commonly used ink-jet technology.

The main issue with the TIJ printhead is that it only lasts a short time due to heater electromigration, bubble cavitation damage, and cracks caused by thermal stress. The lifespan can be extended by making the heater thicker and more shaped (Shah et al., 2021).

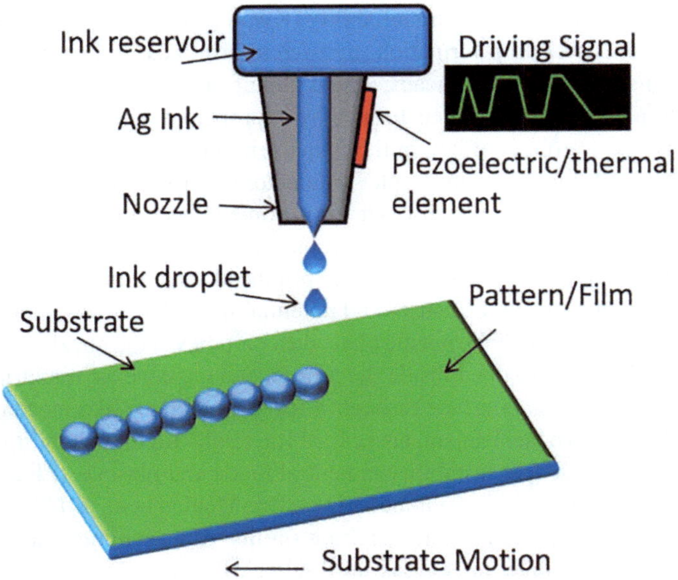

Fig. 7.4 Schematic illustration of inkjet printing technology (Adopted from Uddin et al. (2022))

7.2.3 Electro-Hydro-Dynamic Jet Printing (EHD)

Electro-hydro-dynamic jet (e-jet or EHD jet) printing uses various techniques to create droplets, providing an answer to the low resolution of traditional ink-jet printer systems.

Using pyroelectric effects (detection of temperature changes (Bassani et al., 2005) to deliver materials to surfaces at micro-meter and nanometer length scales without the need for nozzles or electrodes is an interesting idea in electro-hydro-dynamic printing. High-resolution operation is made possible by three features of e-jet printing—(a) the nozzles' inner diameters are smaller than those used in ink-jet printing; (b) the droplets produced by electro-hydro-dynamics may be much smaller than this diameter, and (c) focusing brought on by the distribution of electric field lines reduces lateral variations in the droplet placement (Onses et al., 2015). EHD has various parameters in accordance with the printing process, as represented in Fig. 7.5.

EHD has been successfully used to print a variety of materials on a range of substrates. However, low throughput, particle–substrate interaction, and liquid wetting are still problems. These difficulties can be resolved by altering the electrode's shape, coating the nozzle with a hydrophobic substance, and adjusting the applied voltage and fluid flow rate (Shah et al., 2021).

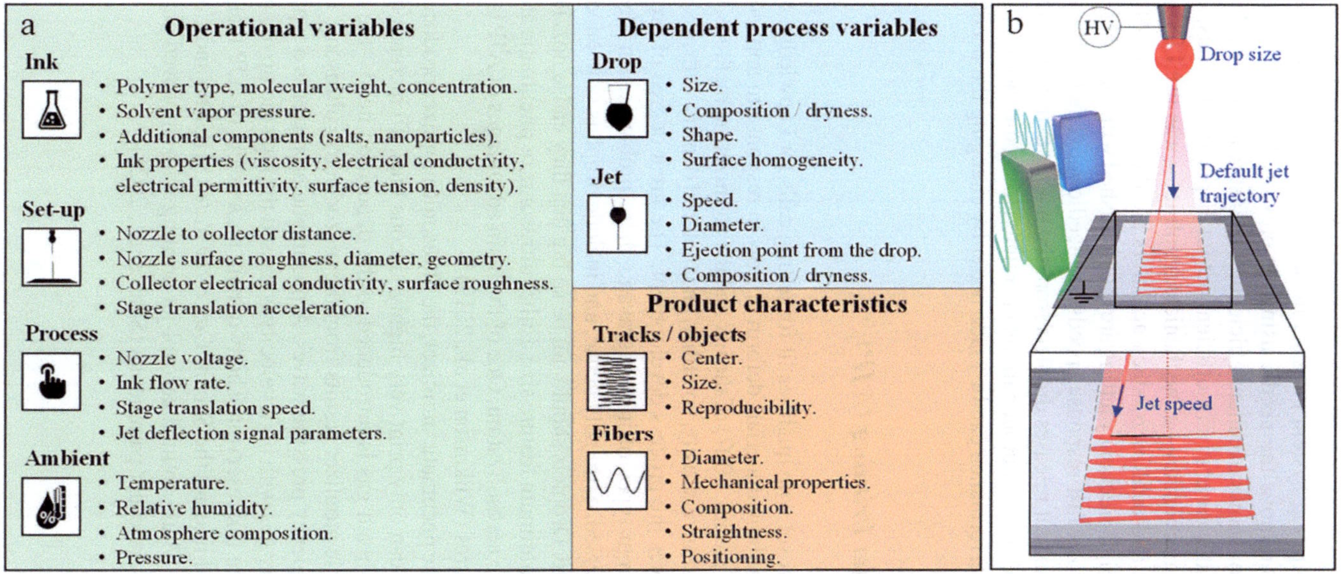

Fig. 7.5 Principle of the EHD jet printing process, along with jet deflection and classification of parameters that make up the operation: **a** parameters involved in the EHD jet printing process; and **b** scheme of an EHD jet printing process with jet deflection electrodes (Adopted from Ramon et al. (2024))

7.2.4 Needle-Based Printing

The needle-based printing system consists mostly of the needle, seat, and nozzle components. One can move the needle using air pressure or a piezoelectric-stack actuator. The actuator shows excellent efficiency and fast reaction. For fluid jetting, the piezo-stack actuator generates insufficient output displacement; thus, a displacement amplification mechanism—a mechanical amplifier—was used for the needle and actuator. Single and dual piezo-stack actuators are the two forms of actuators. High-viscosity and high-frequency jetting are accomplished with the dual piezo-stack actuator. The actuator's fast response speeds the needle movement. The motion of the needle toward the seat releases droplets from the nozzle exit. Fluid flow simulations help one to forecast the output parameters related to needle motion (Shah et al., 2021). The jetting performance can be made better by changing the needle shape.

7.2.5 Aerosol Jet Printing (AJP)

Production of an aerosol with qualities fit for jetting calls for knowledge of both the rheological qualities (how materials deform or flow in response to applied forces or stresses (Figura & Teixeira, 2007) of the ink and the atomization methods. The interaction of surface tension, viscosity, volatility, and material density with aerosolization particularly challenges the process development. The great degree of process variation between different types of inks aggravates these problems. An ideal aerosol would be monodispersed, extremely dense, and include droplets with enough inertia to affect the substrate. Droplets should not be so big that they cause nozzle clogging or compromise the minimum feature size. Either ultrasonic or pneumatic atomization allows one to achieve aerosolization. One of the main advantages of AJP is its ability to produce fine features (Wilkinson et al., 2019).

Compared with the pneumatic atomizer, the ultrasonic atomizer produces uniform aerosol; however, it can only print very highly viscous inks. To increase the performance even more, sheath gas is introduced to the deposition head—the atomizing of ink in an ultrasonic atomizer results from the piezoelectric ultrasonic transducer producing high-frequency pressure waves. A pneumatic atomizer is a type of atomizer whereby a carrier gas with high velocity is passed through the tip of an atomizer nozzle, atomizing the ink reservoir. The two main parameters of AJP that affect the quality of the printed line are the *Sheath Gas Flow Rate* (SHGFR) and *Carrier Gas Flow Rate* (CGFR). Thus, optimal aerosol-jet printing with improved performance requires tuning these relevant parameters (Shah et al., 2021).

7.2.6 *Laser-Assisted Printing*

A laser-assisted printing system consists of a laser beam and a particular type of substrate known as the donor substrate. The top of the donor substrate is covered with an absorbing layer, and the bottom side is covered with an adhesive layer of ink (Brown et al., 2012; Shah et al., 2021). The laser beam's cavitation of the ink layer drives the droplet to the collecting substrate.

Being a nozzle-less method, laser-assisted printing does not suffer from clogging. This technology is used in direct writing and DOD printing. Several materials—such as conductive inks, adhesives, and biomaterials—were printed using this technology (Adhikari et al., 2021).

7.2.7 *Drop Impact Printing*

In a DOD printing method, a sieve replaced the nozzle to produce a single satellite-free (small droplets that form alongside the main droplet during processes like inkjet printing or droplet generation) micrometre-sized droplet. The size of the jetted droplet matched the pore size. Reducing the pore opening helps to drop the droplet size. This method can produce printing for Z values (The Z value is a dimensionless number that characterizes the printability of a fluid in the context of the described printing method), ranging from 3–200. The size can be further enhanced when printing highly viscous inks ($Z < 3$). This method of printed materials is used for both biological and technological uses (Shah et al., 2021).

7.3 Application of Digital Printing

The conventional method for fabricating a semiconductor consists of many steps for the manufacturing of a single device, whereas digital printing projects unmasked lithography [lithography is the transfer of geometric shape patterns in a mask to a thin film of radiation-sensitive material (referred to as resist) on the surface of a semiconductor wafer, and in mask-less lithography, it is done without the mask (https://www.cityu.edu.hk/phy/appkchu/AP6120/5.PDF) and lesser steps, marking its application in printed electronics (Beedasy & Smith, 2020), in Chemiresistive sensors (Hartwig et al., 2018), in supercapacitors (Sajedi-Moghaddam et al., 2020), in pharmaceuticals (Scoutaris et al., 2016) and conductive materials (Shah et al., 2021).

7.3.1 Digital Textile Printing

Nowadays, inkjet printing technology is widely used in textiles and graphic art, replacing conventional printing methods such as rollers or screen printing. Inkjet printing is also widely known as Digital printing. Textile printing has also made its mark following inkjet printing, which is proclaimed to use 30% less water and 45% less electricity than the other conventional methods used.

Ink-jet printing, being a non-contact process, may lead to the spreading of the ink droplets on the substrate surface, affecting the printing quality of the product (Ujiie, 2006). The interaction between the droplets and the textiles in DTP has drawn significant research interest. The image quality of prints on fabrics can be significantly affected by the spreading and coalescing of the droplets (Zhang et al., 2020). A study illustrated the effect of aqueous glycerol droplets on hydrophobic and hydrophilic nylon fabrics. The outcomes illustrated that it penetrated the droplet and created liquid filaments underneath the fabric surface. They also determined that the spherical form of the droplet on the fabric remains unchanged when it comes into contact with permeable substrates of pore size ranging from 100–300 μm during brief time intervals (Zhang et al., 2018). In polyester fabric pores, the penetration of droplets with and without an underlying substrate showed a difference because of the volume loss of the liquid in the spreading ratio of water–glycerol droplets on the substrate with and without the underlying substrate (de Goede et al., 2021). Environment, pre-treatment, and post-treatment influence the colour and its performance. The process has to be done in a controlled environment to enhance the efficiency of the printer. For instance, reactive dyes need a humid environment to form a better bond with the fabric.

Acid and reactive dyes need streaming as post-treatment for fibre bonding. Pre-treatment is usually necessary to enhance the DTP's performance (Shah et al., 2021). Commercially available pre-treatment chemicals were used (DP-300 & DP-302 of Lubrizol Corporation). They exhibited enhanced colour strength and gamut for pigment-type ink grey-scale by pre-treating cotton and polyethylene terephthalate fabrics. They employed chemical padding as the pre-treatment process. Employing acrylic polymers as a pre-treatment chemical enhances the colour of nylon fabrics. Some researchers enhanced the colour quality of a cotton/polyamide fabric substrate by pre-treating it with alpha olefin sulfonate and sodium alginate. In contrast, some also enhanced the colour performance by combining the protease enzyme and sodium alginate for wool fabric ink-jet printing with reactive dyes (Shah et al., 2021).

7.3.2 Display Pixel Printing

Display technology has become ever-present in our lives, and its application is in various fields, such as mobile phones, tablets, desktop monitors, TVs, data projectors, and other display devices (Huang et al., 2020). *Organic Light-Emitting Diodes*

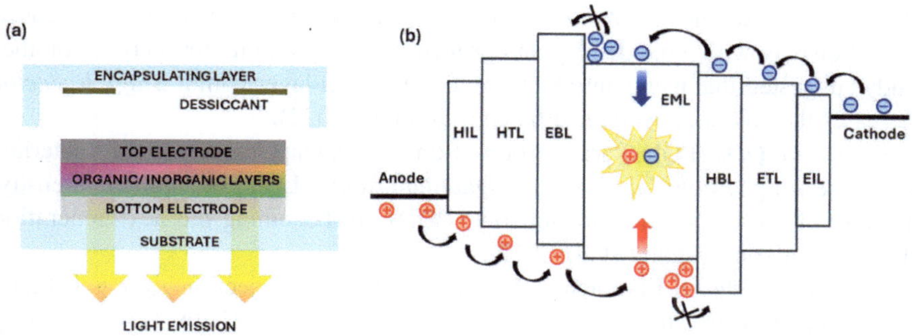

Fig. 7.6 Schematic cross-section of an encapsulated bottom-emitting OLED and Energy diagram of a typical multilayer OLED, highlighting the functions of all possible layers: generally, not all are present as multiple functions are combined in a single layer (Adopted from Liguori et al. (2024))

(OLEDs) are identified as the most potential energy-conserving and environmentally friendly solid-state lighting and full-colour display technology because of the benefits associated with their high efficiency, large viewing angle, high contrast, ultra-thin thickness, flexibility, and transparency. The majority of OLED panels are produced in mass by ultra-high vacuum (<10 − 6 mbar) thermal evaporation processes and consist of a multilayer architecture made up of numerous functional layers, as shown in Fig. 7.6a (Fang et al., 2024). An *Emissive Layer* (EML) between two conductors makes up an OLED. A *Hole Transport Layer* (HTL), *Hole Injection Layer* (HIL), and *Electron Transport Layer* (ETL), and *Electron Injection Layer* (EIL) can be augmented to increase the effectiveness, as shown in Fig. 7.6b (Shah et al., 2021). The advantages of precision patterning quality, efficient material usage, and scalability render ink-jet printing a perfect process for large-area OLED fabrication (Yoon et al., 2021).

An analysis improved the surface's uniformity when a 3-in-blue OLED array was inkjet-printed (Shah et al., 2021). To achieve a uniform film, the emissive layers of the OLED were printed with ink using various solvents. (Interfaces and pattern resolution of inkjet-printed organic light-emitting diodes with a novel hole transport layer). There was an enhancement in the current efficiency and luminance compared to the OLED without the film. Microlenses were printed for pixelated OLEDs to enhance the outcoupling efficiency (Cramer et al., 2018). In a study by multi-nozzle printing, uniform PEDOT: PSS (Poly (3,4-ethylenedioxythiophene) polystyrene sulfonate (https://en.wikipedia.org/wiki/PEDOT:PSS)) film was obtained by combining optimized ink and printing parameters, and they achieved a uniform droplet ejection (Cao et al., 2021).

Quantum dot Light-Emitting Diode (QLED) is another display that can be printed by ink-jet technology. QLEDs have distinct advantages in terms of contrast ratio, response time, and view angle. The blue QLEDs had less device efficiency than the red and green devices (Sun et al., 2019). Highly effective red QLEDs with a QD ink from a mixture of n-tridecane and decalin solvents were printed. Quantum dots also find applications as colour-conversion layers of display technologies (Chen et al.,

2021). To subtract the blue residual emissions, an analysis was fabricated to create a QLED display with a 6.6 display and printed a QD (Quantum dots) layer. Another study suggested that using an inkjet-printed QD pixel layer with a *Bragg Reflector* increases the colour conversion efficiency (Ziming, 2022).

Perovskite QDs (PQDs) are a group of direct-bandgap semiconductor materials that possess high *Photo-Luminescence Quantum Yield* (PLQY) and high PL intensity. Because of these special characteristics, PQDs are becoming the next-generation photonic materials (Sadhu et al., 2022).

The layers of the OLED, such as the emissive layer, hole ejection layer, hole transport layer and electron transport layer, can be deposited with the help of inkjet printing (Sieradzki, 2017).

7.3.3 MEMS (Micro-Electro-Mechanical System) Devices Printing

One of the most adaptive techniques considered in manufacturing MEMS devices (Fig. 7.7) is ink-jet printing because it is environmentally friendly, uses less water, is mask-less, and promotes rapid and multi-material deposition (Shah et al., 2021).

Inkjet printing can deposit various functional materials onto planar and non-planar substrates, such as conductive inks, polymers, and biological solutions. This is useful for printing MEMS devices' sensors, actuators, and interconnects (Ladino-Luna, 2005).

7.3.3.1 Photolithography

One of the most expensive steps in the lithography process is the photoresist coating. However, this coating technique has a disadvantage: photoresist material wastage (Percin et al., 1998). A study showed the deposition of alkanethiol monolayers and DNA oligonucleotides on Au films with the help of ink-jet printing (Bietsch et al., 2004). A study used the EHD technique for printing photoresist lines; the line width was controlled by the voltage supplied to the substrate and nozzle (Qu et al., 2005).

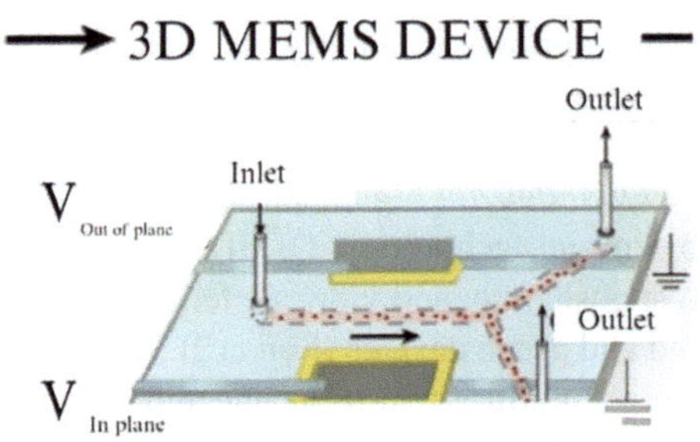

Fig. 7.7 3D Printed MEMS device (Adopted and modified from Blachowicz and Ehrmann (2020))

Micro-lenses and micro-lens arrays play a vital role in various applications, including optical communications, optical storage devices, wavefront sensing, and biomedical instruments. Ink-jet printing is widely used in micro-lens and micro-lens array fabrication. Compared with photolithography, ink-jet printing can deposit various materials at precise locations on any substrate for the fabrication of microlenses (Danzebrin & Aegerter, 2001; Jacot-Descombes et al., 2015).

Inkjet printing does not need a photomask to do direct patterning of functional materials (like photoresists) (Pimpin & Srituravanich, 2005).

7.3.3.2 Etching

An interesting property of ink-jet printing is polymer etching. Microstructures with a concave or convex shape can also be prepared on top of the polymer surface (Shah et al., 2021). Microstructures such as microgrooves, microwells, and hexagonal holes can be well fabricated using inkjet etching (Lee & Park, 2017). The dimensions of these microstructures can be controlled by adjusting the jetted droplet volume, altering the polymer–solvent interaction, substrate temperature, and processing criteria.

7.3.3.3 Deposition

Direct material deposition is the most prevalent use of ink-jet printing in manufacturing MEMS devices. Many materials can be deposited, such as conductive, insulator, sacrificial, and piezoelectric materials. Piezoelectric materials are critical in producing MEMS sensors, actuators, energy storage devices, transformers, and transducers. Several studies have reported the direct deposition of piezoelectric materials using ink-jet printing (Lau & Shrestha, 2017).

Prior to the availability of ink-jet printing, the deposition of functional soluble materials was performed using screen printing and spin coating. Ink-jet printing has since been applied to write electrical interconnects with a wide range of conductive inks onto different flexible and rigid substrates. Direct write technology applications using this method include *Printed Circuit Board* (PCB) production, *Radio Frequency Identification* (RFID) tags, solar cells, batteries, displays and much more (Cummin & Desmulliez, 2012). The entire electrostatic drive motor was fabricated using silver and gold inks by ink-jet printing. The insulator was printed using ink-jet printing technology to isolate the electrode wires of silver and gold electrically (Fuller et al., 2002).

7.3.3.4 Wearable, Flexible, and Stretchable Devices (WFS)

Wearable, flexible and stretchable devices have gained serious importance in the field of human health monitoring and smart devices (Gu et al., 2019). These wearable

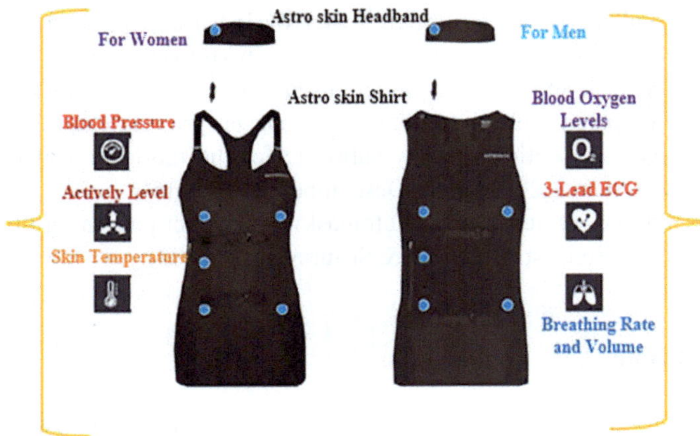

Fig. 7.8 Flexible biosensor (Adopted from Ali et al. (2025))

electronics are integrated into the skin surfaces through the attachment of fibres. There is enormous potential for the fabrication of WFS through ink-jet printing technology, and some of them are low cost, excellent resolution, fewer steps, mask-less working and accurate deposition of small droplets on the surface of the substrate (Yan et al., 2020). Silver nanowires and nanoparticles were used as the ink for ink-jet printing technology in flexible and stretchable electronics (Wang et al., 2022). Inkjet printing has allowed the dielectric, conductor and semiconductor inks to be printed on fabrics and skin. Ink-jet printing can also be applied to print respiratory rate wearable and stretchable sensors, piezoresistive devices, piezoelectric devices, and bio-impedance sensors for electrical impedance tomography imaging. Electronic circuits for WFS devices were also printed using ink-jet printing (Shah et al., 2021). An example of flexible biosensors is shown in Fig. 7.8.

7.4 Influence on Digital Printing Technology

The potential of Digital printing today is implacable. With the enhancement of technologies, products that earlier would not have been produced can now be easily made (Norberg et al., 2001). The widespread adoption of digital textile printing relies on the capability of the technology to enhance production speed and reduce paint costs. Technological advancements, paint and other consumable innovations, and improvements in printheads and machine mechanisms are all important for digital printing (Koseoglu, 2019). Today, the most difficult aspect of mass textile printing is design. Digital printing gives better colour definition with less design time, which cannot be achieved by conventional printing. Moreover, it can be produced in smaller quantities for orders and at a lower investment cost. Lower capital, lower material, lower fabrics and lesser ink reduce the cost. Environmentalists, consumers, and producers are witnessing a significant reduction in electricity consumption and water use in the digital printing of textiles. It can be asserted that more than 30% savings in electricity expenses and nearly 95% in water use are achieved. Another benefit is that the inventory costs are lowered, as the customer's low orders are easily fulfilled (Hu, 2016).

Compared with other printing technologies, digital printing technology has slowly shifted traditional plate-making technology into a digital format, which not only skilfully simplifies the printing process but also enhances printing efficiency (Cheng, 2020).

The finer points to be considered in the case of digital printing are speed to market, design flexibility and new business models. Technological advancements will also significantly reduce the cost in the coming years. There is no colour registration issue, and the dimensional accuracy is very good. The design improvements/revisions are quick, easy, and less costly. Many designs and more colour shades can be achieved in a short span of time. The printing of minute details, the very good softness of half-tone designs and the limitless colour numbers are also some of the major advantages of digital printing. More luxurious and beautiful patterned appearances and home textile designs can be achieved by digital printing (Koseoglu, 2019).

7.5 Environmental Aspects of Digital Printing

In conventional printing, the paste formulation includes dyes (reactive, acid, disperse), thickeners (like sodium alginate, urea (for dye penetration), alkali (sodium carbonate for fixation), and other auxiliaries like a wetting agent and oxidizing/reducing agents. Excess dye, urea, and alkali are washed off during post-treatment (steaming and washing). Urea decomposes to ammonia, a greenhouse gas that increases the carbon footprint. Large amounts of water and energy are required for fixation and washing. In digital printing, the paste formulation comprises pigment or dye-based colourants. Water is used as the primary solvent, along with small amounts of surfactants, salts, and binders (especially for pigments).

There is the usage of no urea, no alkali, and no thickeners (because high-viscosity chemicals can clog print heads) (Tkalec et al., 2022).

Modern ink developments from an ecological perspective have put much consideration into lessening environmental outrage. A principal objective toward this has been to eliminate all forms of waste and emissions and lower production costs in terms of energy. One main upgrade has been the phase out of those inks that emit volatile organic compounds (VOCs), causing air pollution and health threats. Herein, water-based inks became the top choice due to their advantages: cheaper, reduced atmospheric emissions, fire hazards, odour during printing, and facilitating equipment cleaning. Water-based inks, especially pigment-based inkjet inks, have been gaining preference in digital textile printing due to their environmentally friendly formulation and outstanding durability. The inks are carefully designed (stoichiometrically) according to the intended physical and chemical requirements of the various textile surfaces. The inks used for thermal and piezoelectric inkjet systems presently contain little or no VOCs, making the water-based inks a sustainable and healthier alternative to the manufacturers and nature (Tkalec et al., 2022).

Deinkability (the ability of the paper to get the ink removed during the recycling process) is the main environmental impact of digital printing. The biggest drawbacks are paper consumption and, in the use phase, electricity consumption, which are major contributors to the life cycle energy and greenhouse gas emissions of office imaging equipment. Companies using mechanical printing methods consume 500–1200 kWh/tonne of printed products, while the digital EP printing facility consumes 2900 kWh/tonne (Viluksela et al., 2010).

Digital printing produces a lower carbon footprint than traditional screen printing, reducing CO_2e emissions by 2 kg per kg of fabric.

The test drive in screen printing, which is the initial stage, has high energy consumption, but in digital printing, it is negligible. In the after-wash, high-screen printing requires hot water (energy usage), which is negligible or absent for digital printing.

- In comparison to digital printing and dyeing, the carbon footprint for digital printing may be less by one kilogram per fabric.
- For heavier (higher GSM) fabrics, savings can reach ~2.6 kg CO_2e (Guo et al., 2024).

7.6 Fluid Dynamics and Acoustics in the DOD Ink-Jet Printhead

Based on the piezoelectric devices, the most controllable ink-jet printheads are made. Figure 7.9 shows a piezoelectric ink-jet with the actuator. A pressure wave is created in the ink, causing the ink-filled channel to be deformed when a voltage pulse is applied to the piezoelectric element (Antohe & Wallace, 2001). The waveform energy is moved towards the nozzle with the help of the fluid acoustics (Lohse, 2022; Tijdeman, 1975).

Figure 7.10 shows a typical piezo ink-jet channel, which shows the entire process. A commercial printhead consists of thousands of such channels. A thin-layer deposition technique makes up the piezo actuator (Funakubo et al., 2012). The main potential of piezo acoustic devices is that printing precision is good, with small drops of ink, a fast printing rate, and lower production costs. The relationship between the amplitude and frequency of the electrical signal at the piezo actuator and the pressure amplitude, and consequently the droplet volume and velocity, is extremely nonlinear, and resonances are of crucial importance (Wijshoff, 2010). One of the additional benefits of piezoacoustic ink-jet printing is that by varying the actuation amplitude, waveform, and frequency, the pressure amplitude and, consequently, the droplet volume and velocity can be modulated. Generally, each channel and, thus, each nozzle can be controlled independently. One problem that has to be addressed, however, is the

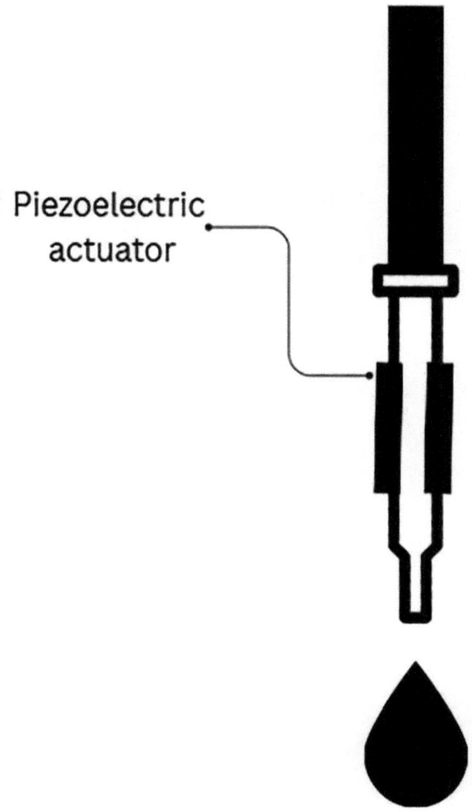

Fig. 7.9 Piezoelectric inkjet with actuator (self-drawn)

so-called cross-talk between adjacent channels, i.e., pressure fluctuations in one ink channel are also partially experienced in the adjacent ink channel, considering that the neighbouring channels should be as close to each other as possible in order to have a compact printhead (Lohse, 2022; Wijshoff, 2010). A piezoelectric element can be used as an acoustic pressure sensor between the pulses, which allows the acoustics in each nozzle to be observed in detail (the process is called *PAINT*) (Wijshoff, 2008) as well as an actuator, which makes it highly advantageous (de Jong et al., 2006a, b; Lohse, 2022).

7.7 Fluid Dynamics in Drop Ejection

The process of drop ejection is controlled by the complex fluid dynamics described by the Navier–Strokes equations. However, factors like inertia, surface tension, and the changing shape of the liquid surface are too complicated to solve using traditional methods. Hence, researchers use a simulation tool named FLOW 3D, which uses a technique called volume of fluid to see how the liquid behaves. The focus of

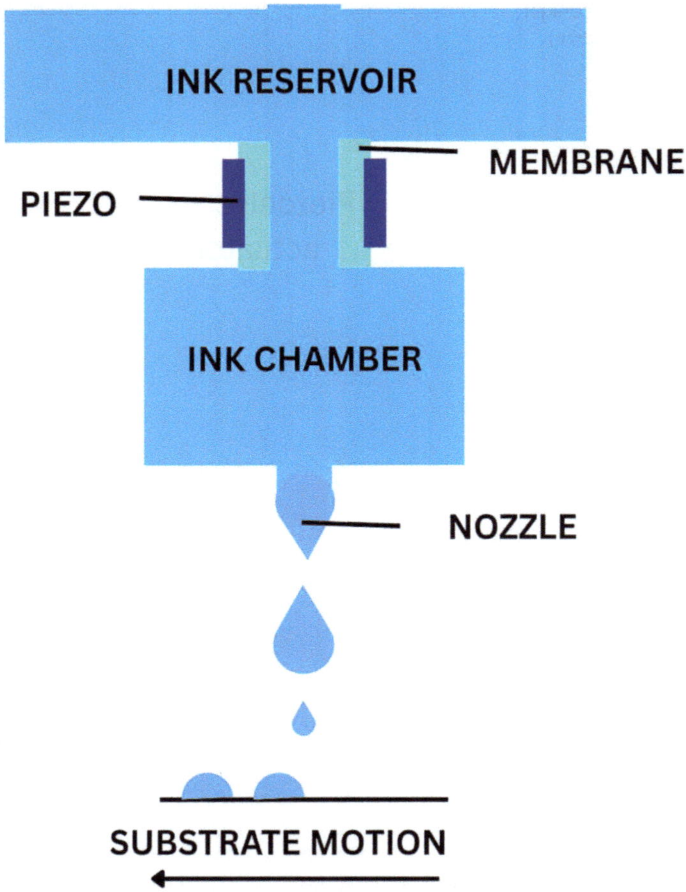

Fig. 7.10 Piezo inkjet channel (Adopted from Lohse (2022))

simulation is to understand practical aspects like drop volume, speed, and shape (Feng, 2002).

7.7.1 *Entrained Bubbles in the Ink Channel*

Using the **PAINT** (*Pressure and Inkjet Nozzle Transient*) technique helps to establish a correlation between changes in the pressure response of the ink channel and deflections of the jetting features, i.e., a decrease in the velocity of droplet jetting. Under some circumstances, such deflections can form following tens of millions of undisturbed cycles. An air bubble entrained into the ink channel was attributed to both an altered acoustical response and a decrease in the droplet jetting velocity. With such a bubble present, the acoustic pressure accumulating near the nozzle is not entirely released, not only during the droplet jetting but also during the compression of the entrained bubble (de Jong et al., 2006a, b). Often, the entrained bubbles grow due to rectified diffusion, but sometimes, after a few cycles, they might be flushed out with the drop (Brenner et al., 2002). For high pressure within the ink channel, the entrained bubble gets compressed and sheds gas to the surrounding ink, but during the low-pressure stage, when the bubble is at its largest size, it may gain gas from the outside since the ink has some dissolved gas (Brenner et al., 2002). Under strong periodic driving, the growth can succeed. The developing entrained bubble increasingly opposes the pressure accumulation in the channel and decreasingly higher droplet velocities during the actuation pulse until such a point as a complete collapse of the jetting process exists (Lohse, 2018). It, therefore, also makes it possible to overcome the technological issue of the bothersome bubbles in the ink channel: When a bubble is sensed, the acoustic driving is shut off.

Meanwhile, other nozzles continue jetting. The bubble can then completely break up, enabling a reactivation of the nozzle. Clearly, it is far preferable to prevent the entrainment of bubbles in the first place (Wijshoff, 2010). Studies directly observed the entrained bubble by inserting a glass nozzle plate between the ink channel and the standard metal nozzle plate. With ultrahigh-speed imaging, they could expose the dynamics of the entrained bubbles, both in position and size. A model was developed to reduce the bubble size from the sensing signal and thereafter correlate it with the direct optical visualization of the bubble (van der Bos et al., 2011).

Subsequently, elucidating further the bubble nucleation, translation, and expansion with an experimental silicon-based printhead with a glass nozzle plate, employing high-speed imaging triggered by variation in the ink channel acoustics (i.e., the PAINT signal).

An illustration of the bubble nucleation process within the ink channel is depicted in Fig. 7.11. It can easily be observed that the impurity in the ink initiates the bubble nucleation on its interaction with the oscillating meniscus. The onset of cavitation on a dirt particle during the rarefaction pressure wave was characterized as a second potential mechanism for bubble formation. Data regarding the extent to which the entrained bubble minimizes acoustic pressure accumulation was also received using the entrained bubble as a pressure sensor (Fraters et al., 2019; Lohse, 2022).

7.7.2 Flow at the Edge of the Nozzle and on the Nozzle Plate

A film of ink at the nozzle plate with a thickness of about 10 nm has a great impact, too (De Jong et al., 2007). The nucleophilic character of the nozzle plate facilitates its occurrence. In this film, a flow can form and carry dirt particles to the meniscus, which can then deform it, with possible bubble entrainment. This nozzle plate flow can have several possible causes. One is merely an air-induced flow as a result of the jetting of the ink. The airflow may pull some of the ink in the film towards the nozzle. A concentration gradient within the film can be another source for such flow since it causes a Marangoni flow (a flow driven by close differences in multiracial

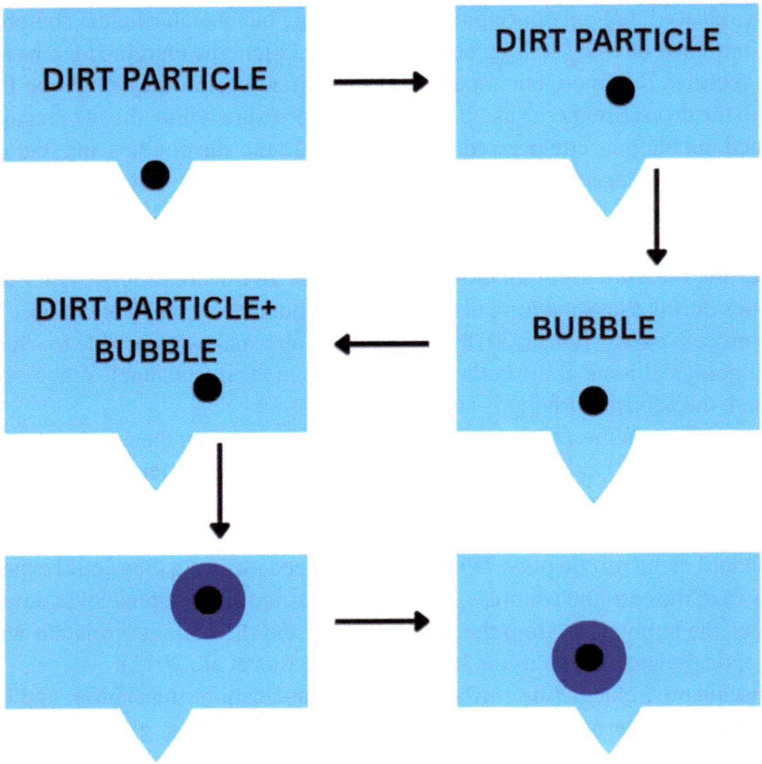

Fig. 7.11 Bubble entrainment in the inkjet channel (self-drawn)

tension caused by temperature or concentration gradients (Schwarzenberger et al., 2014).

The selective evaporation of one component of the ink causes concentration gradients in the ink film, having a thickness of approximately 10 nm (not to scale) on the nozzle plate, resulting in Marangoni flows. In this case, the high surface tension of well-mixed ink within the nozzle results in a Marangoni force on the ink–air interface, drawing liquid from ink film areas or drops resting on the film already exposed to evaporate for a period of time and exhibit lower surface tension (Lohse, 2022).

Such gradients may form between adjacent nozzles that eject different inks or because of the selective evaporation of one or more ink components. The latter situation is shown in Fig. 7.12. There, the ink within the nozzle is properly mixed, owing to the acoustic field-induced flow, and thus possesses the material properties of the initial ink. More distant from the nozzle in the ink film on the nozzle plate or in droplets already resting on the nozzle plate for extended periods, some components of the ink have selectively evaporated. This creates a concentration gradient and, therefore, a Marangoni flow along the film–air interface (Beulen et al., 2007;

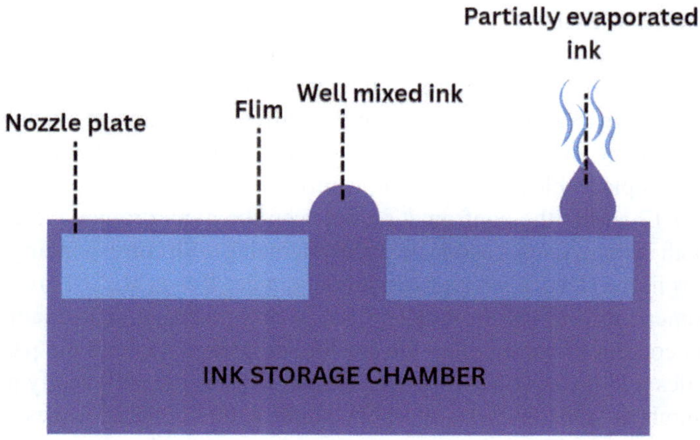

Fig. 7.12 Evaporation of the ink components (Adopted and modified from Lohse (2022))

Lohse, 2022). Thereafter, when the flow is moved towards the nozzle, it leads to the movement of the dirt particles into the nozzle.

The selective evaporation of a single component of a multicomponent ink can lead to other issues, such as an unused nozzle for a few printing cycles. The specific ink in that nozzle is not required. During that period—e.g., several seconds—one or several of the components of the ink can selectively evaporate out of the nozzle. This alters the material properties of the residual ink, such as the surface tension and the viscosity, and thus, the pulse strength required for jetting once the nozzle is triggered again—a significant source of inaccuracy. For instance, when the residual liquid at the meniscus possesses a greater surface tension or viscosity than the well-mixed ink before (partial) evaporation of one or more components, a greater pressure pulse is required to extrude the initial droplets after having suspended the nozzle, which is obviously an adverse effect on the reproducibility of the jetting process (Lohse, 2022).

7.7.3 The Jetting Process, Pinch-Off, and Satellite Formation

Jetting is achieved when the droplet unlinks itself from the ink filament, and this process of droplet detachment is called end pinching. In DOD ink-jet printing, end-pinching is generally the preferred mechanism of droplet formation, but secondary mechanisms include droplet formation by an instability of the filament (the tail of the droplet). This may happen when viscosity is too low, i.e., too low *Oh* (The Ohnesorge number is a dimensionless constant that describes the tendency for a drop to either stay together or fly apart by comparing viscous forces with inertial and surface tension forces (https://www.stevenabbott.co.uk/practical-coatings/ohnesorge.php)) and progressively results in satellite droplets forming. Satellite droplets may be formed when the Oh (Ohnesorge number) is slightly greater, and the threshold relies on the pulse shape (Break-up of diminutive Rayleigh jets). There can be a hierarchy of the sizes and structures of the satellite droplets (A cascade of structures in a drop falling from a faucet). There are four different hierarchies, namely primary, secondary, tertiary and quaternary tails of droplets. However, the tails in the higher orders are unreplaceable and, therefore, controlled by noise.

Figure 7.13 shows the confrontation between the main droplet and the satellite droplet, both in experiments and in the nicely agreeing with corresponding numerical simulations in the slender-jet approximation (van der Bos et al., 2014).

The numerical work (McIlroy & Harlen, 2019) provides certain examples, identifying the conditions for merging the satellite droplets back with the parent drops. These tactics will break down after the satellite droplets are sufficiently minute—*St* (Stokes number) sufficiently small to encounter lateral air drag forces and deflect from the jetting axis. Then, they are a source of contamination.

A last feature of the droplet's journey between the nozzle and the substrate is related to the droplet's shape oscillation: Because elongation forces are applied to the droplet at pinch-off, after pinch-off, the free droplets experience shape oscillations whose frequency (i.e., Rayleigh frequency) (the Rayleigh frequency is the minimum frequency that a finite duration time window can resolve) (https://en.wikipedia.org/wiki/Short-time_Fourier_transform#:~:text=As%20the%20Nyquist%20frequency%20is,signal%20of%20finite%20record%2Dlength) is determined

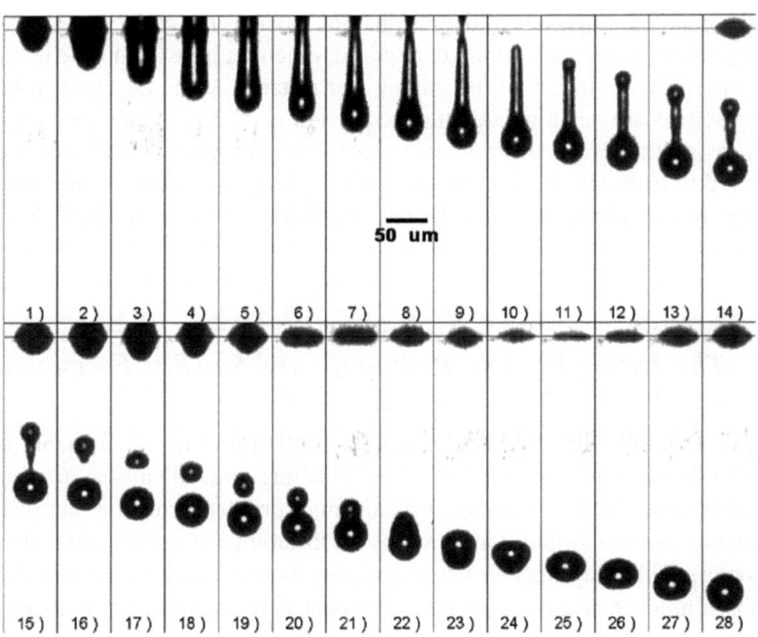

Fig. 7.13 The ink ejection, satellite formation, and recollection process (Adopted from Zhang et al. (2022))

by the surface tension and whose damping is determined by the liquid viscosity or, in dimensionless form, the *Capillary Number* (Ca) (Ca is a dimensionless quantity representing the relative effect of viscous drag forces versus surface tension forces acting across an interface between a liquid and a gas, or between two immiscible liquids (https://en.wikipedia.org/wiki/Capillary_number)). Such shape oscillations have been observed and numerically simulated. As this latter paper demonstrates, the deformation significantly and non-monotonically depends on capillary number (Ca), having a maximum near Ca = 0.7–0.8. Evidently, whether the droplet is elongated along the axis or compressed at the impact moment is critical for the spreading behaviour of the droplet on the receiving substrate (Lohse, 2022).

7.7.4 Drop Impact, Spreading, and Splashing

Weber number (Weber number indicates whether the kinetic or the surface tension energy is dominant) (Moeller & Matyjaszewski, 2012) and *Reynolds number* (The Reynolds number is the ratio of inertial to viscous forces in a fluid and is utilized to describe the laminar or turbulent character of a flow determine the spreading behaviour other than the droplet's shape at the moment of impact. Many different parameters influence a drop impact, such as spreading and splashing, which are strongly impacted by the increasing contact angle between the liquid and substrate (Lohse, 2022), the types and the unevenness of the targeted substrate (Quéré, 2008), the impact angle, the substrate's temperature (Tran et al., 2012), and also the air pressure (Qian & Law, 1997), they are seen in the Strokes number (*St*) (dimensionless number characterising the behaviour of particles suspended in a fluid flow) (https://en.wikipedia.org/wiki/Stokes_number) of the droplet of impact. Figure 7.14 shows the spreading and splashing of the droplet on the substrate.

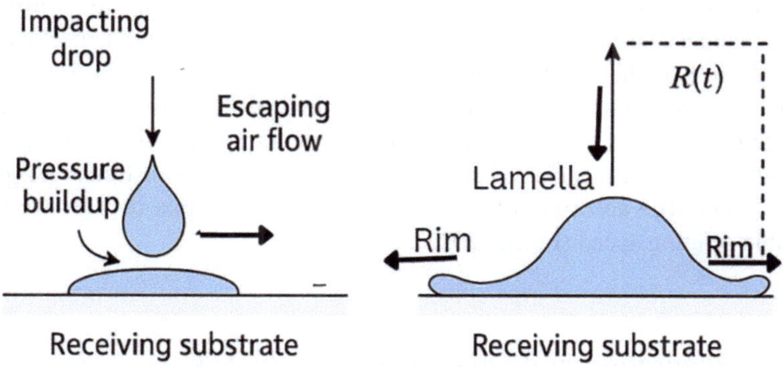

Fig. 7.14 The droplet during impact and spreading on the substrate (Adopted and modified from Lohse (2022))

The contextual air is applicable because, just before the impact, the air between the drop and the receiving substrate is squeezed. Pressure builds up, leading to the formation of a drop called dimple formation (Hicks & Purvis, 2010).

In ink-jet printing, the escaping airflow before impact, the formation of dimples, and the air bubble entrainment under the impacting droplet are undesirable consequences since the airflow improves splashing. The bubble entrainment may alter the optical properties of the drop and minimize the direct contact between the drop and the receiving substrate (Lohse, 2022).

7.7.5 Drop Coalescence, Drop Film Interaction, and Ink Intrusion into Paper

The jetted drops, upon impact, may interact with capillary forces. Our knowledge of the coalescence (or merging) of pure and identical sessile (used to measure the contact angle between solids and liquids (Schuster et al., 2015) drops has become more refined in recent years (Lohse, 2022). In the case of coalescence of various but completely miscible droplets, a tremendously high surface tension gradient across the interface (i.e., a large enough Marangoni number) will result in Marangoni flow. The droplet of low surface tension liquid will be drawn over that with the larger one (Karpitschka & Riegler, 2010). The counteracting force, in general, is of a viscous character. These Marangoni flows may cause undesirable consequences, especially if they carry pigments, thereby influencing the resolution that can be achieved and colour accuracy.

As demonstrated in Fig. 7.15, two drops laid close to one another on such a thin film of viscosity can interact through film deformations. A study demonstrated that the interaction may be attractive or repulsive, depending on the distance between the two drops. The distance at which the interaction switches from attraction to repulsion was found to be a function of film thickness and to increase with time. It showed that the cause of the non-monotonic interactions is the emergence of a visco-capillary wave on the thin film induced by the drops. From the thin-film equation, they determined the scaling law for the spreading of the waves and showed that this determines the range over which the interaction occurs (Hack et al., 2018). Lastly, the ink droplet deposited can also be soaked up by the substrate if it is porous, such as paper. The absorption timescale is one parameter that decides the highest attainable printing speed (Lohse, 2022).

7.7.6 Evaporation and Solidification of the Ink

The last stage of the droplet is the drying or solidification of the ink, which happens through a chemical reaction or cooling process.

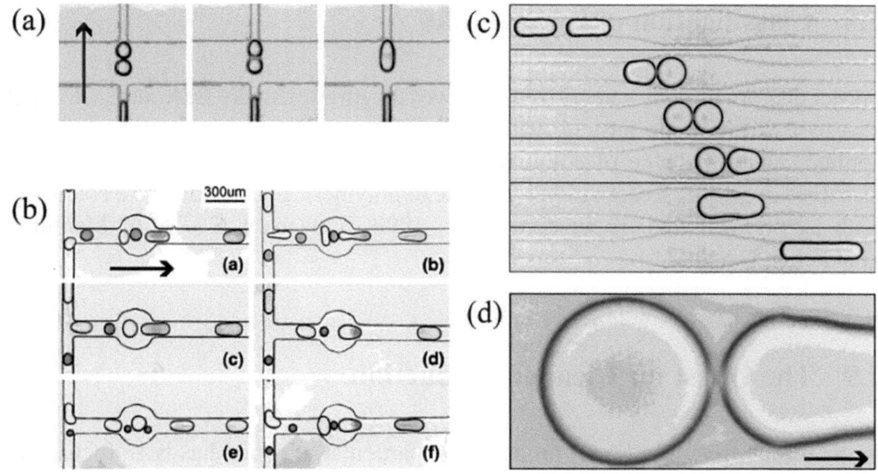

Fig. 7.15 Passive droplet merging achieved by the design of the channel geometry–**a** and **b** indicate the merging of two or more droplets; **c** and **d** illustrate the last moment of droplet merging, known as decompression merging (Adopted from Gu et al. (2011))

Nowadays, the evaporation of a single sessile pure liquid droplet in volumes as small as microliters is reasonably well understood. But much less so is the evaporation of picoliter colloidal droplets, which are the ones that matter in ink-jet printing. While the evaporation of picoliter droplets will be controlled by the vapour diffusion in the air surrounding the droplets, as for microliter droplets, there can be differences because of pinning/de-pinning effects at the surface, since, for picoliter droplets, the contrast between the size scales of usual surface inhomogeneities and those of the droplet itself is much reduced (Lohse, 2022).The behaviour of evaporating binary sessile droplets is also made more complex once surfactants also come into effect. These and the binary droplet structure strongly influence the deposit pattern, as shown with evaporating whisky droplets. A combination of surfactants having an oil-lubricating layer (Kim et al., 2016; Lohse, 2022) might even completely eliminate the coffee stain.

For evaporation from a colloidal droplet on a porous substrate like paper, the residual deposit will be more uniform since the infiltration of the colloidal liquid within the pores may overcome particle motion, providing a chance for the coffee stain effect, i.e., when the combined action of infiltration and evaporation is quicker than the advective and diffusive timescale of particle transport. Lastly, the evaporation of an isolated sessile droplet, of course, is the exception in ink-jet printing applications when numerous droplets are densely laid on the accepting substrate. Hence, their evaporation mechanisms influence one another, typically hindering evaporation due to each other's mutual shielding effects. This also impacts the deposition pattern of each droplet formation. Beyond evaporation and droplet drying, other means of droplet solidification have become very important in ink-jet printing.

7.8 Classification of Ink-Jet Inks

Ink-jet inks are broadly categorized into two categories, namely Base and Colourant. The base is the mode through which the colourant is dissolved or dispersed and applied, while the type of colourant being employed within the medium, hence dissolved dyes, e.g., Reactive dyes, organic or polymeric dyes, or disperse colourants such as pigment and disperse dye (Le, 1998; Tawiah et al., 2016; Madhukar Thakker & Sun, 2024).

7.9 The Quest for Cleaner Production

In the recent drive for cleaner productions, particular attention has been focused on inks that eliminate waste and emissions and reduce energy input during production, and the elimination or reduction of *Volatile Organic Compounds* (VOCs) from inks has gained a tremendous boost (Le, 1998). VOC can reduce emissions by using cost-effective water-based inks that cause less pollution, use fewer solvents, have less potential to fire, emit less print odour, and are easier to wash up on printing equipment. Water-based inks with little or no VOCs are used in piezoelectric and thermal ink-jet printing (Tawiah et al., 2016). These inks have better advantages because fewer bubbles are formed in most digital textile printing thermal ink-jet printers. The water-based ink is simple to wash off print heads and avoids the clogging of nozzles once the ink's rheology is Newtonian. The water-based digital textile ink-jet inks are stoichiometrically (the relationship between the quantities of reactants and products in a chemical reaction or the proportions of elements in a compound) compounded to meet certain substrate-specific needs.

7.10 Recent Advancements in Digital Inkjet Printing

A few of the advancements in digital inkjet printing are discussed below.

7.10.1 AI-Assisted Droplet Control-DMF

The unprecedented advantage of the AI-assisted feedback control system for DMF (digital micro-fluids) is that it can autonomously conduct experimental operations with various reagents, such as protein-rich droplets, on DMF chips without human intervention. Moreover, it can be combined with reinforcement learning to increase the precision and automation of droplet manipulation by recognizing, tracking, and discriminating states, as well as automating droplet control, thereby expanding the capabilities and applications of DMF systems. Certain studies (Guo et al., 2024) showed that digital microfluids (DMF) has become a widely used method for liquid handling due to its distinct capabilities of flexible and discrete fluid control. Specifically, DMF architectures employing electrowetting-on-dielectric (EWOD) mechanisms leverage voltage change to regulate the wettability of fluids on solids, thus enabling the movement of droplets at scales ranging from nL to μL. This method is very convenient because it possesses several features, including small size, programmability, parallelization, and low power consumption (Guo et al., 2024).

7.10.2 Advancement in Printhead

Trident has launched a new product, a 768-jet print head with 256 addressables (the independent firing elements), each with three orifices for testing, making it reliable. Two print heads that work at two different temperatures are used to operate different types of inks that went through a long-term reliability test that jetted over 90 billion drops per orifice. It included the printhead dynamic endurance test, all channels that fired the frequency sweep stress test and the thermal cylindrical printing test. Each print head was seen and characterized for two weeks to examine whether performance changed over time. However, it passed the test (Zhou, 2004).

7.10.3 Closed Loop Systems with Stepper Motor

Stepper motors are usually used in open-loop systems, but using them in a closed-loop controller can greatly improve speed, torque, acceleration, and reliability. Open-loop systems need safety margins to maintain the motor's stability, but closed-loop systems remove those margins or reduce them, making the motor efficient. Closed-loop systems can also control low-cost fused filament fabrication machines (Weiss, 2014).

7.10.4 Theory of Constraints (THC)

The theory of constraints is a management approach developed by Eliyahu M. Goldratt, focusing on identifying and managing bottlenecks or limiting factors that prevent an organization from reaching its full potential. It was initially developed in the 1970s and is an optimized production technology. The core idea is like a chain, depending on the weakest link, where every system has at least one constraint, and that constraint limits the system's overall performance. By improving this, one can increase the overall efficiency. The printing industry faces technical, market, social, and organizational constraints. Technical constraints such as machines, might not be powerful or advanced enough, and there might not be enough space for storage or delays in the production line. Market constraints can be like demand can be unpredictable, or only a few suppliers might be available. Social constraints can be like a lack of workers or poor teamwork, and organizational constraints may be like delays in getting raw materials or weak planning systems (Rogowska, 2024).

7.11 Composition of Ink-Jet Ink

Ink-jet inks are the most important part of the ink-jet printing. The chemistry of the inks determines the quality of the printing, the jetting characteristics of the inks, and the dependability of printing. The composition of digital textile printing differs widely in terms of the type of colourant, the base used to formulate (water, oil, and organic medium), the printing machinery to be utilized, and the curing method. (Tawiah et al., 2016). Inks should have definite properties for the proper functioning of the printhead.

7.11.1 Ink Formulation Components: CIJ Versus DOD CMYK

The comparison of ink formulation components (CIJ and DOD CMYK) is shown in Table 7.1.

From the above content, we get that the composition elements of CIJ and DOD CMYK (Cyan, Magenta, Yellow, and Key) inks are alike in fundamental make-up—both consist of water as the medium, similar proportions of polymer emulsion,

Table 7.1 Ink formulation components: CIJ versus DOD CMYK (Momin, 2008)

Ink component	CIJ CMYK	DOD CMYK
Polymer emulsion	8–14%	8–14%
Water	Balance	Balance
Surfactant	0.05–1%	0.05–1%
Humectants	5–10%	5–10%
Conditioning agent	0.25–0.5%	0.0%
Thickener	0.0%	0.5–5%
Cross-linker	0.5–1%	0.5–1%
Pigment dispersion	2–5%	2.5%
Particle size	1 micron	1micron
Viscosity	3–5cps @ 25 °C	10–14cps @ 25 °C
Surface tension	30–35 dyne/cm	30–35 dyne/cm

surfactants, humectants, and cross-linkers, with the same particle sizes and surface tension. Nevertheless, the crucial differences lie in viscosity and usage of additives: DOD inks are much more viscous (10–14 cPs (centipoise) compared to 3–5 cPs), and thickeners (0.5–5%) are necessary for the stability of jetting, whereas CIJ inks contain conditioning agents (0.25–0.5%) to balance drop formation. These differences stem from the operational conditions each printing technology necessitates—CIJ's requirement of low-viscosity, high-jetting speed inks and DOD's need for more controlled ejection of droplets via accurate nozzle actuation.

7.11.2 Colourants Contents in Digital Textiles Printing Ink-Jet Inks

The main function of the ink is to convert the colourant to the substrate or to simply change the colour of the substrate (Izdebska-Podsiadły, 2015). When the colourant is a dye molecule, it should be present in a concentration lower than the solubility limit, or a little change in storage will lead to precipitation. In these inks, one must identify the dye solubility in the presence of all the components, particularly at low temperatures. The optical properties of dyes are often influenced by slight changes in pH and the existence of electrolytes (in water-based inks), medium polarity, and the existence of surfactants (possible solubilization). To ensure excellent stability and good consistency, a study proposed that the content of dispersed colourants (pigments and disperse dyes) in inks, particularly for textile applications, must not be more than 4% by ink total weight. Others have nevertheless advised that digital textile ink-jet inks, in general, must contain a pigment volume fraction of below 10% of the total weight of the formulation (Pal & Fleming III, 2006).

Effective dispersion of pigments in printing inks is crucial for several reasons. However, the impact of dispersion quality on the rheological performance (the study and measurement of a material's ability to flow or deform under stress) of digital textiles ink-jet printing ink is arguably the most significant factor. Due to the application techniques, flow properties play a significant role in computer textile inks, and this is indeed a challenge that a printing ink must satisfy in order to be considered for potential use (Oyarzún, 2000). Thus, prior to loading pigment inks into a digital textiles ink-jet printer, there is the requirement to stabilize the ink against flocculation; thus, its occurrence kinetically can significantly impact the colour of the printed textiles and lead to nozzle clogging of the printer (Niaounakis, 2014). Conversely, non-graphic ink-jet inks like conductive or ceramic can have pigments at concentrations higher than 50%. A study showed that increased pigment loading in ink-jet inks might lead to poor stabilization, which will, therefore, influence the jetting pattern, print quality and colour luminance, but could then be the case for conductive and ceramic inks since inks for those applications require high viscosity in order to be able to stay stable upon application thermodynamically (Tawiah et al., 2016). Overall, inks that contain dye are more stable than pigment inks, as the ink is thermodynamically stable (all parts are dissolved in one solution). In contrast, in pigment inks, the system is merely kinetically stable. Traditional UV curable inks typically have dye or pigment at below 10% w/w concentration to attain the desired optical density, which can be prepared from a combination of monomeric and oligomeric acrylates or epoxies, which are polymerized with UV light in the presence of a photo-initiator (Jafarifard et al., 2016).

The above details (Table 7.2) speak volumes about how various colourants respond in a distinctive way to each type of fibre, yielding diverse fixation processes and colouristic properties. Pigments are suitable for all fibres but need binder-based surface fixation. Reactive dyes link to the molecules of natural fibres such as cotton, silk, and wool by forming covalent bonds, creating very good wash and rub fastness but poor light fastness. Disperse dyes are polyester-specific, using hydrophobic binding that provides good brightness and very good all-round fastness. Acid dyes attach electrostatically to protein fibres and nylon, creating good brightness and durability. The fixation procedures—ranging from steaming to oven curing—are essential for obtaining the ideal fastness and colour intensity, highlighting the value of choosing the right dye-fibre combination to secure the best textile performance (Tawiah et al., 2016).

7.11.3 Rheology of Ink-Jet Inks

The study of the flow and deformation of matter is called rheology (Oyarzún, 2000). Ink-jet inks should possess Newtonian behaviour, which is guided by examining the shear stress and viscosity of the inks under various shear rates (Jafarifard et al., 2016). Better printing quality, better and speedier processing, environmental acceptance, and cost are the reasons that drive ink technology (Dalwadi et al., 2005). The rheology of ink is an essential parameter since it supports droplet formation through the nozzle in a controlled manner for any given ink-jet technology (Tawiah et al., 2016).

7.11.4 Jetting Behaviour and Droplet Formation

Ink design is mainly influenced by a set of factors like jetting performance, operation reliability, and stability in storage and use. Because of such considerations, inks are typically formulated based on different parameters chosen by the printhead design. According to a study, the general requirements for inks can be said to be purity (to avoid blockage), particle size, viscosity, surface tension, conductivity (the continuous flow of the application), stability, pH Value, foaming/defoaming properties (Tawiah et al., 2016). The tail that comes up with the ink drops during droplet formation is required to be absorbed by the head while falling. Otherwise, many drops may form and settle in various locations on the substrate, which can damage print quality. Still, there can be spatter because the drop falls onto the substrate even though heads and

Table 7.2 Digital textiles ink-jet printing inks, colour fibre-interaction and finishing techniques (Tawiah et al., 2016)

Colourant	Fibre type	Colour fibre-interaction	Colouristic properties	Fixation
Pigment	All fibers	No interaction—complex surface polymer (binder) bonding mechanism	Good washing fastness, Excellent light fastness, good rubbing fastness depending on binder content	Oven curing at 160–180 °C for 30–90 s
Reactive dye	Cotton, silk, wool, linen	Covalent fibre bonding	Bright colours, excellent washing fastness, excellent rubbing fastness, poor light fastness	Steaming for 90–120 °C for 8–30 min, depending on steamer type, washing and drying
Disperse dye	Polyester	Hydrophobic-solid-state mechanism	Excellent light, washing and rubbing fastness, bright colours	Transfer press or "thermosoled" or oven cured depending on the type of disperse ink
Acid dye	Nylon, silk, wool, leather	Electrostatic and hydrogen bonding with fibre	Bright colours, excellent light fastness, good washing and rubbing fastness	Steaming for 20–60 min, depending on steamer type, at 20–120 °C, depending on shade and fibre type, washing and drying (except leathers)

tails coalesce; hence, ink developers need to maintain a balance between controlled dot gains and overspreading. Viscosity, surface tension and drop size are important parameters to ensure this equilibrium. Further, the pressure and the molecular weight of the gas, by which the falling drop plays an essential role, alongside the viscosity (Tawiah et al., 2016).

7.12 Important Properties of Ink-Jet Inks

The important properties of ink-jet inks are discussed below.

7.12.1 Viscosity

Viscosity should be made adjustable in such a way that the flow should be either as high as to avoid starvation at the nozzle or not as high as it flows out in the plate of the nozzle. Viscosity greatly impacts the drop size and drop (Magdassi, 2010). If the drop speed is low due to the deflection of the drop formation on the substrate, it is affected by the air currents, which will lead to poor printing quality. Viscosity modifiers help control quality by modifying the drop formation and break-up process. Since the molecular weights of viscosity modifiers have elasticity, it is difficult for the jet to break up; some examples of viscosity modifiers are Glycerol, Polyethylene Glycol of molecular weight 200 (PEG200) and Carboxymethyl Cellulose (CMC) (Tawiah et al., 2016).

These viscosity modifiers help ink-jet textiles. Printing ink helps maintain constant viscosity, ensuring stable jetting pressure and minimum nozzle clogging (Niaounakis, 2014). The amount of viscosity modifiers used in that particular formulation is subject to the colourant type being utilized, the type of ink-jet printer and nozzle, i.e., DOD or CIJ, and the target fabric material to be printed (Tawiah et al., 2016).

7.12.2 Surface Tension

Surface tension is among the most critical characteristics of ink-jet inks for textile purposes. It has been proposed that ink surface tension is one of the major factors controlling droplet formation and spreading on the substrate upon contact (Magdassi, 2009). Surface tension can be regulated through the use of surfactants and with the proper choice of solvent mixtures. For instance, adding propanol to water will result in a significant reduction in surface tension, from 72.8 dyne/cm to less than 30 dyne/cm, as a function of the concentration of propanol. The surface tension of inks has to be within the range (25–60 dynes/cm) so it will wet capillary tubes, go through the nozzle and shape the droplets properly to provide effective jetting (Tawiah et al., 2016). Surfactants, also referred to as surface-active agents, are substances like detergents that decrease their spreading and wetting character when added to a liquid, thus enhancing their surface tension. Surfactants assist in the dyeing process of textiles; the dye penetrates the fabric evenly (Karsa, 1999). Fluor chemicals, silicones, and hydrocarbons are common surfactants that can reduce the surface tension of any substance and are the most effective wetting agents. Silicones are second-best in effectiveness and cheaper, but some forms of silicone become airborne, creating severe environmental issues as well as contaminating the substrate (Tawiah et al., 2016). While it can be necessary to decrease the surface tension of a coating, it is the opposite for the substrate since the same agent which facilitates the decorating material makes the substrate worthless. Silicone contamination generates the notorious duetting imperfection known as "fish eyes." After surfactant modification, coatings, inks, and paints are typically permanently altered, even upon curing (Tawiah et al., 2016).

7.12.3 Humectants

Humectants are of great importance for optimizing drying characteristics and enabling high-resolution printing (Magdassi, 2010). Humectants like glycols and alcohols function as hygroscopic agents to draw out water from the air during printing or in the idle position of the printer, henceforth avoiding clogging of print heads (Tawiah et al., 2016).

7.12.4 Foaming and Defoamers

The entrapment of the bubbles in the ink is the main problem with ink performance. One solution that is shared in many problems like this is adding a defoamer, a molecule that induces the breakdown of existing foam. The defoamers work by lowering surface tension within a local region to very low levels, thus quickly thinning out these local regions (for example, amyl alcohol) and facilitating the drainage of liquid from the lamellae. A serious issue with ink performance is the occurrence of bubbles in the ink. Foaming can be commonly noticed in inks that have surfactants and polymers. One solution to this issue is adding a defoamer, a molecule that results in the breakdown of foam that has already been formed. The defoamers function by (a) lowering surface tension in a local region to very low levels, leading to these local regions being thinned quickly (e.g., amyl alcohol); (b) facilitating drainage of liquid from the lamellae (e.g., tributyl-phosphate, which, lowers surface viscosity) (Magdassi, 2010).

7.12.5 pH and Electrolytes

pH plays a crucial role in water-based inks and can have a profound influence on the solubility of the different components and on the colloidal stability of the dispersed pigments. The effect of solubility is commonly seen if the ink comprises a polymeric binder like acrylic resin, which is insoluble at low pH. Colloidal stability is influenced by the zeta potential of the particles: the higher the value, the higher the stability. The presence of electrolytes can produce serious stability issues during long storage, owing to the compression of electrical double layers of the particles (which can cause flocculation); thus, the concentration of electrolytes must be extremely low (Tawiah et al., 2016).

7.12.6 Dielectric Properties and Conductivity

These characteristics are required in continuous ink-jet inks, where the droplets are deflected as a result of an electrical field. The charge ability is achieved by incorporating charge control agents—electrolytes and ionic surfactants—that are soluble in the ink medium. The conductivity must be highly accurately controlled; hence, any slight change in conductivity upon storage must be avoided. In this case, a change in conductivity could also result from precipitation of salts, interactions with other parts and the container wall, etc. The conductivity of the ink is also critical for printing systems in which ink recirculation sensors are initiated by conductivity signals obtained by sensor contact with the ink (Magdassi, 2010).

7.12.7 Ink Storage and Stability

All physicochemical properties of ink-jet printing inks of textile, such as pH, viscosity, surface tension, particle size and its distribution if colourant dispersed, and dielectric properties, should be stable during a long duration of time ("shelf life"), usually two years at room temperature. Still there are some exceptions such as UV inks. After the formulation of the ink, it should fulfil certain physicochemical specifications, which will depend on its proposed use (Magdassi, 2010).

The essential parameters taken into account during the preparation of the mentioned ink should remain consistent throughout its shelf life. A stable ink is an ink whose characteristics do not change with time. In most cases, inks do not have un-dissolved matter within them; stability is induced through reactions among components of the ink, for instance, polymerization with UV ink, precipitation as well as separation by phases based on shifts in solubility (faced, for example, when samples are stored or shipped at cold temperatures), and even contact with the walls of the ink container slow (Tawiah et al., 2016).

7.12.8 Fabric Pre-treatment Methodologies for Digital Textiles Ink-Jet Printing

Pre-treatment of the fabric is crucial to obtaining good colour strength, fastness, droplet management, colour spread and penetration to ensure optimal image quality in digital textile printing, particularly with reactive dyes. It is required since ink-jet printing of fabrics has typically been performed on pretreated fabrics, except for carpets.

A Few Pretreatment Chemicals and their Application in Inkjet Printing (Tawiah et al., 2016) are given below:

- *Alginate ester/seaweed* used in reactive dye ink for cotton.

- *Urea/alkali/lyoprint R.G* used in reactive dye ink for cotton/lyocell.
- *Atmospheric pressure plasma (APP)* used in reactive dye ink for cotton.
- *Plasma treatment using Radio Frequency O_2 Plasma (RFOP)* used in pigment ink for polyester.
- *Enzymatic treatment* used in disperse dye ink for polyester.
- *Chitosan* used in reactive dye ink for cotton.
- *Gas plasma* used in pigment ink for cotton.
- *Foam* used in pigment ink for silk.
- *β-cyclodextrin and citric acid* used in pigment ink for polyester.
- *Fluoroacrylate copolymer and Guar Gum* are used in disperse dye ink for polyester.
- *Nano-modification using atmospheric-pressure air plasma (AAP)* used in pigment ink for polyester.
- *Amino compounds* used in pigment ink for silk.
- *Ebecryl 2002 polyurethane acrylate (PUA) based on PEG* used in pigment ink for viscose, nylon, wool, cotton, polyester.
- *Cationization with EPTAC* used in reactive dye ink for cotton.
- *Cationization with DTAB and TMAH* used in reactive dye ink for linen.
- *Cationization with CHPTAC* used in reactive dye ink for cotton.
- *Tertiary amine cationic polyacrylamide* used in reactive dye ink for cotton.
- *Amino-terminated hyperbranched polymer (HBP-NH_2)* used in reactive dye ink for cotton.
- *Cationization with CHPTAC* used in reactive dye ink for cotton.
- *Pretreatment with polyamidoamine (PAMAM) dendrimer* used in reactive dye ink for cotton.

The above details illustrate a high usage of cotton substrates, specifically with reactive dye inks, and the demand for chemical or cationic pretreatments (e.g., CHPTAC, EPTAC, PAMAM, HBP-NH_2) to maximize dye fixation and print quality. Polyester fabrics are treated primarily for pigment or disperse dye inks, sometimes by plasma, enzymatic, or chemical surface modification techniques for improving ink receptivity. Silk and linen also have their treatments represented. In general, the pretreatment type is very specific to the type of fibre as well as to the ink system, further establishing the importance of surface modification in digital textile printing towards the best colour quality, stability, and print accuracy.

7.12.9 Drying Mechanism of Ink-Jet Inks

The different kinds of drying mechanisms are absorption, evaporation, or penetration. Textile inkjet inks are jetted on the substrate; they dry up the substrate through various mechanisms, as mentioned. Figure 7.16 illustrates the different drying mechanisms of the print head (Le, 1998).

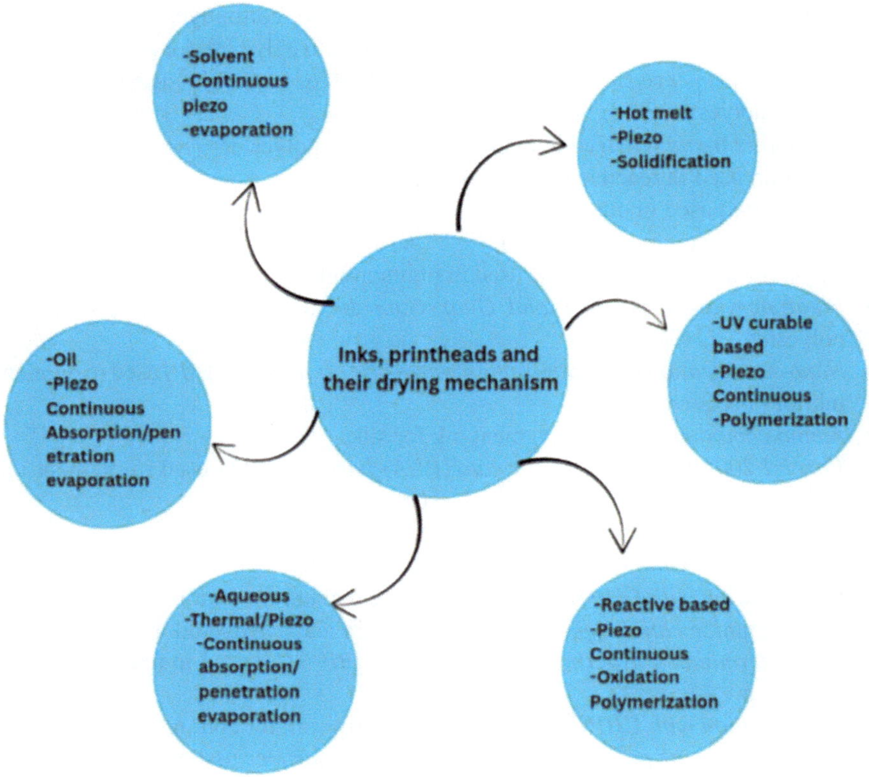

Fig. 7.16 Drying mechanisms of the ink (self-drawn)

7.13 Ink Chemistry

The most important element of ink-jet printing is likely to be the ink. Ink chemistry and formulation determine not only the quality of the printed image but also the drop ejection behaviour and the print system's reliability. Numerous kinds of inks have been designed and utilized in ink-jet applications (Le, 1998). The main function of the ink is to put the functional molecules accurately on the substrate after being jetted out from the printhead. The functional molecules are the colourants (e.g., dyes and pigments), also called graphics art. The ink jetted from the print heads should be in liquid form with a viscosity below 25Cp (centipoises) (Magdassi, 2009). Different colourants behave differently towards light, depending on their opaque nature. The more opaque the colourant is, the more light will be reflected and refracted, depending on the particle size and refractive index (the printing ink manual).

When a droplet of water-based ink hits the surface of uncoated paper like a bond, copy, or plain paper, the ink spreads along the paper fibres and penetrates into the bulk of the paper. The water-based ink indeed relies on penetration and absorption for its drying process. There is some water evaporation, but this drying process is usually extremely slow. Such behaviour of the ink reduces colour density and spot resolution on paper. It has long been recognized that paper or another medium with a coated water-receiving layer could significantly enhance colour density and resolution by regulating ink spreading and penetration through the coated layer. Phase-change ink is also referred to as hot melt or solid ink that is solid at room temperature. The molten liquid is jetted out of the printhead. Upon impact on a recording surface, the molten ink drop immediately solidifies, preventing the ink from spreading or penetrating printed media. The rapid solidification characteristic ensures good image quality on a broad range of recording media (Le, 1998). Ink composition is complex. They contain dyes, pigments, solvents, lubricants, resins, surfactants, particulate matter, fluoresces, and other components. The constituents of inks perform various functions such as dyeing, carrying, flow and ink thickness, and colour when dry. Inks' composition consists of vehicles (a liquid carrier) and additives (analysis of dyes and inks). Inks are intricate blends that can be varied in composition to create the desired writing properties. Consequently, ink-source comparisons are often possible based on the relative amounts of colourants and additives (separation of small mass ions).

7.14 The Technology of Piezoelectric Inks

In the piezoelectric ink-jet, depending on the piezoceramic deformation mode, the technology can be divided into four main types: squeeze, bend, push, and shear.

7.14.1 Bend Mode

A study put forward a piezoelectric ink-jet nozzle and made the piezoelectric ceramic into a buzzer-shaped form. The environment of the piezoelectric ceramic was fixed, and the centre section was deformed to force the droplets to form. In this study, a high-speed camera (CCD) was utilized to capture the droplet formation and separation process. A study performed earlier showed that the process of droplet injection of a typical piezoelectric film nozzle was numerically simulated. The corresponding findings showed that numerous factors, such as the nozzle outlet diameter, injection time interval, surface tension, and fluid viscosity, significantly influenced the injection behaviour. Since then, with the improvement of MEMS technology, some researchers have started to apply MEMS technology to design piezoelectric microjet devices. A study also designed a piezoelectric nozzle with MEMS technology and discovered that nozzle structure, driving voltage, ink viscosity, and meniscus vibration mode all influenced the spray performance of the nozzle. Another study made a piezoelectricity-driven ink-jet printhead on a silicon wafer by using MEMS technology and performed a numerical simulation on this path. It was discovered that the *Helmholtz mode* (a special type of air-spring oscillator, which is an enclosed volume having a small neck and an opening at one end; it is called a Helmholtz resonator) was the dominant resonance mode for controlling the flow oscillation at the nozzle. Since it is one of the essential elements of the piezoelectric nozzle, the piezoelectric vibration plate influences the print head's jet performance directly (Yan et al., 2021).

7.14.2 Push Mode

In push mode design, when the piezoelectric ceramic rods enlarge, they push the ink to emit the droplets from an orifice. Theoretically, the piezoelectric actuators can make contact and push with the ink. In real implementation, a thin diaphragm between the piezoelectric actuators and the ink is employed to avoid the harmful interaction between the ink and the actuator materials. The push mode piezoelectric ink-jet is used effectively in printheads from firms, including Data Products, Trident and Epson (Brünahl & Grishin, 2002).

7.14.3 Squeeze Mode

The actuator of a printhead operating in a squeeze mode consists of piezoelectric ceramic tubing approximately 1 mm in diameter. The radially polarized tube has electrodes on its inner and outer surfaces. Upon application to the transducer of a short rise time voltage pulse when a droplet is to be ejected from the orifice, the polarity is chosen to produce a contraction of the transducer. The sudden reduction in the volume inside the enclosure produces a small amount of liquid being squeezed out from the orifice. The pressure pulse also brings some ink back into the tube, but the quantity is relatively low owing to the high acoustic impedance provided by the length and narrow bore of the tube. The voltage pulse is permitted to decay fairly slowly, and the transducer thus fills up slowly to its original volume. As a result of the low rate of volume change during the decay, the associated decrease in pressure is too insignificant to conquer the surface tension at the orifice. The liquid thus enters the transducer to fill the liquid already pushed out without sucking in air through the orifice (Brünahl, 2003).

7.14.4 Shear Mode

The shear action pushes the piezo plates (structures containing piezoelectric materials (Mechkour, 2022) against the ink to force out the droplets. The *piezo driver* (Piezo drivers are high voltage amplifier type power supplies for the stable driving of the displacement and the vibration for each piezo element based on its purpose) becomes an active wall within the ink chamber. Contact between the piezo material and the ink is one of the design parameters of crucial importance in a shear-mode printhead

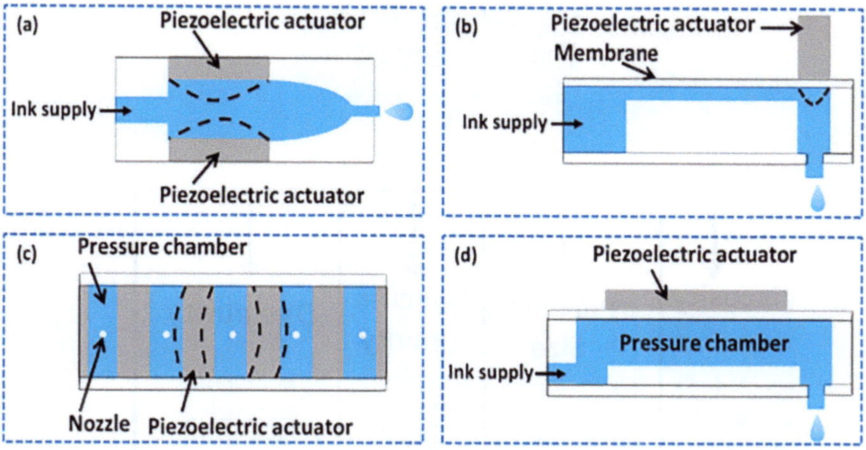

Fig. 7.17 Classification of piezo-driven ink-jet printheads. **a** squeeze mode, **b** push mode, **c** shear mode, **d** bend mode (Adopted from Shah et al. (2021))

design. *Spectra26* and *Xaar27, 28* are the trendsetters of the shear-mode printhead design (Le, 1998).

Figure 7.17 shows different types of piezo-driven ink-jet printheads.

7.15 Fluid Theoretical Printability

The considerable physical parameters of printing fluids are viscosity, density, and surface tension. The travel velocity of the droplet head decreases due to the increased viscosity and increases due to increased velocity (Jang et al., 2009). Research has indicated that physical property variations like viscosity and surface tension induce variations in drop size and drop velocity at the nozzle exit. This leads to inconsistency in drop formation, induces satellite drops and leads to low print quality. Differences in drop size also influence the saturation level of the fabric, thus leading to shade differences. Thus, the physical characteristics of the ink jet inks have to be properly controlled within a certain range for stable drop formation, good jettability and good print quality (Momin, 2008). Inks should have definite properties, as mentioned in Fig. 7.18, for proper functioning of the printhead.

7.16 Effect of Printhead Geometry

Physical geometry, channel sizes and shapes of the printhead, nozzles and drive profile vary according to each printhead type. Because of this, inks are typically adjusted to work for particular printheads with small operating windows and are

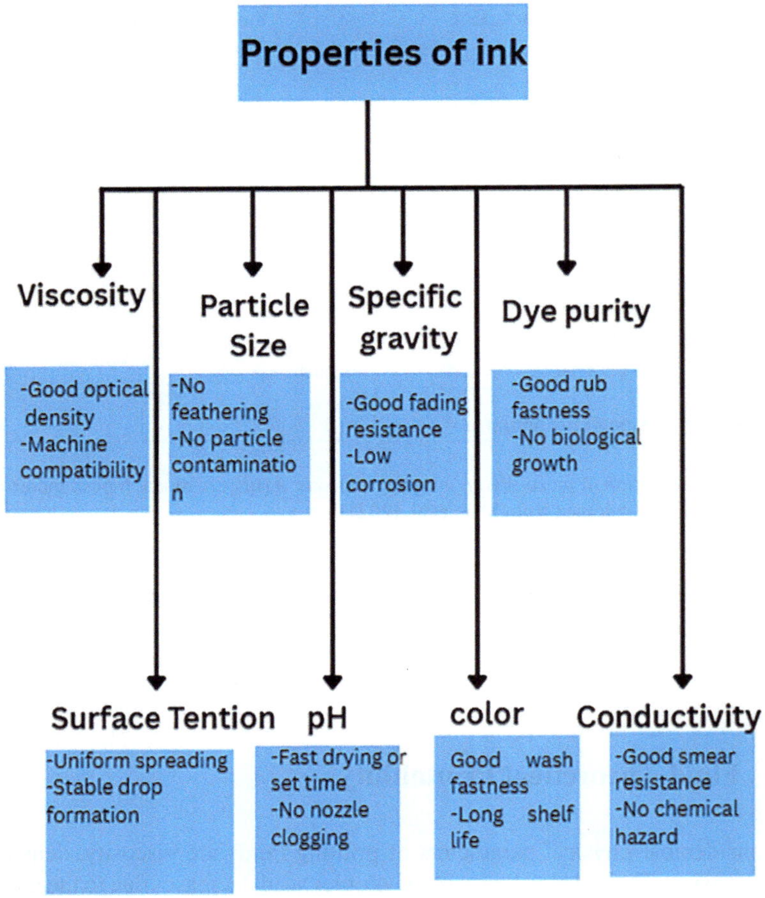

Fig. 7.18 Properties of ink (self-drawn)

capable of managing ink of set properties (Tuladhar et al., 2011). The ink formulation and printing parameters, along with the print head geometry, affect the state of the material upon deposition, which influences film microstructure (Tafoya & Secor, 2020).

7.17 Main Motive of Ink-Jet Printing

The main objective of ink-jet printing is to generate the smallest drops possible and jet at higher frequencies or to operate the device with highly viscous materials. Smaller droplets and increased frequencies not only render piezo acoustic ink-jet printers more accurate and quicker but also provide diverse applications in biology, medicine, microfabrication, and chemistry (Lohse, 2022).

7.18 Future Perspective

Improvements in ink-jet technology have changed the face of textile printing because they have significantly reduced or eradicated the time and human resources consumed by making plates and blending ink in traditional screen textile printing. The transformation will remain incomplete without superior quality, environment-friendly inks, printing all substrates and providing superior colouristic traits. Potential in digital printing fabrics has increased exponentially over the last ten years, with researchers continuing to discover new means to accelerate this growth further. With its low cost, rapid turnaround, and limitless flexibility, digital printing will dominate the sampling and mass customization printing business. With developments in software, hardware, and chemical technology that are realized along with ongoing research by textile chemists to create a universal set of dyes/inks for use in digital ink-jet printing on chemically variant textile materials, it is clear that digital printing of textiles will deliver its potential and take centre stage in textile printing. A common dye/ink set will definitely reduce the complicated inventory of colourants in the industry, minimize machine downtime and reduce dye content in printing effluents (Tawiah et al., 2016).

7.19 Complexities in Ink-Jet Flow

The fluid dynamics issues in ink-jet printing are both on the printhead side—i.e., the piezo acoustic actuation and the intricate flow within the printhead, entrained bubbles, the meniscus dynamics, wetting effects at the nozzle plate, and jet formation—and on the receiving substrate side—i.e., droplet impact, coalescence, the wetting of the substrate, droplet evaporation, and drying. These issues are intertwined since the optimization of the process and the materials regarding either the printhead side or the receiving substrate side is insufficient. The process must be optimized since the same ink is used, and droplet frequency, velocity, and size are critical on both sides. One such example of competing requirements from the printhead side on one side and from the receiving substrate or, more concretely, the paper side on the other side is the ink volatility: It would be better at the nozzle if ink evaporation were avoided to avoid nozzle clogging, but on the paper side, rapid ink evaporation is desired to allow effective printing and avoid paper deformation (Lohse, 2022).

7.20 Limitations and Challenges

Inkjet printing has certain limitations and challenges, which are as follows.

7.20.1 Nozzle Clogging

A reliable ink supply system should be used, or nanoparticles can affect piezoelectric inkjet printing. Water-repellent nanoparticles clog the printhead and must be cleaned thoroughly since they cause repetitive clogging. To achieve a higher resolution, the nozzles must be made smaller, increasing the chances of clogging (Cao et al., 2024).

7.20.2 Effect of Humidity on Inks

Humidity plays a vital role in humidity variation, which causes loss of clarity and image quality due to the movement of the substrate in the dye/pigment (Fricker et al., 2012).

Expanding digital printing for mass production while maintaining cost-effectiveness and sustainability remains challenging. Addressing these challenges will require continued research in materials science, ink chemistry, digital printing technologies, and collaboration across academic, industrial, and regulatory sectors.

7.20.2.1 Limits to Printability

Not all types of inks are compatible with inkjet inks. To become printable, the ink needs to have proper fluid properties such as viscosity, surface tension, and density. If the ink is too viscous, it may not form a drop, and if it is too runny, it may create many droplets, leading to poor-quality printing.

Another limitation of thermal drop-on-demand inkjet systems is that they work under thermal conditions, where the ink is heated to form a bubble that is ejected as a droplet. This method, therefore, limits the types of inks that can be used.

7.20.2.2 Resolution & Surface Definition

In additive manufacturing, resolution does not just mean how sharp or detailed something looks. Resolution directly affects the way the printed material is built up into a solid, functional structure.

In 3D printing, the exact placement of the dot is crucial. Even a slight displacement can lead to weaker parts and rough surfaces, so drop precision plays an important role in determining the surface finish of the final product.

Another challenge is how the material spreads out when the droplet falls, which is directly related to the Marangoni flow (Castrejon-Pita et al., 2013).

7.20.2.3 Printing Speed

Inkjet printing performance is affected by several main factors. Expanding the number of nozzles can significantly increase printing speed, but at the same time, it is more complicated and more expensive, and there are problems with the alignment. The number of drops that can be produced is restricted by how rapidly a nozzle can replenish and eject the ink, usually about 1 μL/sec for each nozzle. Continuous Inkjet (CIJ) systems give higher speeds, but at the same time, they lose resolution because of the droplet deflection. The ink's solid content is also of utmost importance; more solids in the ink mean more material to deposit, which is the case in 3D printing, while low-solids inks only waste the energy and the carrier fluid. Hybrid systems, a combination of powder and a binder, contribute to improving part density and lowering the sintering temperature. Throw distance is another factor that complicates things: Drop-on-demand (DoD) systems have a very small throw distance of less than 5 mm due to low droplet velocity, which reduces the precision. At the same time, CIJ can print from distances as far as 20 mm, but the droplet trajectory and the accuracy of the print still remain (Castrejon-Pita et al., 2013).

7.21 Conclusion

In conclusion, digital printing, especially ink-jet technology, has transformed many industries by providing high-resolution, cost-effective, and eco-friendly alternatives to conventional manufacturing and printing technologies. With fine control over droplet formation, fluid flow, and material deposition, this technology allows for precise patterning, which is critical in applications from textile design and MEMS production to OLED display fabrication and flexible electronics. The evolution of sophisticated inks, waveform designs, and pre-treatment processes guarantees not only higher print quality but also sustainability and compatibility with various substrates. With further advancements in research on ink formulation, printhead design, and printing processes, digital printing is poised to replace traditional lithography-based fabrication further, becoming a key part of the future of smart manufacturing and functional materials. Moreover, progress in waveform tuning, satellite droplet control, AI-assisted droplet manipulation, and printhead design has greatly increased precision, stability, and reliability. Besides, the environmental aspects are also a big step forward since water-based and VOC-free inks, combined with the lowered energy and water consumption, become sustainable alternatives to traditional printing methods. The range of new applications now includes various areas, such as high-resolution textile prints and OLED displays, to MEMS devices and wearable electronics, which illustrate this technology's extensive scope and impact.

In conclusion, the performance and quality of inkjet printing are affected by a complex interaction of factors such as nozzle configuration, drop frequency, ink composition, and throw distance. Even though increasing the number of nozzles can lead to faster printing, it also entails some problems related to alignment and cost. Drop-on-demand (DoD) systems are capable of high resolution; however, their throw distance is small. On the other hand, Continuous Inkjet (CIJ) systems have long throw distances but are less accurate since the droplet deflection causes disturbances.

Declaration The authors declare that they have no known competing financial interests or personal relationships that could have appeared to influence the work reported in this chapter. Generative AI and AI-assisted technologies have been used in the writing process to improve the readability and language of the work only. After using this tool, the authors carefully reviewed and revised the content to ensure accuracy and quality. The authors take full responsibility for the final version of this publication.

References

Adhikari, J., Roy, A., Das, A., Ghosh, M., Thomas, S., Sinha, A., et al. (2021). Effects of processing parameters of 3D bioprinting on the cellular activity of bioinks. *Macromolecular Bioscience, 21*(1), 2000179.

Ali, S. M., Noghanian, S., Khan, Z. U., Alzahrani, S., Alharbi, S., Alhartomi, M., & Alsulami, R. (2025). Wearable and flexible sensor devices: Recent advances in designs, fabrication methods, and applications. *Sensors, 25*(5), 1377.

Antohe, B. V., & Wallace, D. B. (2001, January). Acoustic phenomena in a demand-mode piezoelectric ink-jet printer. In *NIP & Digital Fabrication Conference* (Vol. 17, pp. 885–889). Society of Imaging Science and Technology.

Barker, B., Bell, J. B., & Garcia, A. L. (2023). Fluctuating hydrodynamics and the Rayleigh-Plateau instability. *Proceedings of the National Academy of Sciences, 120*(30), Article e2306088120.

Bassani, F., Liedl, G. L., & Wyder, P. (2005). Encyclopedia of condensed matter physics.

Beedasy, V., & Smith, P. J. (2020). Printed electronics as prepared by inkjet printing. *Materials, 13*(3), 704.

Beulen, B., Jong, J. D., Reinten, H., Berg, M. V. D., Wijshoff, H., & Van Dongen, R. (2007). Flows on the nozzle plate of an inkjet printhead. *Experiments in Fluids, 42*, 217–224.

Bietsch, A., Hegner, M., Lang, H. P., & Gerber, C. (2004). Inkjet deposition of alkanethiolate monolayers and DNA oligonucleotides on gold: Evaluation of spot uniformity by wet etching. *Langmuir, 20*(12), 5119–5122.

Blachowicz, T., & Ehrmann, A. (2020). 3D printed MEMS technology—recent developments and applications. *Micromachines, 11*(4), 434.

Brenner, M. P., Hilgenfeldt, S., & Lohse, D. (2002). Single-bubble sonoluminescence. *Reviews of Modern Physics, 74*(2), 425.

Brown, M. S., Brasz, C. F., Ventikos, Y., & Arnold, C. B. (2012). Impulsively actuated jets from thin liquid films for high-resolution printing applications. *Journal of Fluid Mechanics, 709*, 341–370.

Brünahl, J., & Grishin, A. M. (2002). Piezoelectric shear mode drop-on-demand inkjet actuator. *Sensors and Actuators a: Physical, 101*(3), 371–382.

Brünahl, J. (2003). *Physics of piezoelectric shear mode inkjet actuators.* Doctoral dissertation, Mikroelektronikochinformationsteknik.

Cao, X., Ye, Y., Liu, X., Guo, T., & Tang, Q. (2021, February). 54.3: Realization of uniform OLED pixels based on multi-nozzle by inkjet printing. In *SID Symposium Digest of Technical Papers* (Vol. 52, pp. 395–397).

Cao, T., Yang, Z., Zhang, H., & Wang, Y. (2024). Inkjet printing quality improvement research progress: A review. *Heliyon, 10*(10).

Carey, T., Cacovich, S., Divitini, G., Ren, J., Mansouri, A., Kim, J. M., et al. (2017). Fully inkjet-printed two-dimensional material field-effect heterojunctions for wearable and textile electronics. *Nature Communications, 8*(1), 1202.

Castrejon-Pita, J. R., Baxter, W. R. S., Morgan, J., Temple, S., Martin, G. D., & Hutchings, I. M. (2013). Future, opportunities and challenges of inkjet technologies. *Atomization and sprays, 23*(6).

Chen, M., Xie, L., Wei, C., Yi, Y. Q. Q., Chen, X., Yang, J., et al. (2021). High performance inkjet-printed QLEDs with 18.3% EQE: Improving interfacial contact by novel halogen-free binary solvent system. *Nano Research*, 1–7.

Cheng, F. (2020). Application analysis of digital printing technology in packaging printing. In *E3S Web of Conferences* (Vol. 185, p. 02034). EDP Sciences.

Cramer, M., Verboven, I., Drijkoningen, J., & Deferme, W. (2018, May). Optimizing the outcoupling efficiency and the radiation pattern of organic light emitting devices by inkjet printing lens arrays films. In *Organic electronics and photonics: fundamentals and devices* (Vol. 10687, pp. 11–19). SPIE.

Cummins, G., & Desmulliez, M. P. (2012). Inkjet printing of conductive materials: A review. *Circuit World, 38*(4), 193–213.

Dalwadi, D. H., Canet, C., Roye, N., & Hedman, K. (2005). Rheology: An important tool in ink development. *American Laboratory, 37*(23), 18.

Dressaire, E., & Sauret, A. (2017). Clogging of microfluidic systems. *Soft Matter, 13*(1), 37–48.

Danzebrink, R., & Aegerter, M. A. (2001). Deposition of optical microlens arrays by ink-jet processes. *Thin Solid Films, 392*(2), 223–225.

de Goede, T. C., Moqaddam, A. M., Limpens, K. C. M., Kooij, S. A., Derome, D., Carmeliet, J., et al. (2021). Droplet impact of Newtonian fluids and blood on simple fabrics: Effect of fabric pore size and underlying substrate. *Physics of Fluids, 33*(3).

de Jong, J., de Bruin, G., Reinten, H., van den Berg, M., Wijshoff, H., Versluis, M., & Lohse, D. (2006a). Air entrapment in piezo-driven inkjet printheads. *The Journal of the Acoustical Society of America, 120*(3), 1257–1265.

de Jong, J., Jeurissen, R., Borel, H., van den Berg, M., Wijshoff, H., Reinten, H., et al. (2006). Entrapped air bubbles in piezo-driven inkjet printing: Their effect on the droplet velocity. *Physics of Fluids, 18*(12).

Fang, H., Li, J., Gong, S., Lin, J., & Xie, G. (2024). Inkjet printing of high-color-purity blue organic light-emitting diodes with host-free inks. *Molecules, 29*(9), 2147.

Feng, J. Q. (2002). A general fluid dynamic analysis of drop ejection in drop-on-demand ink jet devices. *Journal of Imaging Science and Technology, 46*(5), 398–408.

Figura, L., & Teixeira, A. A. (2007). *Food physics: Physical properties-measurement and applications.* Springer Science & Business Media.

Foresti, D., Kroll, K. T., Amissah, R., Sillani, F., Homan, K. A., Poulikakos, D., & Lewis, J. A. (2018). Acoustophoretic printing. *Science Advances, 4*(8), eaat1659.

Fraters, A., Van Den Berg, M., De Loore, Y., Reinten, H., Wijshoff, H., Lohse, D., et al. (2019). Inkjet nozzle failure by heterogeneous nucleation: Bubble entrainment, cavitation, and diffusive growth. *Physical Review Applied, 12*(6), Article 064019.

Fricker, A. L., Hodgson, A., Townsend, J. H., & Woods, C. (2012). Humidity sensitivity of inkjet prints. *E-Preservation Science, 9*, 60–66.

Fuller, S. B., Wilhelm, E. J., & Jacobson, J. M. (2002). Ink-jet printed nanoparticle microelectromechanical systems. *Journal of Microelectromechanical Systems, 11*(1), 54–60.

Funakubo, H., Dekkers, M., Sambri, A., Gariglio, S., Shklyarevskiy, I., & Rijnders, G. (2012). Epitaxial PZT films for MEMS printing applications. *MRS Bulletin, 37*(11), 1030–1038.

Gao, D., & Zhou, J. G. (2018). Designs and applications of electrohydrodynamic 3D printing. *International Journal of Bioprinting, 5*(1), 172.

Gruene, M., Unger, C., Koch, L., Deiwick, A., & Chichkov, B. (2011). Dispensing pico to nanolitre of a natural hydrogel by laser-assisted bioprinting. *Biomedical Engineering Online, 10*, 1–11.

Gu, H., Duits, M. H., & Mugele, F. (2011). Droplets formation and merging in two-phase flow microfluidics. *International Journal of Molecular Sciences, 12*(4), 2572–2597.

Gu, Y., Zhang, T., Chen, H., Wang, F., Pu, Y., Gao, C., & Li, S. (2019). Mini review on flexible and wearable electronics for monitoring human health information. *Nanoscale Research Letters, 14*, 1–15.

Guo, K., Song, Z., Zhou, J., Shen, B., Yan, B., Gu, Z., & Wang, H. (2024). An artificial intelligence-assisted digital microfluidic system for multistate droplet control. *Microsystems & Nanoengineering, 10*(1), 138.

Hack, M. A., Costalonga, M., Segers, T., Karpitschka, S., Wijshoff, H., & Snoeijer, J. H. (2018). Printing wet-on-wet: Attraction and repulsion of drops on a viscous film. *Applied Physics Letters, 113*(18).

Hartwig, M., Zichner, R., & Joseph, Y. (2018). Inkjet-printed wireless chemiresistive sensors—a review. *Chemosensors, 6*(4), 66.

He, B., Yang, S., Qin, Z., Wen, B., & Zhang, C. (2017). The roles of wettability and surface tension in droplet formation during inkjet printing. *Scientific Reports, 7*(1), 11841.

Hicks, P. D., & Purvis, R. (2010). Air cushioning and bubble entrapment in three-dimensional droplet impacts. *Journal of Fluid Mechanics, 649*, 135–163.

https://en.wikipedia.org/wiki/Capillary_number).

https://en.wikipedia.org/wiki/PEDOT:PSS.

https://en.wikipedia.org/wiki/Short-time_Fourier_transform#:~:text=As%20the%20Nyquist%20frequency%20is,signal%20of%20finite%20record%2Dlength.

https://en.wikipedia.org/wiki/Stokes_number.

https://www.cityu.edu.hk/phy/appkchu/AP6120/5.PDF.

https://www.stevenabbott.co.uk/practical-coatings/ohnesorge.php.

Hu, Y. (2016, July). The influence of modern digital design on traditional tie-dye. In *2016 International Conference on Sensor Network and Computer Engineering* (pp. 418–422). Atlantis Press.

Huang, Y., Hsiang, E. L., Deng, M. Y., & Wu, S. T. (2020). Mini-LED, Micro-LED and OLED displays: Present status and future perspectives. *Light: Science & Applications, 9*(1), 105.

Izdebska-Podsiadły, J., (Ed.). (2022). *Polymers for 3d printing: Methods, properties, and characteristics*. William Andrew.

Jacot-Descombes, L., Cadarso, V. J., Schleunitz, A., Grützner, S., Klein, J. J., Brugger, J., et al. (2015). Organic-inorganic-hybrid-polymer microlens arrays with tailored optical characteristics and multi-focal properties. *Optics Express, 23*(19), 25365–25376.

Jafarifard, S., Bastani, S., Soleimani-Gorgani, A., & Sari, M. G. (2016). The chemo-rheological behavior of an acrylic based UV-curable inkjet ink: Effect of surface chemistry for hyperbranched polymers. *Progress in Organic Coatings, 90*, 399–406.

Jang, D., Kim, D., & Moon, J. (2009). Influence of fluid physical properties on ink-jet printability. *Langmuir, 25*(5), 2629–2635.

Karpitschka, S., & Riegler, H. (2010). Quantitative experimental study on the transition between fast and delayed coalescence of sessile droplets with different but completely miscible liquids. *Langmuir, 26*(14), 11823–11829.

Karsa, D. R., (Ed.). (1999). *Design and selection of performance surfactants* (Vol. 2). CRC Press.

Kim, H., Boulogne, F., Um, E., Jacobi, I., Button, E., & Stone, H. A. (2016). Controlled uniform coating from the interplay of Marangoni flows and surface-adsorbed macromolecules. *Physical Review Letters, 116*(12), Article 124501.
Koseoglu, A. U. (2019). Innovations and analysis of textile digital printing technology. *International Journal of Science, Technology and Society, 7*(2), 38–43.
Kumar, P., Ebbens, S., & Zhao, X. (2021). Inkjet printing of mammalian cells–theory and applications. *Bioprinting, 23*, Article e00157.
Kwon, K. S., & Kim, W. (2007). A waveform design method for high-speed inkjet printing based on self-sensing measurement. *Sensors and Actuators A: Physical, 140*(1), 75–83.
Kwon, K. S., Rahman, M. K., Phung, T. H., Hoath, S. D., Jeong, S., & Kim, J. S. (2020). Review of digital printing technologies for electronic materials. *Flexible and Printed Electronics, 5*(4), Article 043003.
Ladino-Luna, D. (2005). Van der Waals gas as working substance in a Curzon and Ahlborn-Novikov engine. *Entropy, 7*(1), 108–121.
Lau, G. K., & Shrestha, M. (2017). Ink-jet printing of micro-electro-mechanical systems (MEMS). *Micromachines, 8*(6), 194.
Le, H. P. (1998). Progress and trends in ink-jet printing technology. *Journal of Imaging Science and Technology, 42*(1), 49–62.
Lee, W. H., & Park, Y. D. (2017). Inkjet etching of polymers and its applications in organic electronic devices. *Polymers, 9*(9), 441.
Liguori, R., Nunziata, F., Aprano, S., & Maglione, M. G. (2024). Overcoming challenges in OLED technology for lighting solutions. *Electronics, 13*(7), 1299.
Liu, L., Yi, J., Xu, K., Liu, Z., Tang, M., Dai, L., et al. (2024). High piezoelectric property with exceptional stability in self-poled ferroelectric films. *Nature Communications, 15*(1), 10798.
Lohse, D. (2018). Bubble puzzles: From fundamentals to applications. *Physical Review Fluids, 3*(11), Article 110504.
Lohse, D. (2022). Fundamental fluid dynamics challenges in inkjet printing. *Annual Review of Fluid Mechanics, 54*(1), 349–382.
Madhukar Thakker, A., & Sun, D. (2024). Inks for digital printing of textiles. *The Journal of The Textile Institute, 115*(12), 2377–2390.
Magdassi, S., (Ed.). (2009). *The chemistry of inkjet inks*. World Scientific.
Magdassi, S. (2010). Ink requirements and formulations guidelines. In *The chemistry of inkjet inks* (pp. 19–41).
McIlroy, C., & Harlen, O. G. (2019). Effects of drive amplitude on continuous jet break-up. *Physics of Fluids, 31*(6).
Mechkour, H. (2022). Modeling of perforated piezoelectric plates. *Mathematical and Computational Applications, 27*(6), 100.
Modak, C. D., Kumar, A., Tripathy, A., & Sen, P. (2020). Drop impact printing. *Nature Communications, 11*(1), 4327.
Moeller, M., & Matyjaszewski, K. (2012). *Polymer science: A comprehensive reference*. Newnes.
Momin, N. (2008). *Chitosan and improved pigment ink jet printing on textiles*. Doctoral dissertation, RMIT University.
Niaounakis, M. (2014). *Biopolymers: Processing and products*. William Andrew.
Norberg, O., Westin, P., Lindberg, S., Klaman, M., & Eidenvall, L. (2001). A comparison of print quality between digital and traditional technologies. In *Proceedings of IS&T-SID DPP2001: International Conference on Digital Production Printing and Industrial Applications* (pp. 380–385). Antwerp, Belgium.
Onses, M. S., Sutanto, E., Ferreira, P. M., Alleyne, A. G., & Rogers, J. A. (2015). Mechanisms, capabilities, and applications of high-resolution electrohydrodynamic jet printing. *Small (Weinheim an der Bergstrasse, Germany), 11*(34), 4237–4266.
Oyarzún, J. M. (2000). *Pigment processing: Physico-chemical principles*. Vincentz Network GmbH & Co. KG.

Pal, L., & Fleming, P. D., III. (2006). The study of ink pigment dispersion parameters. *The Hilltop Review, 2*(1), 9.

Percin, G., Soh, T. H., & Khuri-Yakub, B. T. (1998, June). Resist deposition without spinning by using novel inkjet technology and direct lithography for MEMS. In *Advances in resist technology and processing XV* (Vol. 3333, pp. 1382–1389). SPIE.

Phung, T. H., & Kwon, K. S. (2019). How to manipulate droplet jetting from needle type jet dispensers. *Scientific Reports, 9*(1), 19669.

Pimpin, A., & Srituravanich, W. (2012). Review on micro-and nanolithography techniques and their applications. *Engineering Journal, 16*(1), 37–56.

Qian, J., & Law, C. K. (1997). Regimes of coalescence and separation in droplet collision. *Journal of Fluid Mechanics, 331*, 59–80.

Qu, X., Li, J., Yin, Z., & Zou, H. (2019). New lithography technique based on electrohydrodynamic printing platform. *Organic Electronics, 71*, 279–283.

Quéré, D. (2008). Wetting and roughness. *Annual Review of Materials Research, 38*(1), 71–99.

Ramon, A., Liashenko, I., Rosell-Llompart, J., & Cabot, A. (2024). On the stability of electrohydrodynamic jet printing using poly (ethylene oxide) solvent-based inks. *Nanomaterials, 14*(3), 273.

Rogowska, P. (2024). *Constraints in printing industry* (p. 208). Scientific Papers of Silesian University of Technology.

Sadhu, A. S., Huang, Y. M., Chen, L. Y., Kuo, H. C., & Lin, C. C. (2022). Recent advances in colloidal quantum dots or perovskite quantum dots as a luminescent downshifting layer embedded on solar cells. *Nanomaterials, 12*(6), 985.

Sajedi-Moghaddam, A., Rahmanian, E., & Naseri, N. (2020). Inkjet-printing technology for supercapacitor application: Current state and perspectives. *ACS Applied Materials & Interfaces, 12*(31), 34487–34504.

Salary, R., Lombardi, J. P., Weerawarne, D. L., Rao, P., & Poliks, M. D. (2021). A computational fluid dynamics investigation of pneumatic atomization, aerosol transport, and deposition in aerosol jet printing process. *Journal of Micro-and Nano-Manufacturing, 9*(1), Article 010903.

Schuster, J. M., Schvezov, C. E., & Rosenberger, M. R. (2015). Influence of experimental variables on the measure of contact angle in metals using the sessile drop method. *Procedia Materials Science, 8*, 742–751.

Schwarzenberger, K., Köllner, T., Linde, H., Boeck, T., Odenbach, S., & Eckert, K. (2014). Pattern formation and mass transfer under stationary solutal Marangoni instability. *Advances in Colloid and Interface Science, 206*, 344–371.

Scoutaris, N., Ross, S., & Douroumis, D. (2016). Current trends on medical and pharmaceutical applications of inkjet printing technology. *Pharmaceutical Research, 33*, 1799–1816.

Shah, M. A., Lee, D. G., Lee, B. Y., Kim, N. W., An, H., & Hur, S. (2020). Actuating voltage waveform optimization of piezoelectric inkjet printhead for suppression of residual vibrations. *Micromachines, 11*(10), 900.

Shah, M. A., Lee, D. G., Lee, B. Y., & Hur, S. (2021). Classifications and applications of inkjet printing technology: A review. *IEEE Access, 9*, 140079–140102.

Sieradzki, Z. (2017). Inkjet printing technique and its application in organic light emitting diodes. *Display and Imaging, 2*, 339–358.

Sun, Y., Jiang, Y., Sun, X. W., Zhang, S., & Chen, S. (2019). Beyond OLED: Efficient quantum dot light-emitting diodes for display and lighting application. *The Chemical Record, 19*(8), 1729–1752.

Tafoya, R. R., & Secor, E. B. (2020). Understanding effects of printhead geometry in aerosol jet printing. *Flexible and Printed Electronics, 5*(3), Article 035004.

Tawiah, B. E. N. J. A. M. I. N., Howard, E. K., & Asinyo, B. K. (2016). The chemistry of inkjet inks for digital textile printing-review. *BEST, 4*, 61–78.

Tijdeman, H. (1975). On the propagation of sound waves in cylindrical tubes. *Journal of Sound and Vibration, 39*(1), 1–33.

Tkalec, M., Glogar, M. I., & Sutlovic, A. (2022). Ecological sustainability of digital textile printing. In *Economic and Social Development: Book of Proceedings* (pp. 105–115).

Tran, T., Staat, H. J., Prosperetti, A., Sun, C., & Lohse, D. (2012). Drop impact on superheated surfaces. *Physical Review Letters, 108*(3), Article 036101.

Tuladhar, T., Tatum, J., & Drury, P. (2011, January). Influence of printhead geometry, print conditions and fluid dynamic properties on the jetting behaviour. In *NIP & Digital Fabrication Conference* (Vol. 27, pp. 70–73). Society of Imaging Science and Technology.

Uddin, M. J., Hassan, J., & Douroumis, D. (2022). Thermal inkjet printing: Prospects and applications in the development of medicine. *Technologies, 10*(5), 108.

Ujiie, H., (Ed.). (2006). *Digital printing of textiles.* Woodhead Publishing.

van der Bos, A., Segers, T., Jeurissen, R., van den Berg, M., Reinten, H., Wijshoff, H., et al. (2011). Infrared imaging and acoustic sizing of a bubble inside a micro-electro-mechanical system piezo ink channel. *Journal of Applied Physics, 110*(3).

van der Bos, A., van der Meulen, M. J., Driessen, T., van den Berg, M., Reinten, H., Wijshoff, H., et al. (2014). Velocity profile inside piezoacoustic inkjet droplets in flight: Comparison between experiment and numerical simulation. *Physical Review Applied, 1*(1), Article 014004.

Van Der Meulen, M. J., Reinten, H., Wijshoff, H., Versluis, M., Lohse, D., & Steen, P. (2020). Nonaxisymmetric effects in drop-on-demand piezoacoustic inkjet printing. *Physical Review Applied, 13*(5), Article 054071.

Viluksela, P., Kariniemi, M., & Nors, M. (2010). Environmental performance of digital printing. *VTT Research Notes, 2538.*

Wang, S., Wu, X., Lu, J., Luo, Z., Xie, H., Zhang, X., et al. (2022). Inkjet-printed silver nanowire ink for flexible transparent conductive film applications. *Nanomaterials, 12*(5), 842.

Wang, Y., Qi, Y., Lu, L., & Zeng, Q. (2025). Attentive ink MLP droplet detection algorithm based on the satellite droplets threshold domain. *Scientific Reports, 15*(1), 1–23.

Weiss, B. (2014). *Closed-loop control of a 3D printer gantry.* Doctoral Dissertation.

Wijshoff, H. (2010). The dynamics of the piezo inkjet printhead operation. *Physics Reports, 491*(4–5), 77–177.

Wijshoff, H. (2008). Structure-and fluid-dynamics in piezo inkjet printheads.

Wiklund, J., et al. (2021). A review on printed electronics: Fabrication methods, inks, substrates, applications and environmental impacts. *Journal of Manufacturing and Materials Processing, 5*(3), 89.

Wilkinson, N. J., Smith, M. A. A., Kay, R. W., & Harris, R. A. (2019). A review of aerosol jet printing—a non-traditional hybrid process for micro-manufacturing. *The International Journal of Advanced Manufacturing Technology, 105*, 4599–4619.

Wu, S. Y., Yang, C., Hsu, W., & Lin, L. (2015). 3D-printed microelectronics for integrated circuitry and passive wireless sensors. *Microsystems & Nanoengineering, 1*(1), 1–9.

Yan, Q., You, J., Sun, W., Wang, Y., Wang, H., & Zhang, L. (2021). Advances in piezoelectric jet and atomization devices. *Applied Sciences, 11*(11), 5093.

Yan, K., Li, J., Pan, L., & Shi, Y. (2020). Inkjet printing for flexible and wearable electronics. *Apl Materials, 8*(12).

Yoon, D. G., Kang, Y. J., Bail, R., & Chin, B. D. (2021). Interfaces and pattern resolution of inkjet-printed organic light-emitting diodes with a novel hole transport layer. *Journal of Information Display, 22*(2), 91–98.

Zhang, G., Quetzeri-Santiago, M. A., Stone, C. A., Botto, L., & Castrejón-Pita, J. R. (2018). Droplet impact dynamics on textiles. *Soft Matter, 14*(40), 8182–8190.

Zhang, K., Fang, K., Bukhari, M. N., Xie, R., Song, Y., Tang, Z., & Zhang, X. (2020). The effect of ink drop spreading and coalescing on the image quality of printed cotton fabric. *Cellulose, 27*, 9725–9736.

Zhang, Y., Hu, G., Liu, Y., Wang, J., Yang, G., & Li, D. (2022). Suppression and utilization of satellite droplets for inkjet printing: A review. *Processes, 10*(5), 932.

Zhdankin, V., Ripperda, B., & Philippov, A. A. (2023). Particle acceleration by magnetic Rayleigh-Taylor instability: Mechanism for flares in black hole accretion flows. *Physical Review Research, 5*(4), Article 043023.

Zhou, Y. (2004, January). Recent advances in highly durable piezoelectric inkjet print head technology. In *NIP & Digital Fabrication Conference* (Vol. 20, pp. 855–858). Society of Imaging Science and Technology.

Chapter 8
The Role of Designing Tools in Digital Textile Printing

Pravin P. Chavan, T. Pugazhenthi, Pintu Pandit, Ranjeet Kumar Jha, and Subhankar Maity

Abstract The evolution of digital tools has revolutionized textile printing, transforming it from a traditional craft into a technologically advanced and creatively dynamic industry. This chapter explores the pivotal role of designing tools ranging from basic software to cutting-edge 3D modeling systems and AI-integrated platforms in reshaping the landscape of digital textile printing. These tools facilitate seamless integration between artistic expression, engineering precision, and sustainability, empowering designers to expand creative boundaries, enhance production efficiency, and reduce environmental impact. The digital image design and creation process comprising image development or scanning, digital conversion, calibration, and high-resolution inkjet printing is examined to highlight its importance in achieving superior quality outputs. As digital and traditional techniques increasingly converge, the significance of designing tools continues to grow, defining not only the aesthetic future of textiles but also the innovative processes by which they are conceived, produced, and experienced.

Keywords Digital textile printing · Designing tools · Sustainability · Inkjet technology

---

P. P. Chavan (✉)
Department of Knitwear Design, National Institute of Fashion Technology, Kannur, India
e-mail: pravin.chavan@nift.ac.in

T. Pugazhenthi
Department of Textile Design, National Institute of Fashion Technology, Kannur, India

P. Pandit (✉) · R. K. Jha
Department of Textile Design, National Institute of Fashion Technology, Patna, India
e-mail: pintu.pandit@nift.ac.in

S. Maity
Department of Textile Technology, Uttar Pradesh Textile Technology Institute, Kanpur, India

A. Athalye and H. Patil (eds.), *Digital Printing*,
https://doi.org/10.1007/978-981-95-2593-5_8

8.1 Introduction

Digital textile printing has emerged as a ground breaking development in the textile industry, offering unparalleled opportunities for creativity, customisation, and efficiency. Unlike conventional methods, which rely heavily on labour-intensive screen preparation, dye baths, and large-scale production setups, digital textile printing enables the direct application of intricate designs onto fabrics through computer-controlled printing systems. Central to this technological shift is the increasing role of sophisticated designing tools and software applications that streamline the entire printing process—from pattern creation to ink deposition thereby significantly enhancing both the quality and functionality of textile products (Kan & Yuen, 2012; Wang & Memon, 2023).

Designing tools have become the backbone of modern digital textile workflows. These tools include computer-aided design (CAD) software, generative design algorithms, color management systems, and simulation platforms, all of which contribute to a more efficient and accurate translation of digital artwork into physical textile outputs. With the aid of such tools, designers can create, manipulate, and preview patterns with great precision, while also simulating how these designs will appear on various materials. This digital interface enables rapid prototyping and iteration, reducing the time and cost associated with traditional sample-making (Choudhary, 2018). Moreover, these tools support the creation of hyper-personalized designs, responding to the modern consumer's growing preference for unique and customizable products.

A particularly transformative application of design software is the use of generative design, which employs mathematical models and algorithms to create complex, non-repeating patterns. This innovation challenges traditional paradigms of textile design, where patterns were often repetitive and limited by screen size or mechanical constraints. Generative tools allow designers to explore dynamic, evolving motifs that can be fine-tuned based on real-time inputs or user preferences (Russell, 2013). These capabilities enable a high level of creative freedom while also aligning with on-demand manufacturing strategies, which minimize overproduction and waste.

In addition to creative design, software tools play an essential role in optimizing technical parameters critical to the digital printing process. For example, systems like FunTeA CAD assist in managing ink penetration and distribution—two crucial factors that affect the functional and aesthetic qualities of the printed fabric (Ali et al., 2019). These platforms allow designers and engineers to simulate how inks will behave on different substrates, adjusting variables such as ink viscosity, printhead speed, and fabric pretreatment methods. One such technique involves pretreating fabrics with a polyamine and a quaternary ammonium compound, which limits ink penetration and enhances colour brightness and sharpness (Glogar &

Ercegovic Razic, 2024). This synergy between chemical preparation and software-guided printing ensures consistent quality and reduces trial-and-error experimentation. Digital textile printing systems equipped with advanced software offer significant advantages in precision printing. For instance, double-sided inkjet printing techniques, which mirror patterns on both sides of the fabric, demand precise alignment and control—a task that is greatly facilitated by intelligent design and calibration software (Kato et al., 2017; Wang & Zhang, 2020). These advancements are not merely technical conveniences; they are essential for producing high-quality textiles for demanding applications such as fashion, home décor, and technical textiles.

Equally important is the role of software in color management and process coordination. Accurate color reproduction is vital in textile production, where slight deviations can impact product appeal and brand consistency. Designing tools integrated with color profiling and calibration systems ensure that digital outputs match intended color specifications across different devices and materials (Tyler, 2005; Ujiie, 2006). Moreover, these tools enhance collaboration across the textile supply chain by enabling shared digital platforms where designers, printers, and manufacturers can work in synchrony.

Despite its many advantages, digital textile printing still faces several challenges, such as the need for further ergonomic and mechanical improvements in printing machinery and the high operational expertise required to manage complex digital systems (Wang & Memon, 2023). Nevertheless, the trajectory of technological development suggests that these challenges will be addressed through continued innovation in both hardware and software domains.

8.2 Evolution of Digital Tools in Textile Printing

The textile industry has undergone a profound transformation in recent decades, largely driven by the evolution of digital tools. These innovations have revolutionized how fabrics are designed, printed, and produced—ushering in a new era of creativity, efficiency, and sustainability. From the early adoption of inkjet technology to the integration of advanced software and 3D printing, digital tools have redefined the possibilities for textile design and manufacturing. Digital textile printing found its roots in the late 20th century, with the introduction of inkjet technology. Initially limited to small-scale applications due to low resolution and limited color capabilities, early printers were not yet suitable for large-scale textile production (Tippett, 2002). However, as technology emerged, so did the potential of digital textile printing.

Industrial printing systems such as DuPont Artistri and the Mimaki Tx series marked a turning point by offering higher speed, improved colour fidelity, and adaptability to a wide range of fabrics (Ujiie, 2006). These machines laid the groundwork for digital printing to be used in mainstream textile manufacturing, enabling more flexible and responsive production processes.

One of the most critical advancements in this journey was the development of specialized inks designed specifically for textiles. Reactive, acid, and pigment-based inks enhanced the quality of fabric prints, offering greater colour vibrancy, better adhesion, and improved wash fastness (Ujiie, 2006). More recently, the emergence of nano inks has further increased precision and durability in textile prints, pushing the boundaries of what digital tools can achieve (Tendulkar & Goyal, 2024).

The evolution of 3D inkjet printing also represents a significant leap forward, allowing for the creation of textured and layered designs. This not only enhances the aesthetic appeal of textiles but also enables structural innovation in fabric construction (Ng & Zhou, 2013).

8.3 Design Software and Digital Integration

Parallel to advancements in hardware, design software has played an equally transformative role in digital textile printing. Early computer-aided design (CAD) programs laid the groundwork for modern digital textile workflows, and over time, specialized software like Adobe Illustrator, CorelDRAW, Pointcaré, Intarsia by ShimaSeiki, and Canvas have evolved to meet the industry's complex demands. Adobe Illustrator, initially launched in the 1980s, has become a cornerstone for vector-based design, enabling designers to create scalable, intricate patterns with precision and ease (Tippett, 2002). Its robust toolset supports everything from motif creation to repeat pattern design, making it indispensable in textile graphics.

CorelDRAW, similarly, gained popularity as an intuitive graphic design suite that combines vector illustration with advanced colour management. It is widely used in textile studios for its user-friendly interface and powerful layout capabilities, facilitating the preparation of print-ready files with optimized colour profiles (Ujiie, 2006).

Pointcaré, a lesser-known but innovative software, specialises in generating patterns and complex surface designs, enhancing creativity in fabric textures and non-repetitive motifs. Its algorithmic approach allows textile designers to explore new aesthetic territories that traditional manual design cannot easily achieve (Tkalec et al., 2022).

Intarsia, developed by Shima Seiki, integrates CAD with digital knitting and weaving simulations, bridging design with production planning. This tool allows for a detailed preview and adjustment of textile structures before manufacturing, ensuring the final fabric aligns perfectly with the designer's vision (Stellmach et al., 2022). Canvas software complements these by providing versatile illustration and editing functions tailored for textile designers focusing on both surface patterns and technical textiles, supporting a smooth transition from design to digital print output (Tendulkar & Goyal, 2024).

These digital tools have revolutionized the textile design process by enabling seamless integration with production workflows. Asset management systems and automated scheduling tools now allow for faster turnaround times and enhanced flexibility, particularly beneficial for custom or small-batch orders an increasingly important market segment. This advancement supports the rise of Just-In-Time production models, which minimize waste and align with sustainable business practices.

Sustainability is a central concern in the textile industry today, and the synergy between advanced software and digital printing technologies plays a key role. Compared to traditional dyeing and printing methods, digital printing significantly reduces water consumption, chemical use, and waste generation. The adoption of water-based inks and energy-efficient digital printers further bolsters the environmental benefits of this approach (Tendulkar & Goyal, 2024).

Digital design tools also enable more accurate demand forecasting and inventory management, reducing overproduction and supporting on-demand manufacturing. This not only promotes sustainability but also meets growing consumer demands for personalized, fast-delivery products. While the efficiency and precision of digital methods are undeniable, there remains a profound respect for traditional craftsmanship. The most successful textile innovations today balance digital sophistication with the tactile artistry of handcraft, preserving heritage while embracing technology.

Looking ahead, the evolution of textile design software continues to accelerate. Integration with artificial intelligence, machine learning, and 3D printing promises even greater personalization, automation, and creative freedom. These emerging technologies are set to further dissolve the boundaries between design and production, ushering in a new era where textiles are conceived, created, and consumed in profoundly innovative ways (Tkalec et al., 2022).

8.4 Generative Design

Generative design is reshaping textile printing by bringing together the worlds of computer modeling, creative design, and modern manufacturing. At its heart, it's a process where designers use algorithms and sets of rules to automatically create visual patterns. The results are often beautifully complex, unpredictable, and visually captivating—far beyond what traditional, repeat-based methods usually allow (Klooker & Hölzle, 2024).

One exciting approach used in generative design is based on *cellular automata*—a mathematical concept introduced by John von Neumann and later made popular by Stephen Wolfram. Imagine a grid where each square (or "cell") follows simple rules and reacts to the cells around it. When applied to textiles, this system can produce patterns that seem to grow and evolve on their own, creating flowing, non-repeating designs directly on fabric during the printing process (Klooker & Hölzle, 2024; Wolfram & Gad-el-Hak, 2003).

What makes this especially fascinating is how closely it mirrors the thought process of skilled textile designers. Instead of relying on fixed motifs and repeat patterns, the software builds new sections of the design in real time, guided by principles of visual balance and harmony rooted in design theory (Klooker & Hölzle,

2024). This allows for a level of richness and depth that's hard to achieve through conventional techniques.

Generative design isn't just about making things look good—it's also incredibly practical. By linking these design systems with digital printers, designers can create unique, evolving patterns that print continuously. These designs can be saved in parts and sent directly to printers, streamlining the process and reducing unnecessary repetition (Klooker & Hölzle, 2024). It's a perfect example of how creative thinking and technology can come together in interdisciplinary, practice-led ways that are transforming how textiles are made.

Perhaps most importantly, this approach is changing the way the industry thinks about pattern-making. Instead of being limited to a few repeating units, generative design opens the door to endless variation and customization. This supports on-demand production, cuts down on waste, and allows for more sustainable and flexible manufacturing—all while giving designers more creative freedom (Burry, 2011).

Digital textile printing has revolutionized the industry through the integration of advanced design tools and technologies. Key tools include digital image design software, various inkjet printing methods, and specialized printing technologies that enhance design capabilities and production efficiency. These tools not only facilitate intricate designs but also promote sustainability and customization in textile production.

8.4.1 Digital Image Design Software and Its Process

Digital image design allows users to create or scan designs, ensuring compatibility with various operating systems and formats (Wang & Memon, 2023). The process of digital image design and creation of printed images on textile fabrics involves several key steps:

- *Image Design*: The initial step is to create the digital image. Users can either design patterns using a mouse or stylus pen within a software application or scan existing artwork into the computer. Scanning can be performed using a flatbed scanner for smaller images or a rotary scanning device for larger patterns. Colour cameras can also be utilized for image capture, although this method is less common.
- *Data Representation*: Once the image is captured, it is converted into digital data that represents the location and colour of each pixel. This data must be compatible with various operating systems, such as Mac and Windows, and formatted correctly in terms of resolution, colour mode, scale, and size. Different encoding systems have been developed to manage this data, with many tracking the RGB values of pixels. For instance, MS Windows monitors use a bitmap display system that requires three bytes of data per pixel, read line by line from the bottom of the image.

- *Calibration*: To ensure high-fidelity colour measurement and reproduction, the imaging device must be calibrated. This step is crucial for achieving consistent colour output across different prints.
- *Printing Process*: The actual printing of the design onto textile fabrics involves spraying tiny drops of various colour inks onto the fabric surface in a prearranged pattern. This is typically done using thermal bubble inkjet printers, where ink is projected from the nozzle by heating tiny resistors in the print head. The heating creates a bubble that forces a small amount of ink through the nozzle and onto the fabric.
- *Final Output*: The result is a vibrant and detailed printed image on the textile fabric, ready for use in various applications, from fashion to home decor.

8.5 Inkjet Printing Technologies

Includes dye sublimation and pigment printing, which enable high precision and a wide colour gamut (Treadaway, 2009). Ink Jet Printer Structure: Ink jet printers for textiles are similar to those used for paper. They consist of a structure that allows fabric to be fed through while a movable print head sprays ink onto the surface. This process is controlled by a computer that determines how much colour is applied to create the desired image or pattern.

8.5.1 Types of Ink Jet Printing

There are two main types of ink jet printing used for textiles.

8.5.1.1 Direct Ink Jet Printing

This method applies ink directly onto the fabric. It is versatile and can be used on various textile substrates, making it suitable for a wide range of materials. Direct inkjet printing on textiles is a modern, precise method that blends chemistry, engineering, and creativity. At its core, it uses specially formulated water-based inks made of two parts: an aqueous phase (mostly water, some solvents like ethylene glycol, and surfactants to manage flow and wetting) and a dispersed phase that includes colorants—like pigments or dyes—along with ingredients that help form a protective film once the print is complete (Kan & Yuen, 2012).

These inks are designed to flow smoothly through advanced piezoelectric printheads. These printheads release microscopic ink droplets in carefully controlled patterns, enabling detailed, vibrant designs to be printed directly onto fabrics without physical contact (Ali et al., 2019; Ujiie, 2006).

After the ink is deposited onto the fabric, the curing process starts. This is where science takes over again—heat or electron beam energy activates special ingredients in the ink (called initiators), which trigger a chemical reaction. This reaction causes the liquid ink to harden into a thin film that binds the colour tightly to the textile, making it resistant to washing and fading (Glogar & Ercegovic Razic, 2024; Glogar et al., 2025). For delicate fabrics, electron beam curing can be used instead of heat (Wang & Memon, 2023).

A strong bond between ink and fabric is critical for durability. Dyes can form chemical bonds with the fabric fibers, especially on natural materials like cotton, while pigments sit on the surface and are held in place by a polymer film. Pretreatments are often applied to fabrics to help ink stick better and to keep designs sharp and bright (Wang & Zhang, 2020).

Once printed and cured, the fabric goes through drying to remove any leftover moisture and solvents. The result is a soft, breathable textile with long-lasting, colourful designs. Before it's ready for use, every print is checked for colour accuracy, durability, and overall quality.

Thanks to digital tools, designers can now create unique, non-repeating patterns with ease. This technology supports personalized production, reduces waste, and brings creativity and sustainability together in the textile world (Huang et al., 2019; Russell, 2013).

8.5.1.2 Dye Sublimation Printing

This is an indirect method where the dye is first printed onto a special paper. The dye is then transferred to the fabric using heat, which turns the dye into gas, allowing it to penetrate the fibers of the fabric. This method is particularly effective for synthetic fabrics (Boisvert & Adelstein, 2015). Dye sublimation printing is a precise and innovative textile coloration method, widely used on polyester or polymer-coated fabrics. This process uses disperse dyes that can transition directly from solid to gas without passing through a liquid phase—a phenomenon known as sublimation. This enables the dye to penetrate the synthetic fibers, producing vibrant, durable prints (Glogar et al., 2025; Ujiie, 2006).

The process begins with **digital design creation** using software like Adobe Illustrator or CorelDRAW, optimized for CMYK colour formats. The designs are printed onto specialized **sublimation transfer paper** using inks that contain disperse dyes suspended in a glycol- or water-based vehicle (Kan & Yuen, 2012; Russell, 2013). These inks are formulated with humectants to control drying, surfactants to stabilize droplet formation, and preservatives to prolong shelf life (Tyler, 2005).

Next, the printed transfer paper is placed in close contact with the polyester fabric. A **heat press** applies temperatures between 175 and 210 °C, initiating the sublimation process. At this stage, the disperse dye molecules sublimate—transitioning from solid to gas—following the Clausius-Clapeyron thermodynamic principle, which describes the effect of pressure and temperature on vapor pressure (Ali et al., 2019).

The gaseous dye penetrates the **amorphous regions** of the polymer matrix. Under heat, the polyester fibers relax, increasing their permeability. Dye molecules then diffuse into the polymer structure, where they become **physically entrapped** as the fabric cools and returns to its original structure. This results in high wash fastness and durability, as the dye resides within the fiber rather than on the surface (Kato et al., 2017; Wang & Memon, 2023).

Importantly, dye sublimation is considered **eco-friendly**. The process produces no liquid effluent, unlike traditional wet dyeing methods. Since the only waste is recyclable transfer paper, and no post-print washing is needed, its environmental footprint is minimal (Russell, 2013; Tyler, 2005).

Sublimation printing supports complex designs, photographic detail, and seamless all-over prints, making it ideal for apparel, sportswear, and home textiles. Its combination of technical precision, aesthetic quality, and sustainability continues to make it a preferred method in modern digital textile printing.

8.6 Print Heads

The print head is a crucial component that releases tiny droplets of ink onto the fabric. There are different types of print heads, including.

8.6.1 Thermal Drop on Demand

Uses heat to create droplets of ink. Thermal Drop-on-Demand (DOD) inkjet printing is a sophisticated digital printing method widely employed in textiles, utilizing rapid thermal pulses to precisely eject ink droplets onto fabric substrates. The process combines thermal chemistry, fluid dynamics, and digital control to produce high-resolution, vibrant prints.

The inks formulated for thermal DOD printing are aqueous solutions containing water, co-solvents (such as ethylene glycol), surfactants, and colorants including reactive, disperse, or pigment dyes. These inks are optimized for low viscosity (1–5 cP) and controlled surface tension (approximately 30–40 mN/m), critical for stable jetting and prevention of nozzle clogging. Colorants and additives must withstand brief exposure to elevated temperatures (~300 °C) without chemical degradation (Boumegnane et al., 2024; Tawiah et al., 2016).

Upon receiving an electrical trigger, a micro-heating element within the print head rapidly elevates the temperature of ink in the nozzle chamber. This causes a thin volume of ink to vaporise, forming a transient high-pressure steam bubble. The sudden expansion of this bubble exerts a mechanical force, ejecting an ink droplet from the nozzle orifice onto the textile surface. The droplet size and velocity are controlled by modulating the heating pulse's intensity and duration, enabling precise image formation (Lohse, 2022; Wang et al., 2018).

After droplet ejection, the heating element cools, and the vapor bubble collapses, generating a negative pressure that draws fresh ink into the chamber, readying the system for the next droplet. This cycle can repeat thousands of times per second, supporting high-speed and high-fidelity printing (Kan & Yuen, 2012).

On the textile, dye-fibre interactions depend on fibre chemistry: reactive dyes covalently bond with cellulose fibres during curing, disperse dyes sublimate into polyester fibres, and pigment inks adhere via polymeric binders. These chemical fixations ensure wash-fastness and colour vibrancy (Negi, 2025; Saha, 2023).

In summary, thermal DOD printing harnesses thermally induced phase changes in ink to mechanically control droplet ejection, integrating complex chemistry and physics for efficient, detailed textile printing.

8.6.2 Piezoelectric Drop on Demand

Uses an electric charge to push ink through the nozzle, allowing for precise control of droplet size. The **Piezoelectric Drop-on-Demand (DOD)** method is a sophisticated, non-thermal inkjet printing technique widely used for textile applications. It allows for precise control over droplet ejection by using **piezoelectric ceramics**, such as lead zirconate titanate (PZT), that deform under electric voltage (Wang et al., 2019; Wijshoff, 2010). This deformation generates a mechanical pressure pulse in the ink chamber, pushing a droplet through the nozzle with remarkable precision (Derby, 2010).

The **ink formulation** is critical for the system's performance. Textile inks used in piezoelectric printing are typically water-based or solvent-based and include **colorants** (dyes or pigments), **surfactants**, **binders**, and **humectants** (Kan & Yuen, 2012). These ingredients are designed to maintain a stable viscosity (5–20 cP) and surface tension (30–45 mN/m), ensuring clean, clog-free droplet formation (Hoath, 2016; Wang & Memon, 2023; Wang et al., 2018). The ink must also be chemically compatible with the print head materials and fabrics to avoid damage or ink instability.

During operation, when a voltage pulse is applied, the piezoelectric actuator deforms inward, compressing the ink and ejecting a micro-droplet—typically 2–20 picoliters—onto the fabric surface (Dong et al., 2006). The **waveform** of this pulse (rise time, dwell time, and fall time) determines the shape, size, and velocity of the droplet (Liu et al., 2013), often exceeding 3–10 m/s for accuracy.

On the textile surface, ink interaction varies by fabric type and ink chemistry:

Reactive dyes chemically bond with cellulose in cotton during alkaline steaming (Park et al., 2012). **Pigment inks** rely on binder films formed during thermal fixation to adhere to any fiber.

Disperse dyes, typically used on polyester, require sublimation and diffusion at high temperatures (~200 °C) for fixation (Chavan & Langer, 1988). Fabric pretreatments—often cationic polymers—improve ink adhesion, prevent bleeding, and enhance color vibrancy.

This method offers several advantages: no thermal stress on ink, broad ink compatibility, variable droplet sizes (greyscale printing), and fine resolution. It is particularly suitable for **short-run, customized textile production** and aligns well with sustainable goals due to reduced water and chemical use (Jang et al., 2009). Challenges include maintaining ink stability, ensuring uniform droplet shape, and achieving optimal droplet-substrate interaction without bleeding or smudging. As piezoelectric systems become more refined, their integration with **digital textile manufacturing** continues to support high-speed, high-fidelity, and eco-friendly production.

8.6.3 Continuous Ink Jet

Maintains a continuous flow of ink, with some droplets being deflected onto the fabric while others are recycled (Boumegnane et al., 2024). Continuous Inkjet (CIJ) printing is a high-speed, non-contact method widely employed in textile applications for its precision and adaptability across various fabrics. The process involves the continuous ejection of ink droplets from a nozzle under pressure, controlled by mechanical and electrical inputs.

In CIJ systems, ink is pressurized and expelled through a micro-nozzle, forming a continuous stream. A piezoelectric transducer induces high-frequency vibrations, causing the stream to break into uniform droplets via Rayleigh instability. These droplets pass through a charging electrode, where electrostatic charges are applied based on digital image data. Charged droplets are then deflected by high-voltage plates, directing them onto the substrate or into a gutter for recirculation (Vodyashkin et al., 2025). The ink's chemical composition is crucial for optimal performance. CIJ inks typically comprise solvents (e.g., glycol ethers), dyes or pigments, surfactants for wetting control, humectants to prevent nozzle clogging, and pH stabilizers for system longevity. These formulations are engineered to maintain appropriate viscosity and surface tension, ensuring consistent droplet formation and accurate placement (Tawiah et al., 2016).

Upon reaching the textile surface, the ink interacts chemically with the fibers. Reactive dyes may form covalent bonds with cellulose, while disperse dyes are used for polyester via sublimation and physical entrapment. Post-printing, drying and fixation are typically conducted using heat or infrared energy to evaporate solvents and secure colorfastness. CIJ printing offers rapid, uninterrupted production of high-resolution prints on various textiles. Its ability to recycle unused ink and operate with water-based formulations enhances environmental sustainability. However, challenges such as ink evaporation at the nozzle and electrostatic disturbances necessitate chemically robust inks and precise waveform control.

8.6.4 Colorants

The type of ink used depends on the fabric. Common colorants include reactive dyes for natural fibers like cotton and acid dyes for protein fibers like silk. Pigment inks are also used, which adhere to the fabric surface with the help of binders.

Substrate Preparation: Before printing, the fabric may need to be treated with chemicals to help the ink bond effectively. This preparation is essential for achieving vibrant and long-lasting colorr.

Post-Printing Process: After printing, the fabric often undergoes additional treatments, such as steaming or washing, to fix the colors and remove any chemicals used during preparation.

This overview provides a basic understanding of how ink jet printing works on textile fabrics, highlighting the key components and processes involved.

8.7 Digital Journey of Textile Motifs

8.7.1 Evolution of CAD/CAM System

The origins of CAD (Computer-Aided Design) and CAM (Computer-Aided Manufacturing) in the textile printing industry can be traced back to the 1970s (Ikubanni et al., 2022) when early computing technology was first applied to industrial design tasks. Prior to this era, textile printing relied heavily on manual processes—designers hand-drew patterns, calculated repeats, performed colour separations with film, and physically prepared printing screens or rotary cylinders. These manual workflows were time-consuming, labour-intensive, and often error-prone, making the process inflexible and expensive, particularly for small production runs or frequent design changes. The introduction of CAD systems changed this dynamic by allowing designers to create, visualise, and edit patterns on-screen with precision and speed. Initially used for simple layout and colour work, CAD gradually evolved to support more complex tasks such as designing woven structures and generating seamless repeats. In parallel, early CAM tools began to automate aspects of production by generating loom instructions and interfacing with screen engraving systems.

The 1980 and 1990s marked a period of rapid growth for CAD/CAM in textile design, especially as personal computing became more accessible (Jhanji, 2018). Software such as NedGraphics, Pointcarre, and Arahne began to dominate the industry, providing user-friendly graphical interfaces that allowed textile designers to simulate fabric behaviour and colour effects digitally. As colour calibration and printer profiling tools improved, the digital outputs became increasingly reliable, matching physical results with minimal sampling. At the same time, CAM software became more intelligent, capable of directly translating CAD files into formats suitable for rotary screen engraving, flatbed screen printing, and increasingly, digital inkjet printers.

In the 2000's there was an important milestone in the CAD/CAM landscape, the full integration with digital textile printing technologies. Unlike traditional screen printing, which required labour-intensive preparation of physical screens for each colour and pattern, digital printing allowed for the direct application of ink onto fabric via high-resolution inkjet heads (Koseoglu, 2019). This advancement offered significant advantages—designs could be printed in millions of colors, in photographic detail, without the need for physical screens or long setup times. To fully leverage this technology, however, CAD/CAM software needed to evolve. It now had to support colour management systems (CMS), RIP (Raster Image Processing) software, and ICC (International Colour Consortium) profiles to ensure that colours were consistent between screen and print. Applications such as DigiFab's Evolution Textile RIP and Arahne print provided end-to-end control—from the digital artwork to the printer output—making it possible to handle high-speed production and short-run customisation with equal ease.

As digital printing grew in popularity, particularly in fast fashion, sportswear, and home textiles, the demand for flexible, scalable CAD/CAM systems soared. Furthermore, CAD tools were adapted to support pretreated digital fabrics, pigment ink compatibility, and various types of finishing processes (Glogar et al., 2025). This shift transformed the role of the textile designer from an artist working with analog tools into a digital engineer who could control nearly every aspect of fabric creation, from concept to final product. As a result, brands and manufacturers were able to reduce lead times, lower material waste, and produce more sustainable collections with less environmental impact—goals that would have been nearly impossible to achieve with traditional print methods.

Entering the 2010 and 2020s, the textile industry experienced a new wave of innovation with the rise of AI (Artificial Intelligence), machine learning, and cloud computing (Koseoglu, 2019). These technologies brought a new level of intelligence and automation to CAD/CAM systems. AI-powered design tools can now suggest pattern variations, auto-generate colourways based on trend data, and even detect design flaws or production issues before they occur. Companies like Browzwear, CLO 3D, and NedGraphics have integrated 3D garment visualisation into their design suites, allowing designers to preview how printed textiles will look on actual garments or interior products in real-time (Vodyashkin et al., 2025). This not only improves design accuracy but also eliminates the need for multiple physical prototypes, making the sampling process faster, cheaper, and more sustainable.

Another major development is the shift toward cloud-based platforms. With remote collaboration becoming essential in the post-pandemic world, cloud-enabled CAD/CAM systems such as NEDhub, a cloud-based platform from NedGraphics, allow global teams to work on the same project simultaneously, access shared libraries of designs and colour standards, and directly connect with printing service providers (الحلواني & عزة, 2025). These systems also incorporate sustainability metrics, helping designers and manufacturers estimate the environmental impact of different printing methods, ink choices, and fabric substrates. Through advanced simulation and data analytics, companies can now make better-informed decisions about how to reduce their carbon footprint while maintaining design quality and brand identity. These innovations underscore how CAD/CAM technology has evolved from a productivity tool into a strategic asset that supports not only creative excellence but also environmental and operational sustainability.

8.7.2 *Pointcarre*

Pointcarre is a French company that has been a significant player in the textile design software industry for over three decades. Established in 1988, Pointcarre has developed a comprehensive suite of CAD/CAM solutions tailored for the textile industry, encompassing weaving, knitting, printing, and 3D mapping. The company's software is compatible with both Mac and PC platforms, offering flexibility to users worldwide (Halaceli, 2012).

8.7.3 *Ned Graphics*

NedGraphics, founded in the Netherlands during the 1980s, has steadily evolved into a global pioneer in textile CAD/CAM software. What began as a focused effort to streamline woven fabric design soon expanded into a robust suite of tools supporting jacquard, knitting, digital printing, tufting, and advanced color management. In its early decades, the company gained international traction by launching innovative software for weaving and jacquard design. By the 2000s, NedGraphics introduced modular tools such as Texcelle, EasyWeave, and Color Reduction, which allowed designers to manage both creative and technical aspects of textile production more efficiently. As digital workflows advanced, the 2010s saw NedGraphics integrate with Adobe Illustrator and offer photorealistic fabric simulations, making it easier for designers to visualize outcomes and collaborate across platforms, including PLM and cloud systems. More recently, the 2020s have marked the rollout of AI-powered tools, the collaborative NEDhub platform, and Jacquard Connect, which links designs directly to looms. By 2023–2025, NedGraphics had launched specialised studios—Creative, Print, Tuft, and Jacquard—to further streamline workflows and enhance productivity across design disciplines (الحلوانى & عزة, 2025).

8.7.4 *Arahne*

Arahne, established in 1992 in Slovenia, has carved a niche for itself as a specialized software company dedicated to CAD/CAM solutions for woven fabric design. From the very beginning, Arahne focused on addressing the specific needs of textile manufacturers, offering tools that enhance both creativity and production efficiency. In the 1990s, the company launched its flagship software, ArahWeave, which became known for its precision in real-scale fabric simulation and seamless integration with industrial dobby and jacquard looms. As digital design evolved, Arahne expanded its offerings in the 2000s with ArahPaint, a raster graphic editor designed specifically for textile designers. This tool allowed users to create custom repeats, perform color reductions, and develop intricate dobby patterns with ease. Entering the 2010s, the company released ArahDrape, a tool for realistic 3D visualization of fabrics on garments and furniture, while also broadening compatibility with major loom manufacturers like Stäubli and Bonas. In the 2020s, Arahne embraced collaboration and innovation by partnering with Gemell Technology, enabling 3D fabric viewing through FabricViewer. It also introduced automation tools and cost estimation features to support industrial-scale production, while continuing to enhance products like ArahView 3D and ArahCatalog, making visualization, design, and archiving more efficient and accessible for modern textile designers (Sapurina et al., 2020).

8.7.5 *DigiFab*

DigiFab Systems, based in Los Angeles, has been at the forefront of digital textile printing innovation, offering a complete ecosystem of solutions that includes CAD/RIP software, pre-treated fabrics, and high-performance digital printing equipment. Since its inception, the company has focused on making digital printing more accessible, efficient, and high-quality for the textile industry. A major milestone came in 2008 when DigiFab was honored with the DPI (Digital Printing & Imaging Association) Product of the Year Award in the Textile Media category for its inkjet-printable cotton canvas line—an achievement that underscored its expertise in advanced fabric coating technologies. That same year, DigiFab launched the StampaJet, a cutting-edge 8-color digital textile printer capable of delivering high-speed prints on a variety of fabric substrates, reaching speeds of up to 39 linear yards per hour. Building on this momentum, in 2009, the company introduced the Evolution Textile RIP and coloring software, providing designers with a powerful, all-in-one solution for textile design, precise color matching, and streamlined digital workflows (Boisvert & Adelstein, 2015).

8.8 The Way of Converting Design to Print on Textile via Software

Textile printing software supports the full workflow from digital image creation to production-ready output files across various printing techniques. These software is designed to bridge the gap between design and production by converting digital artwork into formats compatible with screen printing, digital (inkjet) printing, rotary printing, and even sublimation or transfer printing methods.

8.8.1 Image Preparation and Design Tools

The process begins with the creation or import of a digital design using raster or vector files. Designers can create motifs or patterns using tools that allow:

- Seamless repeat creation (Glogar et al., 2025)
- Color palette management
- Design layering
- Pattern simulation on real fabric textures or garments.

Once the design is finalised, the software enables image resolution management and scaling based on the required output format (e.g., DPI suitable for digital printing vs. screen printing).

8.8.2 Colour Separation Based on Printing Technique

Colour separation plays a crucial role in adapting designs for different printing techniques. It automatically performs:

- Colour separations for rotary and flatbed screen printing by reducing the number of colour layers, often optimizing to 6–12 colours.
- Indexed colour mapping, essential for generating films/screens in traditional printing.
- Continuous tone (CMYK or RGB) output for digital textile printers, especially inkjet systems that use full-colour processing without separations (Parraman, 2017).
- The designer selects the print method, and the software adjusts the separations accordingly. For example:
- In screen printing, it reduces the design to solid blocks of spot colours.
- In digital printing, it prepares high-resolution output with ICC profiles for colour accuracy.

8.8.3 Simulation and Printing-Specific Output

After separation, some applications support 3D and flat simulation tools that allow the user to preview how the design will appear on fabric or a finished garment under the chosen printing technique. This helps in spotting design inconsistencies, scale issues, or colour mismatches early. Depending on the selected technique, it generates appropriate output files:

- TIFF (Tagged Image File Format), PSD (Photoshop Document), or JPG/ JPEG (Joint Photographic Experts Group) for digital printing.
- 1-bit TIFF or vectorised files for screen engraving, A black-and-white (monochrome) version of the TIFF file, where each pixel is either on or off (1 or 0).
- Separated channel PDF (Portable Document Format) files for rotary or flatbed printers
- The software can also export production-ready colourways, engraving guides, and print parameters directly to compatible machines or RIP software, minimising the need for manual adjustment at the print facility.

8.8.4 Workflow Integration and Machine Compatibility

Printing application supports seamless integration with various production environments and machines:

- Digital Printers: Epson, Mimaki, Mutoh, MS, Kornit, and others.
- Screen and Rotary Printers: Zimmer, SPGPrints, Reggiani
- Color Management Systems: Compatible with Pantone®, ICC profiles, and custom color books.

It also integrates with PLM and ERP systems, enabling full traceability from design through printing to production.

8.8.5 Working and Execution of Software

This diagram as shown in Fig. 8.1 illustrates the sequential workflow involved in preparing an image for screen printing using digital tools. Process Flow Chart of Software Operation and Execution:

1. **Input Image**: The process begins with an image either **created**, **scanned**, or **captured**.
2. **Color Reduction**: The number of colors in the image is minimized to suit the printing process.
3. **Cleaning**: The image is refined to remove any noise or unwanted details.
4. **Repeat Setting**: The design is aligned and adjusted to be seamlessly repeated.
5. **Color Ways**: Multiple color options or variations are created for the same design.
6. **Color Separation**: Each color is separated for individual screen creation.
7. **Screen Generation (B/W)**: Final black-and-white positives are prepared for screen exposure.

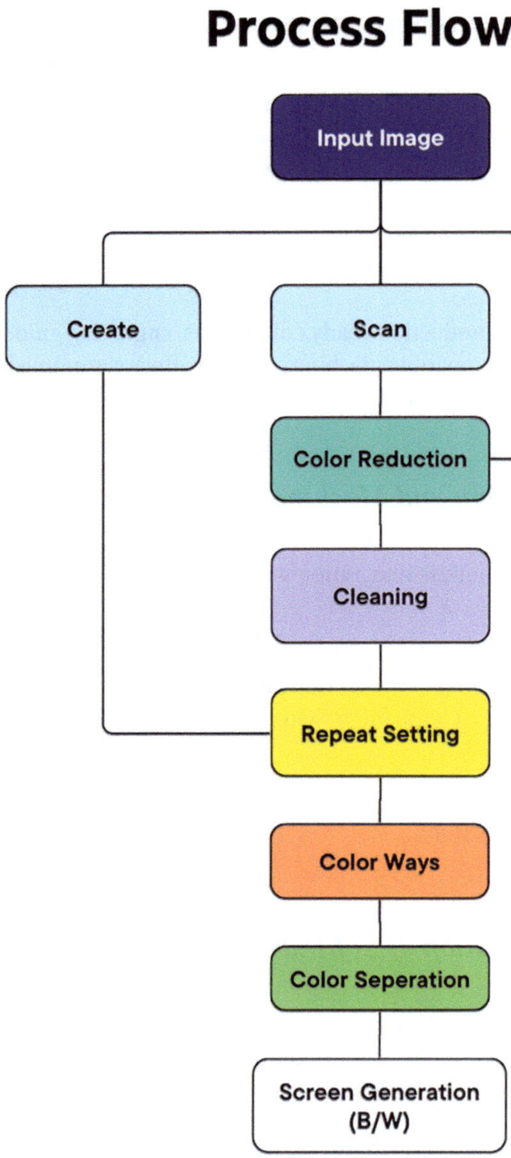

Fig. 8.1 Process flow chart of software operation and execution

The image in Fig. 8.2 visually represents the process of color separation and screen preparation using digital tools like Adobe Photoshop, which is widely used in textile and graphic design industries. The process begins with a Photoshop-generated multicolor design, which undergoes digital color separation a method of dividing the image into its component colors. In this case, the image has been separated into five individual screens, each representing a distinct color used in the final print. This is

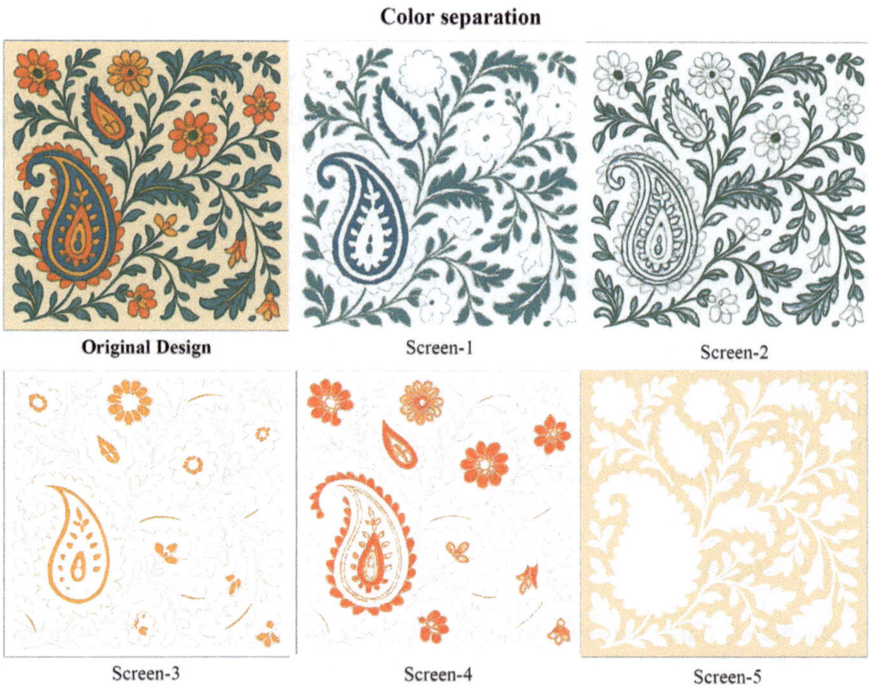

Fig. 8.2 Color separation and screen printing process illustrated using a Photoshop-generated design

crucial for screen printing, where each color is applied separately through different screens.

The key steps depicted in this process include:

Original Design Preparation: The composite image is finalized in Photoshop with all desired colors and details.

Channel Separation: The image is analyzed and separated into different color channels—commonly Cyan, Magenta, Yellow, Black, and an additional spot color if required.

Halftone Conversion (if applicable): Each channel is converted into a halftone dot pattern suitable for screen printing, especially for tonal variations.

Screen Preparation: Each separated color is printed as a black-and-white film positive and used to create individual screens.

Printing Process: During production, each screen is aligned (registered) precisely, and inks are applied one at a time to reproduce the original image accurately.

This method enables layer-by-layer application of colors, ensuring clarity, sharpness, and vibrant reproduction in the final output. It is especially critical in fabric printing where color fidelity and durability are essential.

8.8.5.1 Input Image

The specialised print software is used to import scanned images, digital photographs captured from the camera, or any true colour (RGB) file for processing images and also the user can create a new design directly from the application using the available drawing tools.

8.8.5.2 Colour Reduction and Cleaning

This process is used to reduce and separate the number of colours in a scan to the desired number of colours. Colour Reduction generates flat colours, textures, tones, and watercolour techniques. And automatic cleaning tools eliminate unnecessary colours and replace them with pre-selected colours. The designer can preview the reduced colour image in real time, allowing fine-tuning of colour selections.

8.8.5.3 Repeat Setting

Repeat Setting tools are used to develop new concepts, collections, and coordinating designs by copying motifs and applying textures. Then customise the output to the precise scale and repeat size that is desired. It supports all varieties of pattern repeats, including complex mirroring.

8.8.5.4 Colorways

This feature is to create colour variations for any type of design and automatically save and manage all colour variations in one main file to track and archive our creative ideas. Colour calibration matches colour across monitors and printers, as well as for paper and digitally printed fabrics.

8.8.5.5 Colour Separation

Colour separation processes in one of the most important steps in the printing industry. Except for digital print all other kind of print requires separating the colour. Because the number of colours decides the number of screens required for printing. If there are 5 colours in a design, excluding background, then 5 screens need to be developed. In this process, all the colours are separated into different layers.

8.8.5.6 Screen Generation

This is the last step involved after color separation. All the separated colours are converted into B/W colors to generate the films for further printing process.

8.8.6 Digital Printing Software Advantages for Sustainable Fashion

Digital Printing software plays a critical role in promoting sustainability within the textile printing industry by optimizing production processes, reducing resource use, and enabling environmentally responsible practices.

8.8.6.1 Digital Design Reduces Sample Waste

Design software enables the creation and visualization of digital samples before actual production. This reduces the need for multiple physical prototypes, which often end up as waste. Digital textile design software allows for virtual sample creation, reducing material waste by up to 60% (Glogar et al., 2025).

8.8.6.2 Optimized Color Management and Ink Usage

Advanced design tools include features for precise color calibration and ink control, ensuring exact results with minimal ink waste. Design software integrated with digital printers improves color accuracy and reduces ink consumption by 30% compared to traditional processes (Singholi & Sharma, 2021).

8.8.6.3 Efficient Layout and Pattern Placement

Software like Adobe Illustrator, CLO, or NedGraphics allows for precise pattern placement, which minimizes textile offcuts during printing. Smart patterning reduces up to 20% of fabric waste by optimizing layout efficiency in digital printing (Christina, 2021).

8.8.6.4 Support for On-Demand Production

Design software enables quick iteration and customisation, making small-batch or made-to-order production more feasible. This model prevents overproduction and reduces inventory waste. On-demand textile printing supported by design software helps eliminate unsold stock, a major source of industry waste (Singholi & Sharma, 2021).

8.8.6.5 Integration with Sustainable Materials Libraries

Many design platforms are connected to digital material libraries that highlight low-impact and eco-certified fabrics and inks. Design tools connected to sustainable material databases empower designers to make more eco-conscious choices from the start (Singh et al., 2023).

8.8.6.6 Circularity and Lifecycle Planning

Some software includes features to plan for a product's end-of-life, like recyclability or biodegradability. This fosters circular design thinking in printed textiles. Sustainable design software helps map the lifecycle of printed textiles, encouraging recyclability and reuse (al Faisal, 2024; Ali et al., 2019).

8.9 Conclusion

The transformation brought about by digital tools in textile printing has been remarkable. From basic inkjet printers to sophisticated 3D systems and AI-driven software, each innovation has expanded the creative and operational horizons of the industry. As digital and traditional techniques continue to converge, the future of textile design and production looks not only technologically advanced but also more sustainable, responsive, and artistically rich.

In summary, designing tools are at the core of the digital textile printing revolution, enabling a seamless integration of artistry, engineering, and sustainability. These tools empower designers to push creative boundaries, enhance production efficiency, and minimize environmental impact. As the textile industry continues to embrace digital transformation, the role of designing tools will only grow more significant—shaping not only the aesthetics of future textiles but also the methods through which they are conceived, produced, and experienced.

In summary, the digital image design and creation process for textile printing encompasses designing or scanning images, converting them into digital data, calibrating the imaging device, and finally printing the design onto fabric using advanced inkjet technology. Each step is essential for achieving high-quality printed textiles.

Declaration The authors declare that they have no known competing financial interests or personal relationships that could have appeared to influence the work reported in this chapter. Generative AI and AI-assisted technologies have been used in the writing process to improve the readability and language of the work only. After using this tool, the authors carefully reviewed and revised the content to ensure accuracy and quality. The authors take full responsibility for the final version of this publication.

References

al Faisal, R. (2024). From design to disposal: Sustainable product design and life cycle analysis. *Green Environmental Technology, 1*(2), 67–77.

Ali, M., et al. (2019). Optimisation of screen printing process for functional printing. *Pigment & Resin Technology, 48*(5), 456–463.

Boisvert, S., & Adelstein, M. (2015). DigiFab conference explores how 3D printing and digital fabrication impact the world. *3D Printing and Additive Manufacturing, 2*(1), 2–4.

Boumegnane, A., et al. (2024). Rheological properties and inkjet printability of a green silver-based conductive ink for wearable flexible textile antennas. *Sensors, 24*(9), 2938.

Burry, M. (2011). *Scripting cultures: Architectural design and programming.* Wiley & Sons.

Chavan, R. B., & Langer, M. H. (1988). Sublimation transfer printing of polyester/cotton blends. *Textile Research Journal, 58*(1), 51–56.

Choudhary, K. (2018). Engineered digital textile print design for customized curves. *Journal of Textile and Apparel, Technology and Management,* 10(3).

Christina, C. I. E. (2021). *How can digital print technology prompt an analysis of pattern as a phenomenon, with application to examples in textile design?* RMIT University.

Derby, B. (2010). Inkjet printing of functional and structural materials: Fluid property requirements, feature stability, and resolution. *Annual Review of Materials Research, 40*(1), 395–414.

Dong, H., Carr, W. W., & Morris, J. F. (2006). An experimental study of drop-on-demand drop formation. *Physics of fluids, 18*(7).

Glogar, M., Petrak, S., & Mahnić Naglić, M. (2025). Digital technologies in the sustainable design and development of textiles and clothing—a literature review. *Sustainability, 17*(4), 1371.

Glogar, M., & Ercegovic Razic, S. (2024). Sustainable pre-treatment of cellulose knitwear in digital pigment printing processes. *International journal of clothing science and technology.* (Preprint)

Halaceli, H. (2012). Digital technologies in textile art. In *RMUTP International Conference: Textiles&Fashion.*

Hoath, S. D. (2016). *Fundamentals of inkjet printing: The science of inkjet and droplets.* Wiley & Sons.

Huang, T., et al. (2019). Fabric texture design for boosting the performance of a knitted washable textile triboelectric nanogenerator as wearable power. *Nano Energy, 58*(December 2018), 375–383. https://doi.org/10.1016/j.nanoen.2019.01.038.

Ikubanni, P. P., et al. (2022). Present and future impacts of computer-aided design/computer-aided manufacturing (CAD/CAM). *Journal Européen des Systèmes Automatisés, 55*(3), 349.

Jang, D., Kim, D., & Moon, J. (2009). Influence of fluid physical properties on ink-jet printability. *Langmuir, 25*(5), 2629–2635.

Jhanji, Y. (2018). Computer-aided design—garment designing and patternmaking. In *Automation in garment manufacturing* (pp. 253–290). Elsevier.

Kan, C. W., & Yuen, C. W. M. (2012). Digital ink-jet printing on textiles. *Research Journal of Textile and Apparel, 16*(2), 1–24.

Kato, K., et al. (2017). Tactile element with double-sided inkjet printing to generate electrostatic forces and electrostimuli. In *Adjunct Proceedings of the 30th Annual ACM Symposium on User Interface Software and Technology* (pp. 31–33).

Klooker, M., & Hölzle, K. (2024). A generative design of collaborative innovation space. *R&D Management, 54*(2), 323–346.

Koseoglu, A. U. (2019). Innovations and analysis of textile digital printing technology. *International Journal of Science, Technology and Society, 7*(2), 38–43.

Liu, Y.-F., et al. (2013). Control of droplet formation by operating waveform for inks with various viscosities in piezoelectric inkjet printing. *Applied Physics A, 111*, 509–516.

Lohse, D. (2022). Fundamental fluid dynamics challenges in inkjet printing. *Annual Review of Fluid Mechanics, 54*(1), 349–382.

Negi, A. (2025). Environmental impact of textile materials: Challenges in fiber–dye chemistry and implication of microbial biodegradation. *Polymers, 17*(7), 871.

Ng, F., & Zhou, J. (2013). *Innovative jacquard textile design using digital technologies*. Elsevier.
Park, J.-Y., Hirata, Y., & Hamada, K. (2012). Relationship between the dye/additive interaction and inkjet ink droplet formation. *Dyes and Pigments, 95*(3), 502–511.
Parraman, C. (2017). Colour printing techniques and new developments in colour printing. In *Colour design* (pp. 589–618). Elsevier.
Russell, A. P. (2013). Repeatless: the use of digital technology to extend the possibilities of printed textile design. In *Proceedings of First International Conference on Digital Technologies for Textile Industries*.
Saha, K. (2023). *Ink penetration prediction models for digital textile printing*. North Carolina State University.
Sapurina, I. Y., Shishov, M. A., & Ivanova, V. T. (2020). Sorbents for water purification based on conjugated polymers. *Russian Chemical Reviews, 89*(10), 1115–1131. https://doi.org/10.1070/rcr4955
Singh, P. K., et al. (2023). Eco-conscious creation: Navigating the nexus of sustainability and production design. In *E3S Web of Conferences* (p. 1034). EDP Sciences.
Singholi, A. K. S., & Sharma, A. (2021). Review of materials and processes used in 4D printing. In *Advances in Engineering Materials* (pp. 677–684). Springer.
Stellmach, D., et al. (2022). *Towards a digital workflow solution for cradle-to-gate sustainability information in textile value chains*.
Tawiah, B., Howard, E. K., & Asinyo, B. K. (2016). The chemistry of inkjet inks for digital textile printing-review. *BEST, 4*, 61–78.
Tendulkar, P., & Goyal, P. (2024). Advancements in Digital Textile Printing. *Technological Innovation and Sustainability: Navigating The Future*, 164.
Tippett, B. G. (2002). *The evolution and progression of digital textile printing*. New York, USA: Digital Printing Systems Orangeburg. http://brookstippett.com/docs/Print2002-BGT.pdf. (Pristupljeno 27.04. 2021.) (Preprint)
Tkalec, M., Glogar, M. I., & Sutlovic, A. (2022). Ecological sustainability of digital textile printing. *Economic and Social Development: Book of Proceedings*, 105–115.
Treadaway, C. (2009). Developments in digital print technology for smart textiles. In *Smart clothes and wearable technology* (pp. 300–318). Elsevier.
Tyler, D. J. (2005). Textile digital printing technologies. *Textile Progress, 37*(4), 1–65.
Ujiie, H. (2006). *Digital printing of textiles*. Woodhead Publishing.
Vodyashkin, A. A., Makeev, M. O., & Mikhalev, P. A. (2025). Inkjet printing is a promising method of dyeing polymer textile materials. *Polymers, 17*(6), 756.
Wang, L., & Zhang, K. (2020). Textile-based thermoelectric generators and their applications. *Energy & Environmental Materials, 3*(1), 67–79. https://doi.org/10.1002/EEM2.12045
Wang, L., Hu, C., & Yan, K. (2018). A one-step inkjet printing technology with reactive dye ink and cationic compound ink for cotton fabrics. *Carbohydrate Polymers, 197*, 490–496.
Wang, J., Huang, J., & Peng, J. (2019). Hydrodynamic response model of a piezoelectric inkjet print-head. *Sensors and Actuators a: Physical, 285*, 50–58.
Wang, H., & Memon, H. (2023). Introduction to digital textile printing. In *Digital textile printing* (pp. 1–19). Elsevier.
Wijshoff, H. (2010). The dynamics of the piezo inkjet printhead operation. *Physics Reports, 491*(4–5), 77–177.
Wolfram, S., & Gad-el-Hak, M. (2003). A new kind of science. *Applied Mechanics Reviews, 56*(2), B18–B19.
الحلوانى
& عزة (2025). The role of the textile designer in combining functional and aesthetic properties to produce fabrics for youth bags. *Journal of Heritage and Design*. (Preprint)

Chapter 9
Environmental Impact of Digital Printing Versus Traditional Textile Printing

Akshay Vade and Ashok Athalye

Abstract This case study examines the environmental impact of digital printing compared to traditional textile printing methods. The analysis focuses on four key areas: energy consumption, water usage, chemical usage and waste, and overall carbon footprint. Findings indicate that digital printing is a more sustainable alternative, consuming significantly less energy and water while generating minimal chemical waste. Traditional textile printing, characterized by high energy demands, extensive water usage, and toxic chemical outputs, presents considerable environmental challenges. Nonetheless, while digital printing shows promise in enhancing operational sustainability, it is incumbent upon manufacturers to conduct comprehensive lifecycle analyses and continuously assess new technologies as they emerge. Hybrid strategies that seamlessly integrate digital capabilities with existing traditional frameworks could be an interim solution for facilities with considerable sunk costs in legacy equipment. Ultimately, the transition to more sustainable textile printing solutions depends on coordinated efforts between technology developers, environmental engineers, and policy-makers. As the drive toward sustainability intensifies, digital printing stands as a beacon of innovation that can significantly lower the environmental footprint of the textile industry while ensuring high-quality, efficient production. For industry professionals and environmental engineers, the findings presented in this analysis underscore the critical need for informed technology choices that balance performance with environmental stewardship. Continued investment in digital printing research, coupled with broader regulatory incentives for sustainable practices, will be instrumental in driving the textiles industry toward a more sustainable future. Digital printing supports on-demand production, thereby reducing fabric waste and promoting a lower carbon footprint. The study concludes that, despite the prevalence of traditional methods in mass production, the adoption of digital printing technologies is pivotal in mitigating the textile industry's environmental impact, building a path toward a more sustainable future.

A. Vade (✉) · A. Athalye
Department of Fibres and Textile Processing Technology, Institute of Chemical Technology Mumbai, Mumbai, India
e-mail: txt22an.vade@phd.ictmumbai.edu.in

A. Athalye
e-mail: ar.athalye@ictmumbai.edu.in

A. Athalye and H. Patil (eds.), *Digital Printing*,
https://doi.org/10.1007/978-981-95-2593-5_9

Keywords Carbon foot print · Digital printing · Sustainability

9.1 Introduction

Digital printing has the ability to make dresses and home textile patterns more vibrant and visually attractive. However, like any technology, digital printing comes with a set of its limits. Traditional printing methods also face similar challenges, each of their technical obstacles and physical needs. Since technology grows at a rapid pace, digital printing is continuously improving. He said, progress is going on and technology has not yet reached its full potential. At this stage, it can be challenging for investors to take clear decisions. Investing in digital printing not only requires capital, but also a solid basis of knowledge and experience. Support for this trip can be obtained through professional counseling and intensive research. After acquiring adequate knowledge and proper viability studies, businesses can take informed decisions about investing in digital printing equipment and processes.

The purpose of our article is to offer a comprehensive and comprehensive evaluation of digital printing techniques, expecting researchers and industry professionals who are searching for the field. Digital printing machines are complex systems that bring together many subjects including electronics, mechanics, chemistry, textiles, fashion and art. It is almost impossible for a person to master all these areas. As a result, the operation of digital printing machinery efficiently requires a well - coordinated team. The first stage includes the correct tools to prepare efficient pattern designer and digital-taiyar design files. Next, you need experts for machine calibration and color management. Beyond this core, your team will include machine operators, electronic and mechanical supports.

9.1.1 Traditional Textiles Printing

Traditional textile printing, hand painting, block printing, and dyeing protest refers to the art of applying colors and patterns in fabrics using time-respecting techniques. These methods often employ natural colors and materials, resulting in specific, character-rich patterns. Here are some of the most common traditional techniques look closely (Lee et al., 2023).

Hand Painting/Kalamkari:

The design is painted freehand on the fabric using natural colors, often characterized by complex and highly artistic motifs.

Block printing:

The pattern is carved into wooden blocks. The block is then immersed in dye and cloth is sealing to create frequent design.

Oppose dyeing (Batik, Shibori, Azrakh):

In these methods, a resistance material (e.g., wax or mud) is applied to parts of the fabric. When the fabrics are dyed, the protected areas oppose the dye, making striking patterns.

Mordent printing:

A modern—a chemical that helps the fabric dye bind—is printed in the desired design on the fabric. The fabric is then painted, and the dye only follows the treated areas with the model.

Discharge printing:

This method involves printing a bunning agent on pre-colored fabrics to remove or lighten the color in specific areas, revealing the new pattern.

Examples of traditional textile print.

Kalamkari:

A classic Indian art form that involves either hand painting or block printing, often reflects mythological scenes or motifs inspired by nature.

Ajrakh:

A geometric block-crushed cloth from Gujarat and Rajasthan, built using natural colors and is known for its rich patterns and striking symmetry.

Batik:

A wax-resistant dyeing technique from Indonesia where wax is applied to fabric before dyeing, resulting in detailed and wide patterns.

9.1.2 The Evolution of Traditional Printing Techniques

The origins of traditional printing techniques trace back to ancient civilizations, where they served as powerful tools for communication, artistic expression, and the preservation of knowledge. These early practices laid the groundwork for modern printing methods and remain integral to cultural heritage techniques shown in below Fig. 9.1.

9.1.3 Historical Foundations of Printing

The earliest evidence of printing can be found in China, where woodblock printing emerged during the 7th century. This technique involved carving images and characters onto wooden blocks, which were inked and pressed onto paper or fabric. The

Fig. 9.1 Traditional methods of printing

invention of movable type, attributed to Bi Sheng in the 11th century, introduced clay and porcelain characters, which allowed for the reuse and reconfiguration of text, greatly enhancing printing efficiency.

Among the most notable achievements of this era is The Diamond Sutra (868 AD), recognized as the world's earliest dated printed book. From China, the practice of woodblock printing and movable type spread across Asia, influencing Korea, Japan, and other regions.

The European Renaissance and the Printing Press

The dissemination of paper and printing technologies reached Europe through trade and cultural exchanges, where significant innovations would further transform the craft. In the 15th century, Johannes Gutenberg's invention of the mechanical printing press, which combined movable metal type with a screw press, revolutionized mass communication. Gutenberg's method, known as letterpress printing, produced sharp, durable text and images, enabling the rapid production of books and other printed materials.

This innovation contributed to pivotal social and intellectual movements, including the Renaissance, the Reformation, and the Scientific Revolution, by making knowledge accessible to the origin of traditional printing techniques detects ancient civilizations, where they serve as a powerful tool for communication, artistic expression and conservation of knowledge. These early practices prepared the basis for modern printing methods and remained integral for cultural heritage (Lee et al., 2023).

The first evidence of printing can be found in China, where Woodlock printing emerged during the 7th century. This technique included carving images and characters on wooden blocks, which were ink and pressed on paper or cloth. In the 11th century, the movable inventive inventions, which were responsible for the two -sheng, introduced soil and porcelain vessels, allowing the reuse and re -combination of the text, with great enhancing printing efficiency.

Garden printing (Madhya Pradesh)

Bagh printing, a hand-block printing technique that originates from the garden district, is famous for mainly printed geometric motifs in red and black vegetable colors on a white background. This process begins with the fabric out to remove impurities, ensuring smooth dye absorption. Artisans take inspiration from local architecture, floral motifs and cultural symbols, making suitable designs for household items and dress.

Kalamkari (Andhra Pradesh)

Kalamkari, which means "pen craftsmanship" in Persian, includes freehand drawing and painting on cotton cloth using a sharp bamboo pen. Traditionally, artisans depicted mythology, often from Hindu epics such as Ramayana and Mahabharata. This process includes several stages of waxing, dyeing and hand paintings, producing complex functions that are widely used for both decorative and utilitarian textiles.

Azarak (Sindh, Kutch, and Rajasthan)

Ajrak is a form of iconic resistance-crushed cloth with deep crimson and Indigo Hughes symmetrical. Using carved wooden blocks and natural colors, artisans receive highly wide, reversible design. This technique is believed to have originated from the ancient Indus Valley Civilization, a symbol of regional heritage.

Dabu printing (Rajasthan)

Dabu is a labor-intensive soil-resistant block printing technique that includes selected areas of fabric with paste made of mud, lime and gum. Once dried, the cloth is dyed, and resistance is washed, which reveals the complex pattern. Usually inspired by nature, Dabu Floars are preserved in the cultural practices of Rajasthan.

Gold and Silver Dust Printing (Khari Printing, Rajasthan)

This decorative technique involves applying paste made of cloth with gum and castor oil, followed by stimulating motifs, traditionally after spraying of gold or silver spraying. Often associated with festive and formal clothing, this art form highlights the grandeur of Rajasthani heritage.

Sangneri printing (Jaipur, Rajasthan)

Known for its bright floral designs on a white background, Sangneri printing uses hand carved wooden blocks and vegetable colors. The symmetry and precision are paramount, as artisans have cautiously align the blocks to produce harmonious compositions. The motifs are usually inspired by local flora and organisms.

Bandrani (Gujarat and Rajasthan)

Bandrani, or tie-dai, include binding parts of fabric with thread before dyeing to make dotted patterns. This labor-intensive technique is deeply inherent in regional traditions and is often associated with auspicious occasions including weddings and festivals.

Lehria (Rajasthan)

Leheria is a resistance-dying method that produces stripes like diagonal, wave on bright colored clothing. Traditionally, artisans used natural colors and layered colors to achieve the signature lead (wave) effect. These textiles are usually worn during festive ceremony and as a traditional headgier.

Batik (Rajasthan)

Although primarily at Indonesian origin, the buttics are embraced by Indian artisans, especially in Rajasthan. This process employs a wax-resistant dyeing technique to create complex, fluid design. The application of wax defines the pattern, while the composition is later brought into life by dyeing.

Bagru Printing (Rajasthan)

Bagru is distinguished by the use of natural colors and wooden blocks practiced by the chipped community near Jaipur. The pattern usually features floral and geometric motifs, which is applied to a lighter color background. The technique is laborious and depends on traditional processes including washing, harsh dyeing and sun drying.

Madhubani Painting (Bihar)

The Mithila region of Bihar is a tradition of traditionally practiced by women on the walls of their homes during the origin, Madhubani painting festivals and special occasions. Using natural colors and handmade brushes, artists also portray elements of daily life and nature along with religious and mythological subjects.

9.2 Digital Printing Versus Traditional Textile Printing Which Technique is Preferred?

Digital printing is a modern method that uses ink-jet techniques to print the design directly on the fabric. This technique makes it easier to create a wide, high-resolution print, including complex patterns and even photographs. One of its biggest benefits is flexibility—unlike traditional unlike screen printing, which needs a separate screen for every color, digital printing can create multicolored designs without any extra setup. However, digital printing is not without shortcomings. Issues such as color bleeding, high ink costs, and low durability—especially when it comes to washing, because colors fade over time—can be a concern for both manufacturers and customers are which compared in Table 9.1.

Table 9.1 compares the main printing methods based on their use of a master, ink type, suitable printing surfaces (substrates), and typical applications. Sheet-fed offset printing uses a master and a thick, paste-like ink, making it ideal for printing on paper and plastic for items like publications, stationery, and packaging. Rotogravure, also using a master but with a low-viscosity ink, is commonly used for high-volume jobs like magazines and flexible packaging. Flexography, similar in ink type and materials, is widely used in packaging due to its ability to print on various surfaces. Screen printing also uses a master and medium-to-thick ink, and stands out for its versatility—it can print on paper, plastic, textiles, and more, making it perfect for posters, labels, and fabric prints. On the other hand, electrophotography and inkjet printing are digital methods that do not require a master. Electrophotography uses toner and is mainly used for printing documents, books, and some packaging. Inkjet printing uses liquid ink and is suitable for various materials, often used for documents,

Table 9.1 Comparison of the commercial printing methods

Method	Master	Ink type	Substrates	Main application
Sheet-fed offset	Yes	Paste-like liquid (high viscosity)	Paper, plastic, etc.	Publications, stationery, packaging
Rotogravure	Yes	Low viscosity liquid	Paper, plastic	Magazines, packaging
Flexography	Yes	Low viscosity liquid	Paper, plastic	Packaging
Screen-printing	Yes	Medium to high viscosity	Paper, plastic, textiles, etc.	Posters, labels, textiles
Electrophotography	No	Toner (powder or liquid)	Paper	Documents, books, labels, packaging
Inkjet	No	Liquid	Paper, plastic, etc.	Documents, marking, direct mail, posters, newspapers

direct mail, posters, and newspapers, especially when short runs or variable data printing are needed.

9.2.1 Traditional Textile Printing

High water consumption:

Traditional methods, especially dyeing and screen printing, rining, washing and preparation of clothing which uses large amounts of water, leading to significant water wastage and pollution.

High chemical use:

These methods often include the use of rigid chemicals and colors, which can be toxic and harmful to the environment and human health.

Energy intensive:

Traditional processes such as screen printing and dyeing are energy-intensive, requiring high temperatures and complex machinery, which contributes to high carbon emissions.

Waste generation:

Traditional printing methods often produce a significant amount of waste, including additional fabrics, ink, and chemicals, which or recycle can be difficult to dispose or recycle.

9.2.2 Digital Textile Printing

Reduced Water Consumption:

Digital printing uses significantly less water compared to traditional methods, as it involves precise ink application rather than soaking fabrics in dye baths. Some digital printing processes are even waterless.

Reduced Chemical Usage:

Digital printing uses water-based inks and eliminates the need for many of the harsh chemicals used in traditional methods.

Lower Energy Consumption:

Digital printing processes are generally less energy-intensive than traditional methods, as they require less heating and processing.

Reduced Waste Generation:

Digital printing minimizes waste by only using the amount of ink needed for each print, reducing material waste and chemical waste.

Faster Turnaround Times:

Digital printing is faster and more efficient than traditional methods, allowing for quicker production and customization.

Eco-Friendly Inks:

Digital printing allows for the use of eco-friendly inks and dyes, further reducing the environmental impact of the textile printing process.

Reduced Carbon Footprint:

By using less water, energy, and chemicals, digital printing can significantly reduce the carbon footprint of textile production.

9.3 What Are the Environmental Impacts of Traditional Printing Practices?

9.3.1 Energy Consumption in Printing

Another key environmental concern with printing is the high amount of energy it uses—both in making and running printers, copiers, and other equipment. These machines typically run on electricity, Since it relies on non-renewable sources like coal and natural gas, it ends up adding to greenhouse gas emissions. All this energy use adds to greenhouse gas emissions, which drive climate change.

9.3.2 Enterprises Aspects of the Environmental Impact of Printing the World

To understand the impact of using ink and toner responsibly, consider this: around 500 million ink cartridges are discarded every year. By recycling or refilling cartridges, we can cut down on landfill waste and reduce the energy and raw materials needed to produce new ones. Choosing re-manufactured or eco-friendly cartridges is another great way to lessen the environmental footprint of printing. For more details, you can also visit a toner supplier or retailer (Javoršek & Javoršek, 2011).

9.3.3 Greenhouse Gas Emission

According to ISO 14064, there is a gaseous component of greenhouse gases (GHG) The atmosphere, both natural and anthropological, which absorbs and emit radiation Specific wavelengths within the spectrum of infrared radiation Surface, environment and clouds, and emit radiation on radiation on specific wavelengths within, [. PAS for 2050, [. PAS 2050 for PAS, and the services of services for 2050, the goods and services for the services. How to assess the footprint, defines the word carbon footprint in the form of a special activity or unit of greenhouse gas (GHG) emissions. In view of the movement, carbon footprints of products, services and companies have become popular. There are lots of there too Various versions of carbon calculations in the printing industry. Decided to make one Task Force to develop. European Paper and Printing Industry Association, CEPI and the Intergraph has published carbon footprint guidelines. It is believed that all printed Product groups have paper accounts for 70% of global warming capacity (GWP). Emission data for factories was obtained from two locations: Econwent database (From Swiss Center for Life Cycle Inventory) and Paper Calculator Environmental Protect Fund (EDF). Canonico et al. According to EDF data More comprehensive and shown results that were two to three times more from Eco invent. Echoonuant, using total emissions, based on calculations About 340 million metric tons of equivalent (CO_2e). In particular, six major product groups were responsible for more than 80% of these total greenhouse gas (Ren et al., 2017).

9.3.4 Water Use and Pollution

Printing processes use a significant amount of water—not just for making paper, but also for cleaning and maintaining equipment. Extracting and treating this water can lead to shortages and pollution, especially in areas already facing water scarcity. On top of that, the chemicals used in ink and toner can contaminate nearby rivers and lakes if not handled properly, putting aquatic ecosystems at risk. To put it in perspective, producing just one ton of paper can require between 10,000 and 20,000 gallons of water. The wastewater from printing often contains harmful substances like dyes, heavy metals, and solvents, all of which can damage water quality and harm aquatic.

9.3.5 Waste Production and Disposal

Printing produces a lot of waste—including paper scraps, used cartridges, and Packaging materials. When this waste is not dealt properly, it can crowd Contribution to landfill and environmental pollution. Breaks in the form of paper and ink, they can release methane—a powerful greenhouse gas that runs global warming. In American alone, over 2 million tonnes of paper and paperboard finished in landfill 2020. This is a major lapse opportunity to reduce environmental impact Recycling. Tossing ink and toner cartridge without proper handling.

Leading for soil and water contamination, human health and integrity submission submission Environment Soil and water contamination, which can cause risk to health and environment. Recycling. On top of that, tossing ink and toner cartridges without proper handling can lead to soil and water contamination, posing risks to both human health and the environment. Additionally, improper disposal of ink and toner cartridges can cause soil and water contamination, which can pose risk to health and environment.

9.3.6 E-Waste from Printing Devices

Continuous progress of printing technology causes rapid ambiguity The printing device, resulting in electronic waste or e-waste. E-waste is dangerous Materials like lead, mercury and cadmium, which can pollute soil and water Not managed properly. Recruitment of e-waste and responsible disposal is necessary to reduce its environmental impact and increase stability in action. Aspects of the environmental impact of printing. Enterprise world Global e-West Monitor 2020 reports that the world produced high production. In 2019, 53.6 million metric tonnes of electronic waste, but only 17.4%was rebuilt. Improper handling of e-waste can have serious consequences, as its dangerous Substances can be environment, groundwater, ground water and even inAir. Implementing effective e-waste management strategies, such as recycling and Proper disposal is important to reduce these environmental risks.

9.3.7 Carbon Footprint

It refers to the total amount of greenhouse gases emitted in the printing process Crisis, including extraction, production, transportation and disposal of raw materials. Fossil Fuel dependence for energy and contributes using carbon-intensive materials Carbon footprint of printing. Printing activities have to reduce carbon emissions It is important in reducing climate change. To understand the carbon footprint of printing, consider the production of single. The paper sheet emits about 2.5 g of carbon dioxide. Carbon emissions quickly Connect when printed by billions of pages printed worldwide. And this Transport of printed material and disposal of waste contributes overall Carbon footprints of printing industry.

9.4 Sustainable Printing Practice

Fortunately, many measures can be taken to reduce environmental impact Printing. An effective approach is adopting permanent printing practices. It Recruitment or use of eco-friendly paper made from certified permanent sources. Recycled paper reduces the demand for new pulp, saves and reduces trees Forest harvesting. Additionally, apply double-side printing and optimizing print Settings can help in conservation of paper.

In printing, action involves using vegetable-based ink in stability Petroleum-based ink. Vegetable-based ink is made up of renewable resources and Less volatile organic compounds (VOCs) include, which contribute to air pollution. In addition, the paper waste properly disposal of recycling, ink and toner cartridges, and Encouragement Necessary steps towards a more durable printing industry.

9.4.1 Digital Options and Paperless Solutions

With increasing use of digital technology, now there are options available that can do Low the negative environmental effects of printing. Hugging Digital Options Such as e-book, online newspapers and digital documents can help save paper and Reduce energy consumption. Waste of paper and allied environmental results Paperless solutions in offices, schools can be reduced to a great extent and home.

Aspects of the environmental impact of printing. Enterprise world Consider the benefits of digital options: Saves an e-book instead of a printed book About 25 pounds of CO_2 emissions, paper requirement per year Production, transport and disposal. In addition, the need for printing and Physical document storage can be reduced using cloud storage and digital Cooperation equipment. Person and organization can reduce significantly Environmental footprints by embracing digital infections and paperless solutions Option.

9.4.2 Responsible Ink and Toner Use

The choice of ink and toner cartridge also affects the environmental impact of printing. Non-toxic and renewable materials can reduce environmentally friendly ink and Option for toner. Recycling and refilling ink cartridge can reduce waste production and conservation of resources. To clarify the effects of responsible ink and use of toner, consider that about 500 million. The annual cartridge is left annually. By recycling and refilling not only ink cartridges The number of cartridges sent to landfill can be reduced, but it can also happen. Energy and raw materials required for the construction of new cartridges. Again -Created cartridges or environmentally friendly options can also help reduce Environmental impact of ink and toner production (Wang et al., 2018).

9.5 How Does Traditional Printing Lead to Waste Generation?

In a comparative study by Kadam et al., digital presses were found to produce significantly less paper and chemical waste than traditional sheet-fed offset presses. Paper waste ranged from 5–20% for digital printing, compared to 35–75% for offset. Chemical waste was also much lower, with digital printing generating just 0.2–1.1 g per impression, while offset printing produced 3.2–7.3 g. However, it's worth noting that this study only looked at two relatively small print runs (500 and 3,000 copies), so the results can't be fully generalized.

A Finnish study found that a digital printing plant created 174 kg of waste per ton of printed products, with 136 kg of that being paper waste. Because digital printing uses fewer materials, it also generates fewer types of waste—especially chemical waste compared to traditional methods as shown in Fig. 9.2. Often, equipment suppliers handle the collection, disposal, and recycling of consumable materials, but detailed information about how these materials are actually processed isn't readily available.

Digital presses, especially those using electrophotographic technology, rely on various replaceable components like ink cartridges, OPC drums, belts, and charging coronas. In this context, inkjet technology is seen as a more environmentally friendly

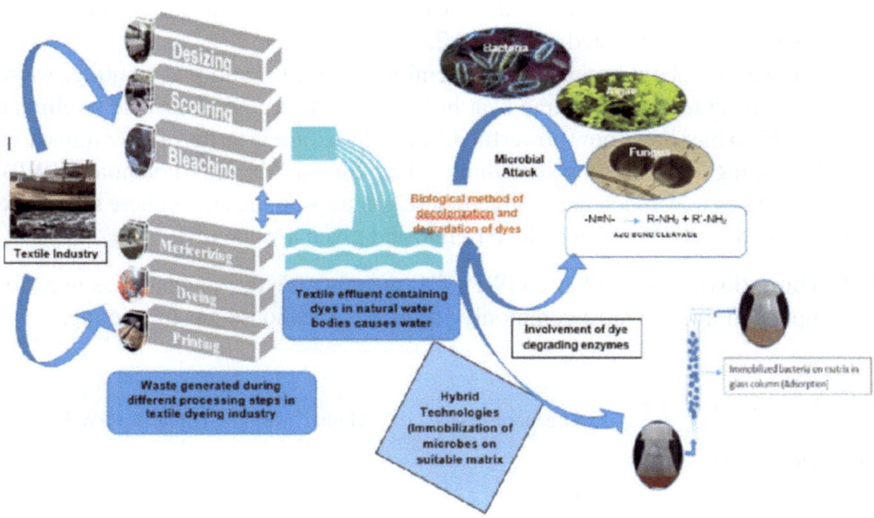

Fig. 9.2 Waste generated by processing steps in textile industries

option. The Finnish study also showed that electrophotographic presses produce far less hazardous, metallic, and landfill waste compared to sheet-fed offset printing. Because digital printing uses fewer materials, it generates less waste compared to traditional printing methods. The process is also simpler and creates less setup (or "make-ready") waste. However, in electroencephalographic printing, several consumable parts—like ink cartridges, OPC drums, belts, and charging coronas—need to be replaced regularly. Typically, the equipment supplier handles the collection, disposal, and recycling of these items. Still, there's limited information available on how exactly these consumables are processed and recycled (Ren et al., 2017).

9.6 What is Carbon Emissions Associated with Traditional Printing?

The carbon footprint is a standard way to measure greenhouse gas emissions throughout a product's life cycle. It includes emissions generated during production, processing, transportation, storage, use, recycling, and disposal (as defined in ISO 14067). However, this study focuses only on the cradle-to-gate carbon footprint—meaning it covers emissions up to the point where the printed fabric is finished. Anything beyond that, like its use or disposal, is not included.

The study follows ISO 14067 guidelines and includes all relevant greenhouse gas emissions. These emissions are converted into carbon dioxide equivalents using a 100-year global warming potential (GWP) scale. Table 9.2 shows the GWP values for the three most significant greenhouse gases: carbon dioxide (CO_2), methane (CH_4), and nitrogenous oxide (N_2O). While other greenhouse gases are considered in the calculations, their impact on the carbon footprint of digital and screen-printed products is minimal are listed in Table 9.2.

When we talk about greenhouse gas emissions from traditional printing, we're mainly referring to gases that trap heat in the atmosphere and contribute to climate change. This table shows how powerful different gases are in terms of their impact on global warming, measured by something called Global Warming Potential (GWP). GWP compares how much heat a greenhouse gas traps in the atmosphere relative to carbon dioxide (CO_2), which is the baseline and given a value of 1.

- **Carbon dioxide (CO_2)** has a GWP of **1**, meaning it's the standard gas used for comparison. It's the most commonly produced greenhouse gas in printing.

Table 9.2 Greenhouse gas emissions associated with traditional printing

Greenhouse gas	Global Warming Potential (GWP)
CO_2	1
CH_4	25
N_2O	298

- **Methane (CH_4)** has a GWP of **25**, meaning it's **25 times more potent** than CO_2 over a 100-year period.
- **Nitrous oxide (N_2O)** is even more powerful, with a GWP of **298**, making it **298 times more effective** at trapping heat than CO_2.

So even small amounts of methane and nitrous oxide can have a much larger warming effect than large amounts of CO_2. This highlights why reducing all types of emissions in the printing process—not just CO_2—is important for environmental sustainability.

9.6.1 The Cost of Traditional Printing Processes on Ecosystems

Although the prepress and imaging process doesn't produce significant VOCs (volatile organic compounds), some chemicals used—like developers and fixers—can release sulfur compounds, acetic acid, ammonia, and unpleasant odors, especially in older or continuous processing systems. The introduction of computer-to-plate (CTP) technology has improved the efficiency of prepress processes across all printing systems and significantly reduced water use. Additionally, harmful chemicals once used in photographic films and developers—such as silver and print developers—are no longer needed. As a result, waste from film, chemical developers, and excess water has been greatly reduced. Another eco-friendly step is replacing acid-based methods with metal engraving for gravure printing preparation, which helps minimize environmental harm in the prepress stage.

9.6.2 Solvent Effect in Printing Environment

The printing environment is often a major source of emissions, largely due to the storage, use, and disposal of chemicals and liquid waste. One of the main concerns is the release of volatile organic compounds (VOCs), which come from various printing and cleaning processes. These VOCs include substances like xylenes, ketones, and aliphatic compounds, which are found in inks, water-based solutions, and cleaning solvents. The chemicals used to clean printing plates, blankets, rubber rollers, and metal cylinders are typically petroleum-based and can include substances such as naphtha, mineral spirits, methanol, toluene, methyl ethyl ketone (MEK), glycol ethers, and trichloroethane (TCA). If not managed properly, these emissions can significantly impact both environmental and human health (Elbadawi et al., 2023).

9.6.3 VOC and HAP Emissions in Printing Performed with Water and Solvent Based Inks

Pigments and binders are the solid, non-evaporating parts of an ink mixture. However, solvent-based inks—commonly used in printing—contain alcohols and esters, which are volatile organic compounds (VOCs) that can make up 50–70% of the ink's composition.

In processes like flexographic, gravure, and screen printing, the solvent used accounts for about 3–4% of the total ink volume. After printing, most of the solvent that isn't absorbed by the material evaporates into the air. While the majority of it is released immediately, some continues to slowly evaporate over time, contributing to long-term environmental exposure.

These VOCs can trigger photochemical reactions in the atmosphere, affecting air quality. VOC emissions occur when organic solvents and thinners used in printing evaporate. About 75–90% of these emissions come from the ink drying stage. Factors such as printing speed, ink formulation, design complexity, and the efficiency of drying systems all influence the volume of emissions.

In the printing environment, most VOC emissions result from solvent evaporation in ink tanks and during the drying of printed materials—especially in areas between printing units where control is limited. The level of evaporation depends on the volatility of the solvent, room temperature, ink tank conditions, dryer performance, and how often the printing machine is stopped and restarted.

The amount of potential VOC (volatile organic compound) emissions during printing is directly related to the amount of solvent used. Common solvents in the printing industry include ethanol, toluene, ethyl acetate, isopropanol, n-propanol, hexane, naphtha blends, methyl ethyl ketone (MEK), isopropyl acetate, glycol ethers, and similar organic compounds. These solvents are typically found in heat-set inks used in web offset, flexographic, gravure, digital (non-impact), and screen printing systems.

Solvent-based inks are widely used in flexographic, gravure, and screen printing due to their fast-drying properties. This is because they contain unstable compounds like aliphatic and aromatic hydrocarbons, alcohols, ketones, and esters, which evaporate quickly and release VOCs into the air during printing. In contrast, water-based inks—which are more environmentally friendly—are less commonly used because they struggle to adhere to certain non-absorbent materials and surfaces with high surface energy.

Despite these limitations, water-based inks are encouraged where possible because of their lower environmental impact. In flexographic and gravure printing, the inks are diluted with solvents and applied to materials through a plate cylinder. After each color is printed, the material passes through a heat-set air dryer, where the solvents evaporate and dry the ink. This process repeats for every color layer.

Unfortunately, the VOCs released during drying often escape into the air without any filtering or treatment—such as afterburners or filtration systems—making them a significant source of air pollution.

9.6.4 *Volatile Organic Compounds (VOCs) and Hazardous Air Pollutants (HAPs)*

The production and use of printing ink can release pigments into the air and contribute to VOC (volatile organic compound) emissions. In sheet-fed offset printing, several materials contain VOCs and hazardous air pollutants (HAPs), including inks, varnishes, lacquers, coatings, roller and blanket cleaners, and ink oils.

Isopropyl alcohol (IPA) is commonly used to control the physical properties of dampening solutions in offset printing. However, this flammable and toxic compound—often used at concentrations of 8–10%—is a major source of VOC emissions, which can degrade indoor air quality and harm the work environment.

VOC and HAP emissions also occur from other materials used in offset printing, such as coatings, moistening solutions, and blanket washes. These emissions are typically released directly into the atmosphere unless properly controlled.

One effective way to eliminate VOCs from IPA use is by switching to waterless (dry) offset printing. This method avoids the use of dampening solutions altogether, significantly reducing emissions—but it comes with high upfront costs.

In heat-set web offset printing, inks contain high-boiling-point mineral oils. These oils evaporate at temperatures between 120–150 °C as the ink dries on coated paper. Unfortunately, the gases released during this process often escape untreated, contributing further to air pollution.

On the other hand, UV printing offers better environmental performance. Because the ink only cures when exposed to UV energy, no chemical reaction (and therefore no VOC release) occurs until that point. UV inks contain minimal or no solvents, making them a strong choice for reducing emissions while maintaining print quality and stability.

9.7 Environmental Comparison Summary

9.7.1 *Time and Cost Comparison: Digital Textile Printing Versus Traditional Printing*

Digital textile printing significantly shortens the production process compared to traditional printing methods. By eliminating complex steps such as color separation, hand drawing, film creation, and screen preparation, manufacturers can handle orders much faster and with lower costs—especially during the proofing stage. Designs can be submitted quickly through modern digital formats like email or CD-ROM, and proofing is typically completed within a single working day. In contrast, traditional proofing can take up to a week, delaying the production timeline.

This streamlined process not only reduces turnaround time but also cuts proofing costs substantially. The absence of a lengthy proofing cycle and the lower cost associated with it provide a major competitive advantage in the market. In today's fast-moving business environment—where time is money and efficiency is critical—digital printing offers companies a faster, more flexible, and cost-effective solution. It's no surprise that digital technology is rapidly becoming the preferred choice for textile manufacturers.

9.7.2 Printing Effects Between Digital Printing Textile and Traditional Printing Textile

How Digital Printing Transforms Textile Design

Digital printing technology has broken past the limitations of traditional textile printing. Unlike conventional methods, it removes constraints like color registration and pattern repetition length—commonly referred to as the "flower back." This allows for much greater freedom in design and production. With digital printing, high-end effects can be achieved directly on fabrics. The process begins with a digital pattern, which is then processed by a computer for color measurement, color matching, and finally, precision inkjet printing. In theory, digital printing can produce up to 16.7 million distinct colors, offering a massive color range that far surpasses traditional methods.

This technique is especially effective for designs that involve complex color gradients, moiré patterns, and other highly detailed elements. Because there's no need to repeat a pattern at fixed intervals, designers have much more creative flexibility. As a result, digital printing not only improves the visual quality of textile products but also expands the possibilities for innovative design.

9.7.3 Production Flexibility Between Digital Printing Textiles and Traditional Printing Textiles

Digital printing offers a major shift in textile production by enabling fast, flexible, and small-batch manufacturing. Unlike traditional printing, digital printing isn't limited by production volume—there are no minimum batch requirements, making it ideal for on-demand or customized orders. The entire process is digitally managed, from design to final output, which significantly enhances production speed and adaptability. In many cases, products can be printed and shipped on the same day, allowing for rapid response to market needs.

One of the standout advantages of digital printing is its seamless integration with computer technology. Designers can easily adjust colors and patterns during production and even print multiple variations of the same design with different tones. This ability to fine-tune and personalize designs in real time sets digital printing apart from traditional methods, making it a powerful tool for modern textile manufacturing.

9.7.4 Energy Consumption Between Digital Printing Textiles and Traditional Printing Textiles

Digital textile printing offers a highly precise printing process that eliminates the need for water and traditional color pastes. Instead, colors are applied "on demand" using computer-controlled inkjet technology, which means there's virtually no waste of dyes or generation of wastewater. This process is not only efficient but also quiet—since it's computer-operated and doesn't involve heavy mechanical equipment, it produces no noise pollution. As a result, digital printing supports a cleaner, greener production method.

By removing the high energy use, heavy pollution, and loud operations typically associated with traditional textile printing and dyeing, digital printing represents a major step forward. It enables low-energy, low-waste, and environmentally friendly manufacturing—marking a true technological breakthrough for the industry.

9.8 Pollution and Chemicals in Printing Industry

The chemical dyeing and printing industries generate three main types of waste: liquid waste, air emissions, and solid waste. Textile production processes like scoring, bleaching, mercerizing, and dyeing are major sources of pollution. One of the most water-intensive stages of textile production is bleaching and dyeing, which require large amounts of water. However, much of this water ends up as waste after the processing is complete. On average, it takes around 0.08–0.15 cubic meters of water to produce one kilogram of fabric. After processing 12–20 tonnes of textiles, about 1,000–3,000 cubic meters of wastewater is produced daily.

This wastewater is often rich in dyes and chemicals, many of which are non-biodegradable and potentially carcinogenic, posing serious risks to both human health and the environment. The waste contains a mixture of organic and inorganic compounds, which can be harmful if not properly treated. Various methods, such as flocculation, trickling filters, and electro-dialysis, are used to treat these waste materials, but the process remains a significant challenge for the industry. Despite efforts to treat wastewater from textile production, current methods are not fully effective at removing all the dyes and chemicals used in the industry. The wastewater not only contains high concentrations of synthetic colors but also various chemicals from different stages of processing. In addition, trace metals like chromium (Cr), arsenic (As), copper (Cu), and zinc (Zn) are often present in these waste streams, posing significant health risks such as bleeding, skin ulcers, nausea, severe skin irritation, and dermatitis.

Organic and microbial impurities are also common in textile waste. Cotton fibers, for example, are primarily dyed with azo dyes—one of the largest groups of synthetic colors used in the industry. Unfortunately, these dyes are difficult to break down with traditional treatment methods. Azo dyes contain nitrogen-nitrogen bonds ($-N=N-$) that make them chemically stable and challenging to degrade. When they decompose, they can release carcinogenic substances.

These dyes have been shown to alter the physical and chemical properties of soil, degrade water quality, and harm plant and animal life. Their toxic nature can even kill soil microorganisms, leading to reduced agricultural productivity and further environmental damage (Elbadawi et al., 2023).

Textile wastewater is highly polluted, with strong colors, high temperatures, turbidity, and large amounts of suspended solids and dissolved materials. It contains a mix of pollutants that are toxic and harmful to both the environment and human health.

Color: One of the biggest issues in textile wastewater is the presence of color, which is easily visible even at low concentrations. This not only makes the water unpleasant to look at but also reduces the ability of aquatic life to photosynthesize, affecting oxygen levels and overall water quality. Traditional treatment methods struggle to remove these colors effectively.

pH Levels: The high pH of textile wastewater can disrupt water quality, impact the biological processes of microorganisms, and even contaminate groundwater by altering soil permeability. This requires neutralizing the wastewater with acidic or basic solutions before it can be safely released.

Hardness (Total Stiffness): Textile wastewater often contains high levels of divalent metals like calcium (Ca^{2+}), magnesium (Mg^{2+}), and iron (Fe^{2+}), which cause water hardness. These metals can lead to turbulence and pollution in water bodies, requiring special treatment before discharge.

Biological Oxygen Demand (BOD): BOD reflects the presence of organic contaminants in wastewater. When ammonia and organic acids in the water undergo anaerobic fermentation, they decrease the water's pH and oxygen levels, leading to hypoxic conditions that harm aquatic life.

Chemical Oxygen Demand (COD): COD measures the level of organic pollutants in the water, such as detergent, softeners, and dye chemicals. High COD values indicate toxic water conditions, reducing oxygen levels and threatening the survival of aquatic organisms.

Total Dissolved Solids (TDS): Textile wastewater tends to have higher TDS values compared to other industries due to the use of bleaching agents, dye fixatives, and other chemicals. High TDS levels make the water unsuitable for drinking or irrigation because the salinity can damage crops and water sources.

Total suspended solids (TSS): High levels of TSS in wastewater indicate an increase in suspended particles, which make the water more turbid (cloudy). This turbidity reduces oxygen levels in the water, disrupting aquatic ecosystems and interfering with the natural food chain.

Turbidity: In many cases, textile effluents are highly turbid due to suspended solids, oily residues, and colloidal matter. This not only gives the water a foul appearance and odor but also blocks sunlight and hinders oxygen transfer. As a result, photosynthesis in aquatic plants is reduced, affecting the entire aquatic ecosystem.

Chloride: Elevated chloride levels often come from the use of bleaching and disinfecting agents like hydrochloric acid, hypochlorous acid, and chlorine gas during textile processing. High chloride concentrations can harm essential microorganisms that support the aquatic food chain. These levels also tend to correlate with other pollutants such as TDS (Total Dissolved Solids), TSS, alkalinity, and sulfates.

Silica: Large amounts of silica in wastewater can make the water more saline and cloudy. This increase in salinity can make the water unsuitable for household use and irrigation.

Oil and Grease: The presence of oil and grease forms a slick layer on the water surface, blocking sunlight and reducing oxygen exchange with the atmosphere. This lowers the amount of dissolved oxygen (DO) in the water, disturbing photosynthesis and harming aquatic life. In animals, this oily coating can damage the insulating properties of fur and feathers. While detergents are often used to remove the oil, they can cause foaming and introduce harmful phosphates, which negatively affect invertebrates and fish. Overall, oil and grease in water bodies pose a serious environmental threat.

9.9 What Harmful Chemicals and Types of Pollution Are Associated with Printing?

Many different chemicals are used in the textile and printing industries, though most are used in small amounts. Some of these chemicals are very specific to textile processing and are not commonly listed in standard life cycle inventory (LCI) databases. In cases where specific LCI data wasn't available, researchers attempted to find equivalent data from existing sources.

According to LCI standards (ISO 14040-44), any input material that makes up less than 1% of the functional unit of the system can be excluded from the analysis—though the total amount excluded should not exceed 5%. Based on this rule, some minor chemicals used in the printing process were left out of the study. However, chemicals used during textile treatment were already accounted for in the cotton production LCI dataset (Kujanpää, 2014).

Digital printing-Greenhouse gas emissions from cradle-to-gate evaluation of the products printed products. Carbon dioxide (CO_2) is a major greenhouse gas that arises from a cradal-legate life cycle of a digitally printed product (cotton). Methane (CH_4) and dinitrogen monoxide (N_2O) emissions arise from cotton cultivation and production Sulfur hexfloride (SF_6) emission comes from the use of grid electricity. In today's world of advanced technology and digital dominance, it might seem like the centuries-old practice of printing doesn't align with sustainability. However, a closer look reveals a major shift in the industry. Innovative methods and smart technologies are helping printing become more environmentally friendly.

As printing evolves, the industry is increasingly embracing sustainability and contributing to a greener future. Drawing on data collected over the past 13 years, we've gained valuable insights into how modern printing technologies are addressing environmental challenges. Micro-technologies now work together to create more efficient, cleaner production processes—ultimately improving the overall environmental impact across the product life cycle. These developments signal a streamlined revolution in printing—one driven by a strong focus on both performance and the planet.

9.10 Rethinking Materials and Processes

Traditional printing has long relied on resource-intensive materials and energy-heavy processes. However, the industry is making impressive strides toward sustainability by adopting eco-friendly alternatives. For instance, some modern equipment now uses ultra-bright LEDs combined with advanced optical techniques, producing display images that are about 2.5 times brighter than older models—while cutting power consumption nearly in half.

Additionally, there's been a shift toward sustainable printing substrates, including recycled paper, FSC-certified materials, post-consumer waste products, and even tree-free alternatives. By rethinking the materials we use in printing, we can dramatically lower our carbon footprint and ease the strain on natural resources.

9.10.1 Embracing the Environmental Benefits of Digital Printing

Digital printing has become a key force in driving sustainable practices, thanks to its efficiency and adaptability. Unlike traditional methods, digital printing doesn't require metal plates, which reduces both energy use and waste. Advances in toner technology—such as highly polymerized toner—now allow for more durable, eco-friendly prints using up to 25% less water compared to traditional toner systems.

Digital printers also use electrostatic rollers or drums for precise toner application, preventing color bleeding and delivering sharp, high-quality images. This method allows printing on a wide range of materials—from packaging and plastic to textiles and metallic foils. By embracing digital printing, we not only open the door to creative flexibility but also make significant strides in reducing our environmental impact (Kujanpää, 2014).

9.10.2 *Energy Conservation: A Paradigm Shift*

Energy conservation is a core principle of sustainable printing. Digital printing technologies offer a much more energy-efficient alternative to traditional methods. For example, industrial inkjet printers can reduce electricity use by up to 57% compared to conventional screen-printing techniques. This drop in energy usage not only lowers operating costs but also significantly contributes to reducing the industry's overall environmental footprint But also has significant positive effects on our carbon emissions. Digital printing can contribute to global efforts to embrace technologies, business climate change.

9.10.3 *Reducing Waste and Embracing Adaptability*

Traditionally, printing was known for generating large amounts of paper waste and inefficient use of resources. But with the rise of digital printing, this has changed significantly. Unlike offset printing—which requires separate aluminum plates for each design—digital printing eliminates the need for these time-consuming and wasteful steps.

In fact, high-quality digital print solutions led to a 61% reduction in CO_2 emissions across the product life cycle in the 2021 financial year compared to 2005. Digital printing also supports on-demand production, reducing the need for large inventories and minimizing the risk of materials becoming outdated. This streamlined approach cuts down waste and makes better use of resources, supporting more sustainable printing practices.

9.10.4 *Smart Technologies for Sustainable Printing*

Smart technologies are playing a key role in advancing environmentally responsible printing. For example, installing gas turbine cogeneration systems—which reuse exhaust heat—can achieve energy efficiency levels of 80–90%, compared to around 40% in conventional thermal power plants. This helps save energy and lower CO_2 emissions.

Industrial inkjet printers offer precise control, eliminate the need for manual color matching between machines, and significantly reduce setup time. These innovations have been recognized globally—such as Konica Minolta's "Three-Star Environmentally Friendly Product" rating from the Japan Federation of Printing Industries. By leveraging technologies like IoT and automation, the printing industry can further boost environmental performance while also improving productivity.

9.10.5 Printing's Role in the Circular Economy

The concept of a circular economy—where materials are reused, recycled, and kept in circulation for as long as possible—is gaining momentum across industries. Printing has a vital role in this transformation. Through responsible sourcing, recycling programs, and the use of biodegradable or compostable materials, the printing industry can support the shift toward a more regenerative model.

By rethinking every stage of the product life cycle—from design and production to reuse and disposal—we can reduce waste and create a truly sustainable printing ecosystem.

Traditional printing is known for high energy consumption from non-renewable sources. It produces a lot of e-waste, including millions of ink cartridges that are left every year, old MFDs, cables, toner cartridges and cords. Additional paper production uses too much water, ranging from 2 to almost 13 L, to produce a single A4 page.

All these factors add an expensive, energy-intensive and disable technology that businesses should reduce when it is reduced. Traditional methods often involve the use of harmful chemicals, energy-intensive equipment and consumption of non-renewable resources, leading to a sufficient environmental footprint. One of the primary concerns with traditional printing is the release of volatile organic compounds (VOCs) during the printing process. These compounds have a harmful effect on air quality, meeting health risks.

Settlement of printed materials such as paper and packaging leads to the accumulation of waste in the landfill and potential contamination of soil and water sources. The manufacture of these materials affects natural resources.

Another concern is the energy-intensive nature of traditional printing devices. The functioning of these machines, from the temperature of the toner to the power of the printing system, results in high energy consumption. This contributes to the carbon footprint of the entire printing process. Recognizing the environmental impact of traditional printing, businesses now understand the importance of infection in environmentally friendly options.

This change is aligned with growing durable practices. This presents an opportunity for many organizations to contribute to a better environment.

9.10.6 Important Benefits of Eco-Friendly Printing

Applications of permanent innovations include three main factors in the printing process: printing technology, ink and print materials. These technologies reduce pollution, waste and energy consumption during printing in production process. They help reduce printing costs while saving the environment.

9.11 Environmental Impact of Traditional Printing

9.11.1 Decreased Pollutants

Environmentally friendly printing methods provide many benefits, the most prominent of the pollutants. Pollutants pose a threat to the environment, which makes it necessary to apply permanent printing methods. It involves durable material, reduces waste, and reduces dependence on non-renewable resources. By adopting these strategies, businesses can greatly reduce their ecological footprints and promote a healthy planet for upcoming generations.

9.11.2 Protection of Natural Resources

Protection of natural resources is an important element of sustainable printing. By applying these permanent printing methods, business can greatly control the environmental footprint produced during printing. There are several ways to reduce the use of natural resources and reduce waste produced by printing activities, such as the use of soy-based ink and recycling paper waste.

9.11.3 Renewable Energy Uses

Another benefit of this method is the use of renewable energy. Printing firms can significantly reduce their environmental impact by choosing renewable energy sources such as solar or wind energy. As the demand for sustainable printing increases, businesses should update their approach with the latest innovations and methods. Including renewable energy in sustainable printing emphasizes the dedication of the area to reduce its ecological footprint and reduce high quality printing standards.

9.11.4 Cost of Permanent Printing Benefits

Eco-friendly printing provides solid cost benefits for businesses, making it an economically viable and compelling option. Nowadays, organizations have improved their production process, contributing to stability.

Generally, printer inquiry also represents about 40–60% IT help desk time, which reduces productivity. According to the data, 65% printed materials are left or recycled on the day they are produced. Keeping all these factors in mind, effective green printing solutions will be led to adopt.

The cost benefits of eco-friendly printing are.

9.11.5 Low Energy Cost

The use of energy-skilled printing devices, such as LED or inkjet printer, leads to significant reduction in energy costs. You can use automatic powering down machines such as energy-saving practices during inaction, thus saving additional expenses.

9.11.6 Paper Reduced and Supply Cost

Refilable ink cartridges or recycled paper selection reduces the cost of these required printing solutions. Modern sustainable printing machines use low ink and toner, which reduces the cost of paper and other supply.

9.11.7 Streamlined Operations

These green practices ensure the systematicity of overall printing functions, which leads to an increase in productivity. The faster the work is done, less resources are used. Streamline operating businesses help improve their productivity on a grand scale (Yıldırım et al., 2020).

9.11.8 Regulatory Compliance

Choosing permanent printing practices enables businesses to stay ahead in this highly competitive market and avoid punishment. Following these rules will help your business to grow better productivity and move towards a permanent future.

9.12 How Can Sustainable Printing Benefit Your Business?

Environmentally friendly printing uses permanent materials and energy-skilled techniques to help reduce their carbon footprints. Adopting these practices supports stability efforts and increases the brand reputation, meets consumer demand for greenery solutions. Even this long -term cost leads to savings, making it an inexpensive option. It provides an opportunity to contribute to a greenery in this highly competitive market.

- **Cost savings:**

Sustainable printing solutions assist companies to reduce their costs as they transfer to more efficient, environmentally friendly processes. By cutting energy use, using advanced machinery, and by reducing waste, business can expect a huge amount of saving on printing expenses.

- **Better public perception:**

Nowadays, consumers understand their impact on the environment and seek companies that align with their principles. Choosing to adopt permanent printing methods can help companies present themselves as environmentally friendly and socially responsible. It can result in potentially increased brand loyalty and customer support. Companies can control their environmental impact by focusing on permanent printing by increasing their reputation among consumers.

- **Increase in product quality:**

Using permanent printing methods, companies can ensure that their products are printed on high quality materials. These materials are known to last more durable and longer than other traditional printing options. This indicates that the items will appear more attractive and professional by reducing waste and encouraging a greenery method of printing.

- **Increased efficiency:**

Eco-friendly printing provides many advantages for businesses, better efficiency is one of the most important. By improving printing methods and processes, business waste can reduce low energy expenditure and increase workflow efficiency.

The Future of 13 printing

When searching for the future of printing, you can discover various permanent functioning and processes to apply in your business. You also need to understand how these changes play an important role in dealing with climate change. It is necessary that we make the right choice in printing methods to promote a healthy ecosystem and run meaningful changes.

Stability is not a tendency, but a revolution that rebuild all industries and business sectors and run its decisions and tasks. Nowadays, companies are actively pursuing innovative solutions to reduce their environmental impact.

The current wave of demand for stability marks a decisive change. Consumers nowadays want to adopt permanent practices, forcing them to focus on stability and reduce their environmental footprint in the supply chain (Jhatial et al., 2023).

- **Innovation in sustainable printing**

As you know, the environment is experiencing an increase in permanent printing technologies. These innovations include the creation of biodegradable ink and void 3D printing techniques. Sprintrs who require low energy and resources are advised to use them.

Ink obtained from algae or plant extracts is more environmentally friendly and offer bright colors that are healthy for you. In addition, introducing renewable energy sources like solar energy, replaces the industry in printing processes, ensuring that your print jobs resulted in low carbon emissions.

- **Role in combating climate change**

Environmentally friendly printing methods are necessary in addressing a dangerous issue of climate change. Companies emphasizing these sustainable practices resonated with your commitment to protect the environment, increase their competitive advantage. Your printing decisions can bring about a cleaner, more durable future. Thus, they strengthen the link between daily behavior and global environmental welfare.

- **Using multifunction printers**

The introduction plays an important role in developing the future of the Multifunction Printer (MFP) printing technology. These printers can add many devices to one, so reduce electronic waste. Introduction to this technique helps reduce waste and increase output, takes a big step forward in furthering environmentally friendly printing practices.

9.13 Ensuring Stability with Environmentally Friendly Printing Solutions with Munter Business Equipment

Permanent printing is necessary for committed organizations to reduce your environmental footprint by providing high quality printed materials. Taking advantage of these procedures and environmentally friendly materials allow businesses to gain more efficiency, reduce costs and create effective solutions without affecting quality.

It is important to select firms that prefer stability with well -managed print services. They try to increase your initiative and ensure that you significantly reduce your carbon footprint. As the owner of a business, you can embrace permanent printing as a fundamental component of your business strategy.

In Munter Business Equipment, we provide ideal printing solutions. If you are wondering which printer to choose to make your business durable, our experts can help. Explore a series of printers from Ireland's leading printer sellers and service providers. Get all your essential office equipment services and have a seamless production process (Jhatial et al., 2023).

9.14 Digital Printing Methods

As highlighted in earlier discussions comparing digital and mechanical printing, digital methods are steadily gaining ground in markets traditionally dominated by mechanical processes. This shift is driven by four key technical advantages:

- **High-quality output** suitable for a wide range of products and applications
- **Broader compatibility** with various paper grades and other substrates
- **Faster production speeds** and improved productivity
- **Greater flexibility** for on-demand, customized, and versioned printing.

These strengths open up new opportunities in areas that demand variable content, such as personalized products and short-run packaging.

The next chapter will explore the specific products and applications that benefit from these advancements. One emerging trend is hybrid production, where digital and mechanical printing are integrated within the same production line. This allows for high-quality digital enhancements to be added to traditional processes—such as in finishing or packaging stages.

Examples include:

- **Kodak Prosper S10** imprinting units paired with Offset Web Presses
- **AGFA**'s combination of inkjet and screen-printing technologies
- **Linoprint**, which merges four-color inkjet printing with packaging systems.

Thanks to the increasing speed, print quality, and compact design of modern inkjet printheads, hybrid solutions are expected to grow in popularity. However, the role of electrophotography in such integrated systems remains more limited (Yıldırım et al., 2020).

9.14.1 Inkjet Versus Electrophotography

Electrophotography is a well-established and mature technology in the digital printing world. In contrast, inkjet printing is still rapidly evolving and offers several notable advantages (Hertwich & Peters, 2009).

Lower cost: Inkjet inks are generally cheaper than toner.

Simpler design: Inkjet printheads have fewer components and consume less power.

Scalability: Inkjet systems can handle larger sheet sizes and wider web formats.

High-quality output: Inkjet technology shows strong potential for delivering excellent print quality.

Versatile ink formulations: Inks can be customized for various materials.

Flexible substrate compatibility: Even thick or rigid materials can be printed using inkjet.

From an environmental perspective, inkjet also seems promising. Large toner-based systems tend to have high power consumption, often requiring substantial energy sources and meeting strict environmental controls. In comparison, inkjet printing consumes significantly less power. However, it's important to note that this claim is not backed by detailed data in the source.

9.14.2 Digital Press

Key Factors in Digital Printing: Quality, Output, and Cost

When evaluating digital printing systems, three critical factors come into play: print quality, output speed, and cost. However, it's important to consider the added value and potential cost savings that digital printing offers over traditional mechanical methods.

For instance, print-on-demand eliminates the need for surplus copies and reduces waste, while personalization can significantly improve the effectiveness of marketing campaigns.

The cost structure also varies depending on the printing model:

In electrophotographic (EP) systems, manufacturers typically provide the printer, toner, and consumables through a "click charge" model—where the cost per page remains constant, regardless of ink coverage.

In inkjet (IJ) systems, manufacturers often source printheads from third-party suppliers, and printing companies purchase ink separately based on their needs. Here, the cost of printing is directly tied to the amount of ink used. This makes cost comparison between the two models challenging. In general, inkjet ink is cheaper than toner, but for high-coverage print jobs, EP systems may offer better cost efficiency due to the flat click charge (Yıldırım et al., 2020).

Print Quality and Substrate Compatibility

The literature does not offer a clear consensus on print quality. Often, resolution is used as a benchmark, but factors such as the type of paper or substrate significantly affect tone reproduction, Madhukar Thakker and Sun (2024)

- Adapt print methods to different paper types, and
- Optimize processes to allow the use of standard (offset) paper grades in digital printing.

New inkjet systems also have the ability to vary droplet size, enabling smoother gradients and grayscale effects—offering a level of nuance that goes beyond basic binary printing.

As a result, comparing print quality across systems is complex. That said, digital printing today often claims to achieve "offset-quality" results—though this term is usually used more as a benchmark than a precise technical standard. As Frank Romano puts it, "*Quality is coming into alignment with industry needs.*"

Output Speed

Unlike print quality, output speed is more straightforward to measure. Manufacturers typically report speeds in A4 pages per minute (ppm) for simplex (single-sided) printing. Based on data from manufacturer websites and product datasheets, some of the latest digital presses—even those still in pilot testing—are capable of printing up to 14,000 A4 pages per minute (Madhukar Thakker & Sun, 2024).

9.15 Waste Management

There are some waste flows from the waste management printing plant. Waste is taken to dangerous waste treatment plant, which is 100 km from AITOO (with a truck of 25t). There is a small amount of waste going into landfill, but the amount of landfield waste is considered negligible and thus excluded from the study. 1.9 Waste Water Treatment Printkorpio Textile Printing House has a self -water purification plant, which purifies waste water coming from printskorpio and also has wastewater from another company nearby. Water is purified and its quality is measured regularly. Thus, with the value used for water quality -the average water quality is based on the average water quality measurement to estimate the emission load, which is performed by the conservation organization of the Cocamenzoki River. Textile printing manufacturing, outgoing waste water in relation to incoming groundwater is measured in relation to water going out of purification plant with purification plant. The average water use amount on daily locations is around 14 m^3. Wastewater sample is manually made from sewer. This report is based on the average of the year 2013 for data measurements used in the data measurement. The total efficiency of the purification system is for total phosphorus (P) and total nitrogen (N).

What are some permanent printing options and environmentally friendly practices?

Clothing stability: a path for a green future

Stability in various industries has become a paramount concern, and the fabric textiles sector is no exception. The textile industry, which is known for its important environmental impact, is now undergoing a change towards more durable practices. This change is important not only to reduce environmental damage but also to meet the growing consumer demand for environmentally friendly products. This blog faces the importance of stability in fabric textiles, challenges, and innovates the innovative solutions being adopted to create a greenery future.

Importance of stability in clothes of clothes

The textile industry is one of the largest pollutants in the world, which contributes to water pollution, carbon emissions and significant waste production. The production of traditional textiles involves the use of large amounts of water, energy and chemicals, causing environmental decline. Permanent practices in fabric textiles aim to reduce these negative effects by adopting environmentally friendly procedures and materials. Stability in textiles involves various aspects, including the choice of raw materials, manufacturing processes, and the choice of life management. By focusing on these areas, the textile industry can significantly reduce its environmental footprint. Additionally, permanent textiles can increase the welfare of workers and communities involved in production, promote fair labor practices and improve economic condition.

Challenges in Achieving Sustainability

Despite the increasing awareness and demand of sustainable garments, the industry faces many challenges in achieving true stability. One of the primary challenges is the complication of the textile supply chain, which includes several stages from the extraction of raw materials to the product disposal. Each stage has its own environmental effect, making it difficult to apply wide sustainable practices.

Availability of durable raw materials is another important challenge. Natural fibers such as cotton and wool, although the use of renewable, adequate water and pesticides is required. On the other hand, synthetic fibers such as polyester are obtained from petroleum, contribute to plastic pollution and are not biodegradable. It is necessary to develop and source permanent options, but often involves high cost and technical challenges.

In addition, infection for permanent practices requires significant investment in research and development. Many textile manufacturers, especially in developing countries, may lack financial resources and technical expertise required to adopt new technologies and processes. Consumer awareness and demand for permanent products also play an important role. Without adequate market demand, business stability may be reluctant to invest in initiatives (Valli Nachiyar et al., 2023).

Innovative solution for durable clothes

Despite the challenges, the textile industry is seeing an increase in innovative solutions with the aim of promoting stability. These innovations are spread in various stages of textile life cycle, ranging raw materials to manufacturing and recycling.

Sustainable Raw Materials

One of the most promising areas of innovation is the development of durable raw materials. For example, organic cotton is grown without synthetic pesticides and fertilizers, reduces its environmental effects. Similarly, cannabis and bamboo are being discovered as a permanent option due to their rapid growth rate and low water requirements. Recycled fibers, such as consumers, are also gaining popularity, which are providing a way to reduce plastic waste.

Environmentally friendly manufacturing procedures

Progress in manufacturing technologies is enableing more environmentally friendly production processes. For example, waterless dyeing techniques, water consumption and waste water production significantly reduces the production. Digital printing techniques also reduce the use of water and chemicals than methods of traditional dyeing. Additionally, adopting renewable energy sources in textile manufacture may reduce dependence on carbon emissions and fossil fuels.

Circular economy and recycling

The concept of a circular economy is receiving traction in the textile industry. Unlike the traditional linear model of "Tech, Make, Disposion", the purpose of a circular economy is to keep products and materials as long as possible. This approach involves designing textiles for durability, recurrence and recycling. Companies are investing in technologies to recycle textiles at the end of their life cycle, converting them into new fiber and clothing. Brands like Patagonia and H&M are proceeding by offering recycling programs for old clothing (Ghoreishi & Happonen, 2022).

Sustainable Fashion and Consumer Awareness

Consumer behavior plays an important role in running stability in textiles. The rise of durable fashion, also known as "slow fashion", encourages consumers to buy low, high quality textiles that last long. The brands are rapidly transparent about their supply chains and stability practices, allowing consumers to make informed options. Education and awareness campaigns are also necessary in promoting sustainable consumption and encouraging responsible disposal of garments.

9.16 What Are Some Technological Progress in Printing?

Screen printing is generally better suited for large-scale print runs, while digital printing is more ideal for smaller batches. This is largely because screen printing requires extensive setup and test runs, which consume resources. For small print jobs, this setup effort isn't practical or efficient. In contrast, digital printing is more flexible and precise, and it doesn't require test runs—making it far more suitable for short or custom batches. From an environmental perspective, especially when looking at greenhouse gas emissions (and excluding product lifespan), digitally printed items tend to have a lower climate impact, regardless of the quantity printed.

A sensitivity analysis also suggests that digital printing remains the more environmentally friendly option, even for larger print volumes. However, this analysis focuses solely on environmental impact and doesn't account for other important factors like cost-effectiveness, print quality, **or** specific production requirements. It's also worth noting that the results apply not just to greenhouse gas emissions, but to other environmental impact categories studied—except for water availability, where the findings differ (Wang et al., 2018).

9.16.1 Characterizing Display, Input and Output Devices

To achieve frequent performance, acquisition and reproduction of textile Design, it is necessary for both different pieces of both calibrate and characteristic To repeat these controls at the equipment and then regular intervals Digital printing of 168 textiles For example, sensor reactions, light sources, for example For emissions and printers, in the standardization of ink supply and Properties of textile substrate especially. Three types of device the way they are required by some different approach Controlled (Ghoreishi & Happonen, 2022).

9.16.1.1 Control of Color Performance Monitor

Understanding CRT Monitor Gamma and Color Calibration

A CRT (Cathode Ray Tube) monitor displays images using a screen made up of tiny red (R), green (G), and blue (B) elements. The image is created by scanning the screen with electron beams controlled by each electron gun (one for each color). The intensity of the light produced on the screen depends on the drive voltage sent to these guns.

This voltage comes from digital RGB values processed by a Digital-to-Analog Converter (DAC) and a video amplifier on the computer's video card. However, the relationship between the digital RGB input and the resulting brightness on the screen is not linear. Instead, it follows a mathematical curve known as a gamma function, which includes an initial offset and a power-based transformation.

The term "gamma" refers to this nonlinear relationship. While the native gamma of most CRT monitors typically falls between 2.5 and 3.0, computer operating systems often apply a different gamma value to correct for this. For example, macOS traditionally uses a gamma of 1.8, while Windows uses 2.2. These settings help standardize how colors appear across different systems.

To manage this mismatch, gamma correction can be applied at various stages—such as during scanning, image processing by a video card, or when storing an image file. Some image file formats like TIFF, TGA, and PNG can even store gamma values directly within the file (Obele et al., 2024). The need to calibrate monitors for accurate color reproduction has led to the development of tools like Adobe Gamma, Praxisoft Wiziwyg, ColorWizzard, and GretagMacbeth Websync. These tools allow users to adjust contrast, brightness, and color balance visually on the screen.

Additionally, many monitor manufacturers offer generic color profiles. More advanced users can fine-tune gamma settings individually for red, green, and blue using video card software—though this process is often tedious and trial-and-error based.

For most users, basic adjustments provide a decent level of accuracy. However, in professional color-critical environments, companies often rely on specialized color management software (such as Imagemaster by Datacolor International) to ensure accurate color communication across multiple systems. This requires complete profiling and calibration of every monitor involved in the workflow.

Before calibration, a monitor typically needs to warm up and stabilize. The calibration process usually includes adjusting contrast (to ensure that a digital input of zero results in zero brightness) and tweaking brightness and glow levels to achieve an accurate display baseline (Javoršek & Javoršek, 2011).

9.16.1.2 Colour Communication

In today's fast-paced commercial textile industry, efficient and reliable communication systems are more important than ever. These systems are needed not just for internal coordination—often across multiple locations—but also for seamless communication with customers, suppliers, and retail partners.

There's a growing trend for local sourcing from suppliers and major retailers rather than global sourcing, which makes fast and dependable data exchange even more essential. Ideally, these systems should use open (non-proprietary) standards to ensure compatibility and ease of use across different platforms and companies.

Some examples of color-related data that may need to be shared include:

Large, high-resolution image files, especially uncompressed formats like TIFF or BMP in 48-bit color. Color measurement data, which may be formatted using XML—for instance, in CXF (Color Exchange Format) developed by GretagMacbeth.

Color management data, such as device calibration information and ICC profiles that help ensure color consistency across devices.

Challenges of Sharing Color Data Online

Sharing and accurately displaying color images over the internet presents specific challenges—especially in fields like online fashion retail, where accurate color representation is critical. Ideally, what the sender sees should look the same to the receiver, but that's not always the case.

Historically, web designers used "web-safe" colors (a limited palette of 256 colors) to ensure consistent display across different systems and browsers. These were often 8-bit GIF images optimized for fast loading. However, with the rise of broadband internet and 24-bit "true color" displays, this practice has become less common.

Still, some systems occasionally fall back on 16-bit color, which can't display the full range of colors and often creates mismatches—especially in designs with large blocks of solid color, like fabric patterns. In such cases, browsers or operating systems (like Windows, macOS, or Linux) may approximate colors by blending nearby shades or using color dithering, which can lead to unwanted visual effects.

To address these issues, several companies offer web-based color management tools, such as: NetProfiler by GretagMacbeth, Online color exchange tools by Ewarna.

These tools help ensure that digital color specifications are accurate and consistent, even when shared across different systems and networks.

9.17 Future Prospects

Recent literature shows that inkjet printing is rapidly advancing in key areas like productivity, print quality, and substrate flexibility. Thanks to improvements in print-head design and ink formulations, inkjet technology is now suitable for a wide range of both graphic and non-graphic applications. It's even expected to make inroads into traditional offset printing markets.

In fields like transactional and transpromo printing, inkjet continues to compete with electrophotography. Meanwhile, in sectors such as outdoor advertising and wide-format printing, inkjet has already carved out a strong position, **as** screen printing (serigraphy) steadily loses market share.

Until recently, achieving high-quality inkjet printing on paper required specialty papers with specific surface properties to control ink absorption and dot gain. However, new innovations—like HP's bonding agent technology—are making it possible to use standard newspaper and bulk-grade papers, significantly lowering costs and broadening potential applications.

The versatility of inkjet inks and the simplicity of the printing process open up numerous possibilities for non-graphic uses as well. According to the literature, many of inkjet's previously identified weaknesses are being addressed or replaced by newer business applications. For instance, HP's T300 Inkjet Web Press uses a bonding agent and dual hot-air dryers to make water-based inks compatible with a wide range of standard papers.

Efforts are also underway to improve data handling and print speed to meet commercial demands. Even concerns about adopting new technologies seem less relevant—Kodak's Versamark Press, for example, has been on the market for over a decade, showing that inkjet technology is both mature and reliable.

One of the most pressing concerns remains the high cost of ink. As one industry expert put it:

> The company that dares to break the cartel on ink pricing will gain a major advantage in the digital print revolution

Interestingly, while many environmental considerations are being explored, deinkability—the process of removing ink during recycling—is still seen as a major environmental challenge in digital printing, yet it is not included in the common SWOT analyses of inkjet printing as shown in Table 9.3 (Yıldırım et al., 2020).

The findings of this study offer valuable insights, but they are based on limited available data, much of which comes from equipment manufacturers, material

Table 9.3 SWOT analysis of inkjet printing, according to Pira International Ltd.

Strengths	Weaknesses
Non-impact	High quality on coated paper not here in 2008;
Ink cost and productivity dependent on coverage	Concept shown
Wide variety of applications	New technology risk
Wide variety of inks	Complex data transfer for high-speed variable
Easily used in a hybrid line	Print
Wide format	Water-based ink drying
Irregular shapes can be printed directly	
Opportunities	Threats
Faster, wider fixed-head single pass Packaging Coating for decoration and security Material deposition	Electronic bill presentation and payment

suppliers, or commercial print operations. As a result, there's a strong need for independent scientific research to validate these observations and expand our understanding—particularly in terms of environmental impacts at both the process and company levels.

Below is a summary of the key areas where further research is most needed:

Energy Use and Material Content in Digital Printing Technologies

Examine energy consumption across various digital printing techniques.

Study how different toner compositions in electrophotography (EP), as well as solvents in liquid EP and inkjet (IJ) inks, affect energy usage.

Compare the energy requirements for ink drying and toner fixing across different systems.

Investigate the energy used by individual components like imaging systems, paper transport, data communication, and the role of paper or other substrates, including how bonding agents influence overall energy and material use.

Energy Consumption throughout the Entire Production Chain

Assess energy usage in all stages of digital printing, including:

Computers, servers, and data transmission systems.

RIP (Raster Image Processor) systems.

Printers.

Online and offline finishing equipment.

Building and Facility Energy Impacts

Evaluate how building services contribute to energy use in digital printing environments, including:

- Space requirements
- Ventilation systems
- Air humidity and temperature control.

Explore the potential for recovering waste heat from digital presses to improve energy efficiency.

By addressing these topics through thorough and independent research, we can gain a more complete understanding of digital printing's environmental footprint and find new ways to make the process more sustainable (Robert, 2015).

9.18 Conclusion

The environmental impact of digital textile printing reduces significantly when compared to conventional printing approaches. The method requires lower water and energy consumption and generates fewer dangerous substances which makes it an environmentally preferable solution for sustainability-focused times. Standard printing techniques continue to use substantial resource quantities particularly water while generating significant environmental pollution throughout numerous regions.

Digital printing emerges as a strong environmental solution when the textile industry searches for alternative green approaches. The method delivers environmental sustainability benefits while preserving excellent product quality alongside unrestricted design possibilities. Digital printing technology together with improved ink formulations and energy-efficient solutions shows promise to establish a more sustainable production approach for textiles.

Declaration The authors declare that they have no known competing financial interests or personal relationships that could have appeared to influence the work reported in this chapter. Generative AI and AI-assisted technologies have been used in the writing process to improve the readability and language of the work only. After using this tool, the authors carefully reviewed and revised the content to ensure accuracy and quality. The authors take full responsibility for the final version of this publication.

References

Elbadawi, M., Basit, A. W., & Gaisford, S. (2023). Energy consumption and carbon footprint of 3D printing in pharmaceutical manufacture. *International Journal of Pharmaceutics, 639*, Article 122926. https://doi.org/10.1016/j.ijpharm.2023.122926.

Ghoreishi, M., & Happonen, A. (2022). The case of fabric and textile industry: The emerging role of digitalization, internet-of-things and industry 4.0 for circularity. In X. -S. Yang, N. Dey, & A. Joshi (Eds.), *Proceedings of Sixth International Congress on Information and Communication Technology* (pp. 189–200). Springer. https://doi.org/10.1007/978-981-16-3071-2_16.

Hertwich, E. G., & Peters, G. P. (2009). Carbon footprint of nations: A global, trade-linked analysis. *Environmental Science & Technology, 43*(16), 6414–6420. https://doi.org/10.1021/es803496a.

Javoršek, D., & Javoršek, A. (2011). Colour management in digital textile printing. *Coloration Technology, 127*(4), 235–239. https://doi.org/10.1111/j.1478-4408.2011.00304.x.

Jhatial, A. K., Gianchandani, P. K., Adamu, B. F., Babar, A. A., & Yesuf, H. M. (2023). Colorants for digital textile printing and their chemistry. In *Digital textile printing* (pp. 119–135). Elsevier. https://doi.org/10.1016/B978-0-323-90966-4.00006-7.

Kujanpää, M., & Nors, M. (2014). *Environmental performance of future digital textile printing* (VTT Research Report No. CR-04462-14). VTT Technical Research Centre of Finland.

Lee, W., Sung, E., Moon, J., Ahn, I., Yoon, K., Park, Y., & Kim, J. (2023). Performance evaluation of ink and digital textile printing fabric using natural indigo. *Fibers and Polymers, 24*(4), 1309–1319. https://doi.org/10.1007/s12221-023-00137-4.

Madhukar Thakker, A., & Sun, D. (2024). Inks for digital printing of textiles. *Journal of the Textile Institute, 115*(12), 2377–2390. https://doi.org/10.1080/00405000.2023.2289975.

Obele, C. M., Ugwu, S. C., & Chukwuneke, J. L. (2024). Development of textile printing ink using expanded polystyrene waste as a binder. *Progress in Rubber, Plastics and Recycling Technology, 40*(1), 48–61. https://doi.org/10.1177/14777606231186635.

Ren, J., Chen, G., & Li, X. (2017). A fine-grained digital textile printing system based on image registration. *Computers in Industry, 92–93*, 152–160. https://doi.org/10.1016/j.compind.2017.08.003.

Robert, T. (2015). "Green ink in all colors"—printing ink from renewable resources. *Progress in Organic Coatings, 78*, 287–292. https://doi.org/10.1016/j.porgcoat.2014.08.007.

Valli Nachiyar, C., Rakshi, A. D., Sandhya, S., Britlin Deva Jebasta, N., & Nellore, J. (2023). Developments in treatment technologies of dye-containing effluent: A review. *Case Studies in Chemical and Environmental Engineering, 7*, 100339. https://doi.org/10.1016/j.cscee.2023.100339.

Wang, L., Hu, C., & Yan, K. (2018). A one-step inkjet printing technology with reactive dye ink and cationic compound ink for cotton fabrics. *Carbohydrate Polymers, 197*, 490–496. https://doi.org/10.1016/j.carbpol.2018.05.084.

Yıldırım, F. F., Yavaş, A., & Avinc, O. (2020). Printing with sustainable natural dyes and pigments. In S. S. Muthu & M. A. Gardetti (Eds.), *Sustainability in the textile and apparel industries* (pp. 1–35). Springer. https://doi.org/10.1007/978-3-030-38532-3_1.

Chapter 10
Practical Challenges and Remedies in Sustainable Digital Textile Printing

Shriyasha Tari and Ashok Athalye

Abstract Digital textile printing (DTP) has become a revolutionary technology in the textile industry providing benefits such as high-resolution design capabilities, reduced water usage, low-waste and on-demand production. The ability to quickly customize and minimize inventory waste makes it relevant to the current needs for sustainable textile processing. Achieving sustainability in DTP presents several challenges. Major issues include the reliance on synthetic inks, pigments and chemical additives, high energy demands, limited fabric compatibility and concerns related to effluent treatment. This chapter explores these practical obstacles and investigates the innovative strategies aimed at enhancing the sustainability of DTP. It discusses the advancements in eco-friendly ink systems, energy-efficient fixation techniques, sustainable pretreatment processes and closed-loop water management practices. Further sections of this chapter discuss the industrial practices that must be adopted and the future prospects for environmentally conscious digital printing. In a nutshell, this chapter discusses the practical challenges in the sustainability aspect of digital textile printing and the alternatives that can be employed to make digital textile printing greener and safer.

Keywords Binder-free · Self-dispersing · Sustainable inkjet · Volatile organic compounds

10.1 Introduction

The textile printing process has revolutionized in 4 stages: Block printing was considered as the first generation, flat-bed screen printing the second, continuous rotary printing the third and lastly the digital printing/inkjet technology. Although

S. Tari (✉) · A. Athalye
Department of Fibres and Textile Processing Technology, Institute of Chemical Technology Mumbai, Mumbai, India
e-mail: tch22sm.tari@phd.ictmumbai.edu.in

A. Athalye
e-mail: ar.athalye@ictmumbai.edu.in

A. Athalye and H. Patil (eds.), *Digital Printing*,
https://doi.org/10.1007/978-981-95-2593-5_10

the rotary screen-printing technology offered advantages such as high speed and cost-effectiveness, it also has some disadvantages. Hence, the shift to digital textile printing became necessary. Textile digital printing evolved in in the 1990s (Tyler, 2005). This transition further allowed designers to use unlimited colors and patterns (Chen et al., 2015). In digital textile printing, inkjet printers are specifically modified to print designs directly onto the fabric (Javoršek & Javoršek, 2011). They offer advantages in terms of process efficiency, cost effectiveness and environmental impact. The main advantages of digital textile printing are production of photographic images and intricate designs, enabling the printing of extensive fabric lengths with minimal setup costs. Industry 4.0 involves the integration of digital technologies for the purpose of efficient manufacturing (Ghoreishi & Happonen, 2022). Therefore, digital printing is also referred to as 'Printing 4.0'. Digital textile printing is a relatively simple and hassle-free process in comparison to conventional textile printing. It begins with a fabric pre-treatment process for better ink absorption followed by designing a pattern using graphic software. The design is then uploaded to the printer which uses inkjet technology to deposit precise ink droplets. The printer controls ink flow for sharp and distinct prints. Finally, the fabric is dried or cured to ensure ink adhesion and long-term durability of printed fabrics. Digital printing and inkjet technology offers various sustainable advantages and it has become a preferred choice of textile processors over the last few decades (Wang & Memon, 2023). The advent of digital printing technology has reduced the consumption of chemicals, water, time and resources required for printing fabrics (Kumelachew et al., 2023). However, with the growing environmental awareness and consumer needs the sustainability concerns of the digital textile printing technology are also increasing rapidly. This chapter highlights the practical challenges in the sustainability aspect of digital textile printing. It also discusses the alternatives that can be employed to make digital textile printing greener and safer. In a nutshell, this particular chapter discusses the practical challenges in digital textile printing and the potential remedial measures that can be taken to make it more sustainable.

10.2 Digital Printing: Challenges in Pretreatment and Recent Research Initiatives

Pretreatment process prepares fabrics to adhere to ink effectively, ensuring vibrant hue, good fixation and optimal fastness properties. It has been observed in previous studies that processes like mercerisation greatly influence the print-quality (Zhao et al., 2022). The pretreatment processes differ significantly depending on the type of ink and the fibre used (Ding et al., 2019). For instance, pigment inks do not have direct affinity for fabric substrates so a binder (typically polyurethane or acrylic based) is padded onto the fabric prior to printing. On the other hand, Reactive inks which are mainly used for cellulosic fibers like cotton, viscose, linen require a more elaborate pretreatment. The objective of pretreatment before printing with reactive inks is to create an alkaline and humectant environment that promotes covalent bonding between the dye molecules and the fiber. This is typically achieved by padding the fabric with a mixture of wetting agents, alkali, urea and thickeners (Ji et al., 2024). In the case of acid dye inks applicable for protein fibers such as wool, silk or polyamide fibers like nylon, the pretreatment focuses on maintaining an acidic pH to enable ionic bonding between the dye and the fiber. The fabric is pretreated with acids (like acetic acid), wetting agents followed by gentle drying (Choi et al., 2005). For disperse dye inks used on synthetic fibers like polyester, pretreatment mainly involves scouring to remove any hydrophobic impurities like oils, finishes that may block dye absorption. Sometimes, a light padding with dispersants or lubricants is done. The fabric is then heat-fixed using a calendar press or hot air. In sublimation printing, which also uses disperse dyes but via transfer paper, no pretreatment is needed on the fabric itself, as the design is printed on coated paper and then transferred using heat (Stojanović et al., 2021). Vat inks are rarely used in digital printing as they require a chemically intensive process. Pretreatment may involve using reducing agents and alkalis to convert the dye into a soluble form. After printing with vat inks an oxidation step is necessary to revert the dye to its insoluble form followed by extensive soaping and washing. Solvent and UV inks are used mainly for technical or coated textiles like PVC or PET require minimal pretreatment. These fabrics are given only surface cleaning treatments such as brushing or corona discharge treatment to ensure ink adhesion. No wet chemical processes are involved (Ahmed & Ahmed, 2025). In summary each ink type demands a tailored pretreatment strategy to suit its chemistry and the fabric's characteristics. Effective pretreatment enhances print quality, color performance, and process reliability, making it an indispensable component of digital textile printing workflows. Recent research efforts have focussed on avoiding the need for pretreatment step before digital textile printing. Pretreatment steps before digital printing with reactive inks is a major challenge. The Table 10.1 mentioned below summarises the various types of pretreatments done before digital textile printing depending on the fibre type.

10.3 Inks and Colorants Used in Digital Textile Printing

Digital textile printing inks are similar to dyes but in aqueous forms, overall substrate affinity remains same. Digital printing inks can be either be dye based or pigment inks (Jhatial et al., 2023). Digital textile printing employs various ink types, each designed for specific fabric substrates (Madhukar Thakker & Sun, 2024). Pigment inks are popular due to their ease of use and adaptability. Especially use of pigment and sublimation inks reduces the water consumption substantially (Rainert et al., 2024). They are applicable on all types of fibres like cotton, polyester, and blends. Fixation requires only heat curing process thus eliminates the need for water-heavy steaming or washing. This makes them eco-friendly and suitable for streamlined processing. However, pigment prints lack the color intensity of dyes and can make the fabric feel slightly harsh stiffer. Reactive inks are ideal for cellulosic fibers like cotton, viscose,

Table 10.1 Summary of pretreatments done for each fibre type

Ink class	Fibre type	Pretreatment purpose	Pretreatment steps
Pigment inks	Cotton, polyester, blends	– Improve ink adhesion – Prevent ink bleeding – Enhance color strength	– Pad with binder (usually acrylic or PU-based) + thickener + softener – Dry at ~100–120 °C
Reactive inks	Cellulosic fibers (cotton, viscose, linen)	– Promote dye–fiber reaction – Enhance color yield – Ensure uniform dyeing	– Pad with alkali (e.g., sodium bicarbonate or carbonate), urea, wetting agent, thickener – Dry without curing
Acid inks	Wool, silk, nylon (polyamide)	– Maintain acidic pH for dye fixation – Improve dye penetration and leveling	– Pad with mild acid (e.g., acetic acid), wetting agent, softener – Dry gently to avoid fiber damage
Disperse inks	Polyester and blends	– Clean the surface – Open up the fiber for better dye uptake (*for direct disperse printing only*)	– Scour with detergent or surfactant – Sometimes pad with dispersant and lubricant – Dry before printing
Sublimation inks	Coated polyester (polymer-coated fabric or paper)	– No pretreatment for paper transfer – For direct-to-fabric, coating may enhance results	– For coated fabrics: apply polymeric coating – For paper transfer: no pretreatment needed
Vat inks	Cotton, viscose	– Assist in reducing the dye for better penetration	– Pad with reducing agent (e.g., sodium dithionite), alkali, wetting agent – Dry under inert atmosphere
Solvent/UV inks	Coated synthetic fabrics (PVC, PET, etc.)	– Ensure adhesion to the coated surface – Remove dust or grease	– Usually no wet chemical pretreatment – Surface cleaning (dry brushing or corona treatment if needed)

and linen. They form strong chemical bonds with the fibers thus resulting in bright colors and excellent wash and rub fastness. The reactive ink fixation process involves steaming and complex washing thus making it resource-intensive in terms of water and energy use. Acid dye inks are mainly used for protein-based fibers such as wool and silk, as well as polyamide (nylon). They provide rich, saturated colors and good overall fastness, but their application requires precise pH control and fabric pretreatment to ensure proper dye uptake, adding to the process's complexity and requiring careful technical oversight. Disperse dye and sublimation inks are specifically crafted for synthetic fibers, especially polyester and its blends. Sublimation transfers the ink to the fabric using heat and pressure, resulting in highly vibrant, detailed prints with excellent fastness and no need for post-washing. Vat inks, known for their exceptional wash and light fastness and are less commonly used in digital textile printing due to their complex chemical fixation process, which involves reduction and oxidation steps. Their use is typically limited to niche or industrial applications where durability is crucial.

Recent studies have reported the developments in self-dispersing inks which avoid the need for pretreatment. It's an effective strategy to reduce chemical waste and consumption and time. Pigments with self-dispersing properties are used to create binder-free inkjet inks that eliminate the need for fabric pretreatment or post-print washing when applied to cotton textiles (Nie et al., 2023b). In digital pigment printing process, there have been research studies reported for avoiding the use of binders. A digital direct inkjet printing technique was devised to enhance the precision of printing on Polyester fabrics without requiring any pretreatment. This method involved creating a specific type of inkjet printing ink that responds to changes in temperature by altering its viscosity. The increased viscosity helps prevent ink from spreading on the fabric, thus enhancing printing accuracy. A thermosensitive disperse dye ink for direct inkjet printing was formulated by incorporating cetyltrimethylammonium bromide (CTAB) and 3-methylsalicylic acid (3MS) into the ink. By assessing the variations in shear viscosity, ink particle size, and temperature-dependent viscosity it was determined that this thermosensitive ink possesses an optimal average particle size and exhibits unique viscosity changes with rising temperature. When this heat-sensitive ink is applied to polyester fabric, there is no need for pretreatment to achieve clear printing and the resulting printed fabric demonstrates excellent color fastness to both rubbing and washing (Jiang et al., 2024). Traditional pigment-ink formulations require dispersing agents to stabilize pigment particles and binders to adhere the pigments to fabric surfaces. However, binders can make textiles stiffer and heavier, while dispersants may become unstable under certain conditions. A recent study reported development of a novel dye-like pigment ink, synthesized using a chemical grafting method. The resulting pigments had self-dispersing properties, allowing the formulation of inks to be free from both dispersants and binders for textile printing. These modified pigments behaved similarly to dyes in ink formulations but retained the desirable properties of pigments. This approach involves the preparation of thermodynamically stable pigment particles designed for high-performance digital printing on textiles. The process includes grafting hydrophilic functional groups from *p*-aminobenzene sulfonic acid (PASA) onto the surface of pigment particles through covalent bonding, which is carried out during the milling stage. This chemical modification enables the pigment particles to become self-dispersible in water, allowing them to behave similarly to dye molecules during the fabric printing process. As a result, the formulated pigment inks achieve excellent stability in aqueous systems without the need for additional dispersing agents or binders. The pigment particles range in size from 100 to 200 nm, with a zeta potential of approximately −40 mV. These inks were applied on cotton, wool, and polyester fabrics and compared to conventional dye-based inks (Nie et al., 2023a).

Reducing the number of steps in textile wet processing by combining them is another effective way to save chemical consumption, water, time, energy and effluent treatment costs. Especially in digital printing of reactive inks, research attempts have focused on eliminating the pretreatment step. A recent research case study introduced a one-step inkjet printing method for cotton using reactive dye ink and a cationic ink containing CHPTAC, applied simultaneously on a cotton substrate followed by steaming (Wang et al., 2018).

Apart from the main colorant there are many ingredients in an ink formulation which affect the digital print quality. A research study has examined how alcohols with different boiling points can affect the flow behavior, drying speed, and shelf life of reactive inks. This study also reported how the presence of alcohols influences printed shade results in terms of color strength, dye fixation, ink spread, and edge sharpness. Inks with glycerol, diethylene glycol, and triethylene glycol evaporated more slowly but had lower storage stability and reduced dye fixation. Adding alcohols generally helped the ink soak into the fabric more but made printed lines less sharp. Among them, 1,2-propanediol, which has a medium boiling point gave the best results by improving dye fixation and overall print quality (Hou et al., 2019).

Another study reported a novel reactive dye-copolymer nanosphere system developed by adsorbing C.I. Reactive Red 218 (RR218) onto cationic polystyrene-butyl acrylate-vinylbenzyl trimethylammonium chloride (PSBV) nanospheres. The developed ink system was tested for digital printing of woven cotton fabric. The resulting ink nanospheres exhibited uniform particle size and good stability making them suitable for ink formulation. Compared to conventional RR218 dye solution, the fabric printed using RR218 and PSBV showed a 1.4-fold increase in color depth and approximately 45% reduction in dye residue in the printing effluent. Additionally, the consumption of sodium carbonate and urea was reduced by around 3.3 mg/cm^2 and 22.8 mg/cm^2 respectively thus contributing to 30% energy savings (Song et al., 2020).

10.4 Biodegradability Issues in Digital Textile Printing

The biodegradability of inks utilized in digital textile printing can differ significantly based on their chemical makeup and the substrates they are designed for. Reactive inks which are primarily used for cellulosic fibers like cotton, contain reactive dyes and solubilizing agents such as urea. Synthetic dyes are usually not easily biodegradable. Additionally, the presence of urea leads to a high chemical oxygen demand (COD) in wastewater, resulting in a considerable environmental impact (Benkhaya et al., 2022). Acid inks, typically used for protein-based fibers like wool and silk also contain synthetic acid dyes and dispersants that resist microbial breakdown. Both reactive and acid inks necessitate extensive post-treatment processes such as steaming and washing which produce large volumes of dye-laden effluents that are challenging to treat biologically (Rane & Joshi, 2021). Synthetic polymeric binders used in pigment printing are often acrylic-based that are not readily biodegradable. These inks tend to remain on the fabric surface and can break down into microplastics during washing thus raising environmental concerns (Zhang et al., 2012). Disperse inksconsist of hydrophobic azo or anthraquinone-based dyes that are not only resistant to biodegradation but can also decompose into toxic aromatic amines under anaerobic conditions. In sublimation printing, disperse dyes are suspended in a solvent medium like glycol ethers and transferred onto fabric using heat. Although this method avoids generating wastewater, the dyes themselves still pose long-term persistence and toxicity challenges in the environment (Valli Nachiyar et al., 2023).

In contrast the natural or bio-based inks which use plant-derived dyes like indigo, turmeric or madder and biodegradable binders such as starch, gum arabic or modified cellulose present a more environmentally friendly alternative (Yıldırım et al., 2020).These formulations are much more biodegradable and less eco-toxic compared to their synthetic counterparts (Lee et al., 2023). But limitations in reproducibility, color fastness and compatibility with high-speed digital print heads have so far hindered their commercial scalability (Repon et al., 2024). The increasing regulatory pressure and consumer demand for sustainable goods are driving innovation towards the development of bio-based, non-toxic, biodegradable ink systems which can transform the environmental impact of digital textile printing in the near future.

10.5 Toxicity of Inks and Additive Chemicals Used in Digital Textile Printing

The digital textile printing process can significantly influence indoor air quality by releasing various chemical pollutants notably total volatile organic compounds (TVOCs) and suspended particulate matter (SPM). In this context a study was conducted to analyze the concentration levels of SPM and TVOCs. These two pollutants were quantified using the gravimetric method and gas chromatography, respectively. The analysis revealed that neither the digital printing equipment nor the cutting machine generated fine particulate matter in the particulate matter 2.5 (PM2.5) range. But notable levels of PM10 particles were detected with cumulative mass concentrations measured at 9.91 $\mu g/m^3$ for the single-colour digital printer, 4.45 $\mu g/m^3$ for the four-colour digital printer, and 2.02 $\mu g/m^3$ for the cutting machine. When evaluating TVOC emissions, the data indicated that the four-colour digital printer released significantly higher concentrations i.e., approximately 4–12 times greater than the single-colour printer. The four-colour printer exhibited TVOC levels ranging from 28.69 to 57.30 mg/m^3 while the single-colour printer ranged from 2.31 to 13.51 mg/m^3 (Adamović et al., 2022).

A key priority in the textile printing sector today is the shift towards more eco-friendly pigment formulations and auxiliaries especially those aimed at reducing or eliminating formaldehyde residues on fabrics. In this context, curable oligomers have been introduced as alternatives to traditional organic solvents, functioning as effective pigment binders. Recently developed aqueous polyurethane acrylate binders, synthesized using either polyethylene glycol or glycerol ethoxylate-co-propoxylate and free from volatile organic compounds (VOCs) have been employed in pigment printing (Elmolla & Schneider, 2006). The adoption of water-based and UV-curable ink systems is gradually replacing solvent-based inks, significantly lowering the emission of volatile organic compounds (VOCs) during the printing process (Robert, 2015). UV-curing inks offer environmental and economic benefits like fast printing, low VOC emissions, and strong adhesion, but drawbacks include residual monomers, odor, and allergenic potential. Few UV-curing inks are bio-based. A recent research study introduced acrylate-free, 100% renewable polyester itaconates as binders for UV-curing offset inks, synthesized using $Zn(OAc)_2$ to avoid side reactions. These resins showed slightly lower reactivity than commercial alternatives but delivered promising printing performance (Robert et al., 2019). VOCs and particulate matter emitted during processing steps is still a major challenge the digital textile printing workers face (Pongboonkhumlarp & Jinsart, 2022). All these harmful chemicals released during digital textile printing increase the occupational hazards in the digital textile printing industry (Kiurski et al., 2016). Another research paper introduces for the first time, a low-energy inkjet printing technique using biodegradable pigments derived from spalting fungi whose synthesis by-products are also fully degradable. Using acetone and ethanol-based biopigment inks, vibrant prints were achieved on both polyester and cotton without any use of synthetic dyes, mordants, or water-intensive processes thus eliminating liquid effluent completely. The study examines how substrate type, solvent system, and number of print passes affect color shade, depth, uniformity, pigment crystallization, and fastness. Compared to drop-casting, inkjet printing enhances pigment crystallization and color sharpness. pH interactions between fabric, solvent, and pigment lead to tunable color outcomes (e.g., bluish to reddish tones) and fabric-dependent intensity. This method offers a sustainable, resource-efficient alternative for textile coloration (He et al., 2022). Apart from this, developing eco-friendly inks by recycling waste materials has also gained impetus. A research study showcased, development of sustainable textile printing ink utilizing recycled expanded polystyrene as a binder (Obele et al., 2024).

10.6 Challenges in Scale-Up

The trade-off between printing speed and resolution is a critical challenge in scaling up digital textile printing technology. High-resolution designs in fashion, home textiles require the printer to deposit more ink droplets per inch (i.e., higher DPI). This necessitates slower carriage movement and more passes per area. While this ensures superior image clarity, fine detailing, and color gradients, it significantly reduces throughput. In an industrial setting, where large volumes and quick delivery schedules are a demand, slow printing speeds can become a production bottleneck. This discrepancy hampers the cost-efficiency and scalability of digital textile printing, especially when printing complex patterns or photographic imagery across wide-width fabrics. Additionally prolonged machine run times increase costs, wear and tear of print heads, and energy consumption. Overcoming these challenges requires investment in advanced printhead technology, optimized RIP (Raster Image Processing) software, and strategic segmentation of high- and low-resolution jobs within production workflow. In pigment ink digital printing, color mismatches can occur due to device color gamut limitations and fabric properties. To enhance color accuracy in digital textile printing wo methods were reported in research studies: the Full-Process Optimized (FPO) method and the Neural Network Aided Original Adjustment (NNAOA) method. The FPO method improves color management by adjusting printing parameters based on fabric and machine characteristics, reducing color difference from 20.8 to 5.254. The NNAOA method uses a neural network to refine original colors for 20 patches with high color variance, lowering the difference from 8.54 to 6.60. Both approaches significantly improve color reproduction (Li et al., 2023).

10.7 Water Consumption and Residual Waste Water Treatment

Digital textile printing (DTP) is generally more water-efficient than traditional textile printing methods, but certain steps like cleaning print heads and pretreatment processes, still consume considerable amounts of water and generate wastewater. Print head cleaning is a crucial task to maintain nozzle functionality and print quality, on an average it uses around 1–5 L of water per cycle. This process removes residual inks, solvents, and other contaminants from the print heads thus resulting in wastewater that often contains pigments, cleaning agents, and chemical solvents. Such wastewater requires specialized filtration, such as activated carbon or reverse osmosis, and chemical treatments like pH neutralization to prevent environmental contamination. Water recovery systems that recycle treated wastewater for subsequent print head cleaning or pretreatment steps are crucial to minimizing overall water consumption in digital textile printing. Improper or insufficient cleaning of print heads in digital textile printing can result in a range of technical issues that negatively impact both the print quality and the overall performance of the printing system. Clogging and nozzle blockages are the most immediate consequences, as residual ink, especially from highly pigmented or viscous inks, can dry and accumulate in the fine channels of the print head. Over time, this build-up prevents the smooth flow of ink, leading to issues such as uneven ink distribution, missing prints, or distorted images. This results in reduced print resolution, color fidelity, and image sharpness, severely affecting the final output. Additionally, ink cross-contamination becomes a significant problem if the print heads are not properly flushed between different print runs. Residual ink from one print can mix with the ink used in the subsequent job, leading to inaccurate color reproduction or unwanted color shifts, especially when transitioning between distinct color pallets or ink types, such as switching from reactive inks to acid dyes. Furthermore, clogged nozzles increase the frequency of manual interventions to clear blockages, leading to greater downtime and inefficiencies in production workflows.

Another overlooked consequence is the environmental impact of improper cleaning. To clear blockages, printers may consume excessive amounts of cleaning solutions or solvent-based cleaners, often requiring larger quantities of ink during cleaning cycles. This increases the overall waste generation, with more cleaning agents and solvents being used, which could contribute to harmful environmental effects if not disposed of properly. The wastewater generated from insufficiently cleaned print heads may contain residual solvents, pigments, and chemicals that require specialized treatment to avoid contaminating water systems.

In conclusion, proper cleaning of print heads is crucial not only to preserve the integrity of print quality and machine longevity but also to reduce operational costs and the environmental footprint of digital textile printing. Implementing effective cleaning protocols, such as automated cleaning systems, using appropriate solvents, and maintaining a routine maintenance schedule, can prevent the accumulation of ink residues, ensure consistent color accuracy, and maximize the efficiency of the printing process. Advanced treatment technologies such as UV disinfection and nanofiltration can further purify water for reuse. By incorporating closed-loop water recycling, waterless pretreatment solutions, and advanced filtration systems, digital textile printing can reduce its ecological footprint, improve water-use efficiency, and contribute to sustainable textile manufacturing. The implementation of sustainable technologies like laser, nano-technology, plasma, ultrasonic, microwave irradiation, ultrasound, has promising solutions for the textile industry in terms of its waste management (Al-Sayed & Abdelrahman, 2021).

10.8 Fabric Compatibility Issues in Digital Textile Printing

Digital textile printing faces several challenges related to fabric compatibility, ink interaction, and overall print quality. Different fabrics, such as natural fibers (e.g., cotton) and synthetics (e.g., polyester), require specific inks and treatments. Natural fabrics typically work well with reactive inks, but may need pre-treatment to improve adhesion, while synthetics require disperse inks and heat fixation for vibrant prints. Fabric blends present additional challenges due to the differing properties of the fibers.

Pre-treatment (e.g., urea for cotton) and post-treatment (e.g., heat fixation) are essential for achieving durable prints, but they vary by fabric type and can complicate the process. Ink saturation can also be problematic, with some fabrics absorbing too much ink, causing over-saturation, while others may absorb too little, leading to pale prints.

Issues such as fabric movement, shrinkage, and misalignment may occur, particularly with stretchy or delicate fabrics, affecting print precision. The speed and efficiency of the process can be reduced by fabrics that require special handling or slower printing speeds. In an attempt to solve fabric compatibility, issue a novel method was proposed to address the challenge of digital printing on woolen fabrics. The demand for digitally printed wool is often hindered as the print quality is often hindered due to wool scales. To address this issue a study had reported In the pretreatment process for wool fabrics used in inkjet printing, two eco-friendly agents—hydrogen peroxide (H_2O_2) and papain—were employed. H_2O_2 oxidized and disrupted the disulfide bonds in keratin, softening the edges of the wool scales and making them more accessible for papain to break down the protein macromolecules in the outer layer. This combined action of H_2O_2 and papain resulted in a smoother fiber surface with improved wettability and lower zeta potential. As a result, it eased dye penetration into the fibers and minimized the electrostatic repulsion between wool and reactive dyes (Shi et al., 2021). Non-planar fabrics are seen as luxurious due to their rich textures and shine. However, printing directly on them often causes distortion. A recent study reported a fine-grained digital printing system that combines digital printing with non-planar fabrics. It uses a hardware-software setup to capture the fabric's shape, digitally adjust the print design using image registration, and then accurately print the modified design, ensuring better alignment and visual quality (Ren et al., 2017). Additionally, the environmental impact of **synthetics**, which are non-biodegradable, and the need for efficient **waste management** of fabric scraps and ink waste pose sustainability challenges. In summary, addressing these challenges requires selecting the right inks for each fabric type, optimizing treatment processes, and managing printing conditions to ensure consistent, high-quality prints across various textiles.

10.9 Need for Technically Skilled Labour

The role of technically skilled labour in the transition from conventional screen printing to digital textile printing is very crucial. Digital textile printing is a complex process. It involves multiple interlinked processes like Digital design preparation, Ink formulation and management, Fabric pre-treatment, Precise color calibration and profiling, post-treatment and finishing processes, Waste management and energy optimization. Each of these steps requires specialized knowledge, and poorly trained operators may lead to significant production inefficiencies, quality issues, and environmental burdens. For instance, in the accurate design file preparation skilled technicians ensure that files are optimized for printing resolution, color modes (RGB vs CMYK). Sustainable DTP also requires precise color calibration: ICC color profiling to minimize ink wastage and ensure right-first-time printing. Technicians need to match on-screen colors with fabric output, reducing the need for physical sampling Digital printing technicians should possess adequate subject knowledge. Skilled labour can abide by the needs of careful handling, storage, and formulation. Operators must ensure that inks and auxiliaries comply with environmental standards like GOTS, OEKO-TEX, ZDHC for which technical know-how is important. Awareness about correct pre-treatment is crucial for dye fixation, penetration and color fastness. Skilled workers also monitor fabric humidity and surface properties to ensure consistent printing performance. Digital printers are highly sensitive. Skilled technicians can prevent nozzle clogging, head strikes, and downtime. Operating RIP

(Raster Image Processor) software requires an understanding of print parameters, ink limitations, and fabric profiles. Technicians must regularly inspect and calibrate machines to reduce waste and improve energy efficiency. Skilled operators optimize temperature and duration to minimize energy use and prevent fabric damage. Sustainable DTP demands knowledge of water recycling, filtration systems, and chemical neutralization processes. Technicians must identify and correct problems such as color bleeding, poor penetration, or sharpness loss. Sustainable production also requires troubleshooting to investigate root cause of technical errors to avoid unnecessary resource consumption.

Skilled staff must manage ink waste, water usage, and air emissions. Life cycle assessment (LCA): Technicians should understand how each process impacts the overall sustainability footprint. Without trained staff, companies may face legal penalties, reputational damage, or financial loss due to non-compliance. Skilled operators reduce rework, reject rates, and downtime. Proper operation extends the lifespan of expensive digital printers. Technically skilled workers can quickly adapt to emerging technologies, such as pigment-based digital printing, 3D printing on textiles, and AI-based print optimization.

10.10 Existing Challenges in Sustainable Alternatives

The major challenges of digital textile printing and their potential solutions are mentioned in Table 10.2.

Sustainable alternatives such as water-based and bio-based inks in digital textile printing offer reduced environmental impact, but they present notable technical and chemical limitations. Water-based inks, though low in VOCs, are less compatible with synthetic or hydrophobic fabrics like polyester due to poor wettability and require fabric pre-treatment (e.g., cationic agents or polymer coatings). They also need high-temperature drying or steaming, increasing energy consumption and operational complexity. Their high-water content can lead to microbial growth and nozzle clogging, while limited penetration and lower pigment concentration reduce print vibrancy and fastness. Bio-based inks, derived from natural sources such as vegetable oils or algal pigments, often suffer from inconsistency in molecular structure, leading to variations in viscosity, dispersion stability, and shelf life. They also exhibit lower color strength, lightfastness, and wash durability, requiring additional synthetic binders or crosslinkers, which may compromise sustainability. Additionally, both ink types face challenges in achieving fine resolution and reliable jetting performance in commercial piezoelectric printheads, limiting their scalability for high-speed, high-resolution textile applications. Volatile organic compounds (VOCs) are widely used in textile digital printing as solvents like polyethylene glycol, ethanol, methanol, and toluene. Additionally, humectants such as urea are utilized during fabric pretreatment and ink formulation. These substances contribute to environmental concerns including greenhouse gas emissions, ozone layer depletion, and human health risks like DNA mutations. Therefore, it is crucial to explore circular materials and adopt

Table 10.2 Summary of challenges in digital textile printing and potential solutions

Challenge	Description	Solutions/innovations
1. Printing speed	Slower than screen printing, limiting scalability	High-speed continuous inkjet (CIJ), multi-nozzle arrays, pulsed jet systems
2. High costs	Expensive equipment and ink; higher per-unit cost	Greyscale heads reduce ink use; pigment ink systems eliminating pre/post steps
3. Ink-fabric compatibility	Ink must suit specific fiber types (cotton, polyester, wool, etc.)	Reactive, acid, disperse, and pigment ink chemistries tailored to fibers
4. Colour management complexity	Limited colour gamut; mismatch with screen print expectations	RIP software, CMYK + (Hexachrome) systems, digital colour calibration tools
5. Print quality issues	Problems with droplet tailing, misting, and fabric absorbency	Piezo heads, variable drop sizes, waveform control, pre-treatment of fabrics
6. Resistance to change	Industry accustomed to screen printing workflow	Targeted training, showing small-batch advantage, hybrid systems for sampling
7. Substrate handling and fastness	Poor fixation or wash/light fastness; stiff fabric handle	UV-curable inks with freezing tech; novel polymer dispersants in pigment inks
8. Pre/post treatment needs	Steaming, washing, drying needed for many inks	Single-step pigment and UV systems; integrated print-steam-wash machines
9. Limited productivity	Narrow width, low-speed machines limit mass production	Development of wide-format machines (e.g., Osiris, Virtu RS) with inline curing
10. Software and file limitations	Raster/vector file handling and pixelation issues	Advanced RIPs, greyscale imaging, scalable vector design workflows

low-energy, eco-friendly technologies such as plasma surface treatment and waterless inkjet printing to help mitigate the environmental impact of textile coloration processes (Faisal et al., 2022). In another research study various biodegradable organic acids containing different numbers of carboxylic acid groups were added to ink formulations, both in the presence and absence of sodium hypophosphite (SHP). Inkjet printing was performed on cotton fabrics that were either untreated or pretreated according to standard industrial procedures, and the resulting samples were analyzed for their dye fixation and antibacterial properties. The results revealed that, at an optimized pH, untreated fabrics printed with inks formulated using multicarboxylic acids like BTCA and without SHP achieved greater reactive dye fixation and enhanced antibacterial activity than pretreated fabrics printed with inks that did not include biodegradable organic acids (Soleimani-Gorgani & Karami, 2016). The major problems in digital textile printing, its cause and effects are listed in Table 10.3.

Table 10.3 Digital textile printing: problems, causes and effects

Problem	Causes	Effects
Shade variation	Inconsistent fabric pre-treatment; Variation in absorbency/GSM; Incorrect color profiling; Ink viscosity/temp/humidity changes; Print head variability	Uneven shades; Customer complaints; Poor reproducibility
Sharpness of design	Ink spreading due to over-wetting; Ink migration; Low-res artwork; Print head misalignment	Blurred edges, bleeding; Loss of detailing
Fabric surface irregularities (Piles, Neps, Hairiness)	Poor fiber prep; Low-grade/recycled fibers; Inadequate singeing/shearing	Uneven ink deposition; 'White dots', print defects; Nozzle clogging
Print head clogging	Unstable inks; Undissolved particles; Dust/temp/humidity; Long idle time	Banding, line breaks; Ink/substrate wastage; Maintenance cost
Uneven drying	Wrong dryer settings; Fabric thickness variation; Uneven hot air	Tone differences; Uneven curing; Water marks, waviness
Dye penetration (excess/less)	Poor pre-treatment; Wrong fabric/ink match; Improper ink laydown; Moisture variation	Over: dull front, show-through; Under: poor fixation, dullness
Foil adhesion failure	Binder layer too thin/thick; Incomplete curing; Low-quality binder; Surface unevenness	Patchy foil; Low wash durability; Product rejection
Ink consistency issues	Viscosity changes; Pigment settling; Poor stirring/mixing	Shade deviations; Dull/brilliant color loss; Color mismatch
Fabric pH impact	Poor neutralization; pH fluctuations in bath	Poor bonding; Shade shifts; Low fastness
Post-washing issues	Improper washing; Uneven thickener removal; Poor fixation	Back-staining on whites; Patchy dull shades; Bleeding on unprinted areas
Absorbency and whiteness impact	Pretreatment variation; Presence of OBAs	Shade inconsistency; Loss of vibrancy; Yellow cast
Fixation chemicals	Incorrect dosage; Chemical impurities	Uneven fixation; Harsh feel; Color shift
Fabric structure effects (Weave, GSM, Packing)	Loose weave; High GSM; Dense or open structure	Sharpness/shade variation; Design distortion; Ink usage differences
Fixation time/temperature	Low temp/short time; High temp/over-fixation	Poor fastness; Color degradation; Fabric yellowing or weakening

10.11 Conclusion

Digital textile printing offers cost-effective sampling, rapid production, and the flexibility to create intricate and diverse designs without limitations on color or repeat patterns. It delivers high print quality while minimizing production costs and effluent waste. The final print quality is influenced by several factors, including the type of fiber, yarn, fabric, pretreatment, post-treatment, dyes or inks used, and the printing equipment (Adamu et al., 2023). In order to strive for sustainability, researchers and textile processors are making constant efforts towards innovation in digital printing technologies. Predictive technology powered by AI can assist in identifying flaws and optimizing ink usage to increase productivity, cut waste, and enhance print quality. AI-powered features have been incorporated into the digital textile printers. Nowadays, digital textile printer's clogged nozzles are automatically checked and cleaned by the AI-based Nozzle Check Unit (NCU). In the meantime, good nozzles are replaced selectively by the Nozzle Replacement Unit (NRU). These AI-assisted technologies are beneficial in achieving sustainable development goals of responsible consumption. Not only digital printing machinery but constant innovation is seen in manufacturing inks as well. Water based and toxic solvent free ink formulations are increasingly in demand. Digitalization and automation has played a great role in bringing about sustainable changes across various sectors and the textile industry is no exception.

Recent advancements in digital textile printing (DTP) have transformed the industry by improving sustainability, speed, and print quality. High-speed single-pass inkjet systems and refined piezoelectric printheads now offer higher resolution and faster production rates, making digital methods more competitive with traditional rotary printing. The development of eco-friendly inks, such as waterless pigment-based formulations and bio-derived reactive and disperse inks, has significantly reduced the environmental impact of printing. Additionally, innovations in pre- and post-treatment such as enzymatic surface treatments and plasma technology have enhanced ink adhesion and color vibrancy, expanding the range of compatible fabrics. The integration of artificial intelligence in raster image processing (RIP) software has further improved color management and ink optimization while automated workflows have streamlined production, reduced errors, and lowered costs. Hybrid printing systems that combine digital and screen printing are also gaining popularity for their flexibility in mass customization.

Despite these advancements, several challenges persist in DTP. The high cost of specialized inks and equipment continues to be a barrier for small and medium enterprises. To address this, open-source ink formulations and affordable leasing models are being explored. Fabric compatibility also remains an issue, particularly with natural fibers, but innovations in universal pre-treatment coatings and multifunctional binders show promise. Pigment-printed fabrics often suffer from poor wash and rub fastness, which researchers are tackling through nano-binder systems, cross-linkers, and improved post-treatment protocols. Environmental concerns related to chemical-laden pre- and post-treatment processes are being mitigated through enzymatic and bio-based alternatives, along with the development of closed-loop water and energy systems. Color consistency across different print runs is another challenge, now being addressed with spectrophotometer-based color matching systems and standardization efforts. Furthermore, limited ink penetration on thick or textured fabrics is being countered with multi-layer printing and penetration-enhancing additives. Looking ahead, digital textile printing is poised for further innovation through the development of 3D printing technologies that add tactile textures. These advancements indicate a promising future for digital textile printing, balancing technological innovation with ecological responsibility.

Declaration The authors declare that they have no known competing financial interests or personal relationships that could have appeared to influence the work reported in this chapter. Generative AI and AI-assisted technologies have been used in the writing process to improve the readability and language of the work only. After using this tool, the authors carefully reviewed and revised the content to ensure accuracy and quality. The authors take full responsibility for the final version of this publication.

References

Adamović, S., Rajs, V., Adamović, D., Mihailović, A., Samardžić, S., Banjanin, B., & Stojanović Bjelić, L. (2022). Potential chemical stressors emitted during the operation of machines in the digital printing process. *International Journal of Environmental Analytical Chemistry, 102*(16), 3841–3854. https://doi.org/10.1080/03067319.2020.1776859.

Adamu, B. F., Gebeyehu, E. K., Wagaye, B. T., Kumelachew, D. M., Tadesse, M. G., & Jhatial, A. K. (2023). Quality of digital textile printing. In *Digital textile printing* (pp. 185–206). Elsevier.

Ahmed, K. A., & Ahmed, E. A. (2025). Fluorescent inks and their potential applications in textile printing. *Journal of Industrial Textiles, 55*, 15280837241310890. https://doi.org/10.1177/15280837241310890.

Al-Sayed, W., & Abdelrahman, S. H. (2021). Sustainable chemistry in textile processes (pretreatment, coloration and chemical finishing). In *Green chemistry for sustainable textiles* (pp. 93–111). Elsevier.

Benkhaya, S., M'rabet, S., Lgaz, H., El Bachiri, A., & El Harfi, A. (2022) Dyes: Classification, pollution, and environmental effects. In: S. S. Muthu, & A. Khadir (Eds.). *Dye biodegradation, mechanisms and techniques* (pp. 1–50). Singapore: Springer Singapore.

Chen, L., Ding, X., & Wu, X. (2015). Water management tool of industrial products: A case study of screen printing fabric and digital printing fabric. *Ecological Indicators, 58*, 86–94. https://doi.org/10.1016/j.ecolind.2015.05.045.

Choi, P. S. R., Yuen, C. W. M., Ku, S. K. A., & Kan, C. W. (2005). Digital ink-jet printing for chitosan-treated cotton fabric. *Fibers Polym, 6*(3), 229–234. https://doi.org/10.1007/BF02875647.

Ding, Y., Shamey, R., Chapman, L. P., & Freeman, H. S. (2019). Pretreatment effects on pigment-based textile inkjet printing–colour gamut and crockfastness properties. *Coloration Technology, 135*(1), 77–86. https://doi.org/10.1111/cote.12377.

Elmolla, M., & Schneider, R. (2006). Development of ecofriendly binders for pigment printing of all types of textile fabrics. *Dyes Pigments, 71*(2), 130–137. https://doi.org/10.1016/j.dyepig.2005.06.017.

Faisal, S., Ali, M., Naqvi, S., & Lin, L. (2022). Statistical optimization and bulk scale validation of the effects of cationic pre-treatment of cotton fabric for digital printing with reactive dyes. *Journal of Natural Fibers, 19*(13), 6737–6747. https://doi.org/10.1080/15440478.2021.1932671.

Ghoreishi, M., & Happonen, A. (2022). The case of fabric and textile industry: The emerging role of digitalization, Internet-of-Things and Industry 4.0 for circularity. In: X. -S. Yang, S. Sherratt, N. Dey, & A. Joshi (Eds.), *Proceedings of Sixth International Congress on Information and Communication Technology* (pp. 189–200). Singapore: Springer Singapore.

He, Y., Cao, Y., Hwang, H., Gutierrez, S. M. V., Li, S., Chen, H.-L., et al. (2022). Inkjet printing and in-situ crystallization of biopigments for eco-friendly and energy-efficient fabric coloration. *International Journal of Precision Engineering and Manufacturing-Green Technology, 9*(3), 941–953. https://doi.org/10.1007/s40684-021-00360-5.

Hou, X., Chen, G., Xing, T., & Wei, Z. (2019). Reactive ink formulated with various alcohols for improved properties and printing quality onto cotton fabrics. *Journal of Engineered Fibers and Fabrics, 14*, 1558925019849242. https://doi.org/10.1177/1558925019849242.

Javoršek, D., & Javoršek, A. (2011). Colour management in digital textile printing. *Coloration Technology, 127*(4), 235–239. https://doi.org/10.1111/j.1478-4408.2011.00304.x.

Jhatial, A. K., Gianchandani, P. K., Adamu, B. F., Babar, A. A., & Yesuf, H. M. (2023). Colorants for digital textile printing and their chemistry. In: *Digital textile printing* (pp. 119–135). Elsevier.

Ji, Y., Li, X., Jin, K., Fan, Z., Hou, K., Du, P., et al. (2024). Research progress and development trend of textile auxiliaries. *Fibers and Polymers, 25*(5), 1569–1601. https://doi.org/10.1007/s12221-024-00545-0.

Jiang, C., Yu, Y., Zhu, J., Nie, L., Xu, Y., Liu, X., et al. (2024). Direct printing of disperse dye inks to polyester fabrics without chemical pretreatment. *Langmuir, 40*(40), 20897–20905. https://doi.org/10.1021/acs.langmuir.4c01586.

Kiurski, J. S., Marić, B. B., Aksentijević, S. M., Oros, I. B., & Kecić, V. S. (2016). Occupational hazards in printing industry. *International Journal of Environmental Science and Technology, 13*(3), 955–972. https://doi.org/10.1007/s13762-016-0937-z.

Kumelachew, D. M., Wagaye, B. T., & Adamu, B. F. (2023). Digital textile printing innovations and the future. In: *Digital textile printing* (pp. 241–259). Elsevier.

Lee, W., Sung, E., Moon, J., Ahn, I., Yoon, K., Park, Y., & Kim, J. (2023). Performance evaluation of ink and digital textile printing fabric using natural indigo. *Fibers and Polymers, 24*(4), 1309–1319. https://doi.org/10.1007/s12221-023-00137-4.

Li, Y., Zhang, X., Yang, L., Zhang, R., & Li, R. (2023). Study of color reproduction in pigment digital printing. *Textile Research Journal, 93*(11–12), 2718–2737. https://doi.org/10.1177/00405175221147725.

Madhukar Thakker, A., & Sun, D. (2024). Inks for digital printing of textiles. *Journal of the Textile Institute, 115*(12), 2377–2390. https://doi.org/10.1080/00405000.2023.2289975.

Nie, L., Dong, Y., Chen, Y., Chang, G., & Li, R. (2023a). A study for self-dispersing pigment-based inks printing on various fabrics. *Colloids and Surfaces A: Physicochemical and Engineering Aspects, 658*, Article 130689. https://doi.org/10.1016/j.colsurfa.2022.130689.

Nie, L., Xu, X., Chen, Y., Dong, Y., Chang, G., & Li, R. (2023b). Development of binder-free pigment inks for direct inkjet printing on cotton fabric without pretreatment. *Langmuir, 39*(17), 6266–6275. https://doi.org/10.1021/acs.langmuir.3c00573.

Obele, C. M., Ugwu, S. C., & Chukwuneke, J. L. (2024). Development of textile printing ink using expanded polystyrene waste as a binder. *Progress in Rubber, Plastics and Recycling Technology, 40*(1), 48–61. https://doi.org/10.1177/14777606231186635.

Pongboonkhumlarp, N., & Jinsart, W. (2022). Health risk analysis from volatile organic compounds and fine particulate matter in the printing industry. *International Journal of Environmental Science and Technology, 19*(9), 8633–8644. https://doi.org/10.1007/s13762-021-03733-0.

Rainert, K. T., De Oliveira, C. R. S., Felipe, B. H. S., Herpich, H., Valle, R. D. C. S. C., & Valle, J. A. B. (2024). Textile printing: An integrated view of processes, properties, and future prospects. *Coloration Technology, 140*(4), 556–570. https://doi.org/10.1111/cote.12746.

Rane, A., & Joshi, S. J. (2021). Biodecolorization and biodegradation of dyes: A review. *Open Biotechnology Journal, 15*(1), 97–108. https://doi.org/10.2174/1874070702115010097.

Ren, J., Chen, G., & Li, X. (2017). A fine grained digital textile printing system based on image registration. *Computers in Industry, 92–93*, 152–160. https://doi.org/10.1016/j.compind.2017.08.003.

Repon, Md. R., Islam, T., Paul, T. K., Jurkonienė, S., Haji, A., Shukhratov, S., & Toki, G. F. I. (2024). Natural dyes in textile printing: Parameters, methods, and performance. *Environmental Science and Pollution Research, 31*(35), 47552–47583. https://doi.org/10.1007/s11356-024-34424-1.

Robert, T. (2015). "Green ink in all colors"—printing ink from renewable resources. *Progress in Organic Coatings, 78*, 287–292. https://doi.org/10.1016/j.porgcoat.2014.08.007.

Robert, T., Eschig, S., Biemans, T., & Scheifler, F. (2019). Bio-based polyester itaconates as binder resins for UV-curing offset printing inks. *Journal of Coatings Technology and Research, 16*(3), 689–697. https://doi.org/10.1007/s11998-018-0146-4.

Shi, F., Liu, Q., Zhao, H., Fang, K., Xie, R., Song, L., Wang, M., & Chen, W. (2021). Eco-friendly pretreatment to the coloration enhancement of reactive dye digital inkjet printing on wool fabrics. *ACS Sustainable Chemistry & Engineering, 9*(30), 10361–10369. https://doi.org/10.1021/acssuschemeng.1c03486.

Soleimani-Gorgani, A., & Karami, Z. (2016). The effect of biodegradable organic acids on the improvement of cotton ink-jet printing and antibacterial activity. *Fibers and Polymers, 17*(4), 512–520. https://doi.org/10.1007/s12221-016-5865-5.

Song, Y., Fang, K., Bukhari, M. N., Ren, Y., Zhang, K., & Tang, Z. (2020). Green and efficient inkjet printing of cotton fabrics using reactive Dye@Copolymer nanospheres. *ACS Applied Materials & Interfaces, 12*(40), 45281–45295. https://doi.org/10.1021/acsami.0c12899.

Stojanović, S., Geršak, J., Trajković, D., & Ćirković, N. (2021). Influence of sublimation transfer printing on alterations in the structural and physical properties of knitted fabrics. *Coloration Technology, 137*(2), 108–122. https://doi.org/10.1111/cote.12508.

Tyler, D. J. (2005). Textile digital printing technologies. *Textile Progress 37*(4), 1–65. https://doi.org/10.1533/tepr.2005.0004.

Valli Nachiyar, C., Rakshi, A. D., Sandhya, S., Britlin Deva Jebasta, N., & Nellore, J. (2023). Developments in treatment technologies of dye-containing effluent: A review. *Case Studies in Chemical and Environmental Engineering, 7*, Article 100339. https://doi.org/10.1016/j.cscee.2023.100339.

Wang, L., Hu, C., & Yan, K. (2018). A one-step inkjet printing technology with reactive dye ink and cationic compound ink for cotton fabrics. *Carbohydrate Polymers, 197*, 490–496. https://doi.org/10.1016/j.carbpol.2018.05.084.

Wang, H., & Memon, H. (2023). Introduction to digital textile printing. In: *Digital textile printing* (pp. 1–19). Elsevier.

Yıldırım, F. F., Yavas, A., & Avinc, O. (2020). Printing with sustainable natural dyes and pigments. In S. S. Muthu & M. A. Gardetti (Eds.), *Sustainability in the textile and apparel industries* (pp. 1–35). Springer International Publishing.

Zhang, J., Li, X., Shi, X., Hua, M., Zhou, X., & Wang, X. (2012). Synthesis of core–shell acrylic–polyurethane hybrid latex as binder of aqueous pigment inks for digital inkjet printing. *Progress in Natural Science: Materials International, 22*(1), 71–78. https://doi.org/10.1016/j.pnsc.2011.12.012.

Zhao, H., Zhang, K., Fang, K., Shi, F., Pan, Y., Sun, F., et al. (2022). Insights into coloration enhancement of mercerized cotton fabric on reactive dye digital inkjet printing. *RSC Advances, 12*(17), 10386–10394. https://doi.org/10.1039/D2RA01053D.

MIX
Papier aus verantwortungsvollen Quellen
Paper from responsible sources
FSC® C105338

If you have any concerns about our products,
you can contact us on
ProductSafety@springernature.com

In case Publisher is established outside the EU,
the EU authorized representative is:
Springer Nature Customer Service Center GmbH
Europaplatz 3, 69115 Heidelberg, Germany

Printed by Libri Plureos GmbH
in Hamburg, Germany